Primates and Their Relatives in Phylogenetic Perspective

ADVANCES IN PRIMATOLOGY

Primates and Their Relatives in Phylogenetic Perspective

Edited by

ROSS D. E. MacPHEE
American Museum of Natural History
New York, New York

PLENUM PRESS • NEW YORK AND LONDON

Library of Congress Cataloging-in-Publication Data

Primates and their relatives in phylogenetic perspective / edited by
  Ross D.E. MacPhee.
       p.   cm. -- (Advances in primatology)
    Includes bibliographical references and index.
    ISBN 0-306-44422-4
    1. Primates--Phylogeny.   2. Cladistic analysis.   3. Anatomy,
  Comparative.   4. Primates--Evolution.   I. MacPhee, R. D. E.
  II. Series.
  QL737.P9P677   1993
  599.8--dc20   93-4969
                                                                  CIP

Cover: Order primates as defined by Carolus Linnaeus in the 10th edition of *Systema Naturae* (1758).

ISBN 0-306-44422-4

© 1993 Plenum Press, New York
A Division of Plenum Publishing Corporation
233 Spring Street, New York, N.Y. 10013

Printed in the United States of America

For Veronica: *Omnia praeclara rara*

Contributors

Ronald M. Adkins
    Department of Wildlife and Fisheries
    Texas A&M University
    College Station, Texas   77843

S. K. Babcock
    Department of Biological
        Anthropology and Anatomy
    Duke University Medical Center
    Durham, North Carolina   27710

Wendy J. Bailey
    Department of Anatomy and Cell
        Biology
    Wayne State University School of
        Medicine
    Detroit, Michigan   48201

K. Christopher Beard
    Section of Vertebrate Paleontology
    Carnegie Museum of Natural History
    Pittsburgh, Pennsylvania   15213

John Czelusniak
    Department of Anatomy and
        Cell Biology
    Wayne State University School of
        Medicine
    Detroit, Michigan   48201

Morris Goodman
    Department of Anatomy and
        Cell Biology
    Wayne State University School of
        Medicine
    Detroit, Michigan   48201

Rodney L. Honeycutt
    Department of Wildlife and Fisheries
    Texas A&M University
    College Station, Texas   77843

J. I. Johnson
    Anatomy Department, and
        Neuroscience Program
    Michigan State University
    East Lansing, Michigan   48824

J. A. W. Kirsch
    Zoological Museum and Department
        of Zoology
    University of Wisconsin-Madison
    Madison, Wisconsin   53706

Traute K. Kleinschmidt
    Max Planck Institute for Biochemistry
    Munich, Germany

Spencer G. Lucas
    New Mexico Museum of Natural
        History
    Albuquerque, New Mexico

W. Patrick Luckett
    Department of Anatomy
    University of Puerto Rico
    Medical Sciences Campus
    San Juan, Puerto Rico   00936
    Current address:
    Institute des Sciences de l'Evolution
    Laboratoire de Paléontologie
    Université Montpellier II
    Montpellier, France

Ross D. E. MacPhee
Department of Mammalogy
American Museum of Natural History
New York, New York 10024

James R. Martin
Department of Anatomical Sciences
and Neurobiology
School of Medicine
University of Louisville
Louisville, Kentucky   40292

John Nickerson
Department of Ophthalmology
Emory University
Atlanta, Georgia   30322

Todd M. Preuss
Department of Psychology
Vanderbilt University
Nashville, Tennessee   37240

John G. Sgouros
Max Planck Institute for Biochemistry
Munich, Germany

Jing-Sheng Si
Department of Ophthalmology
Emory University
Atlanta, Georgia   30322

Nancy B. Simmons
Department of Mammalogy
American Museum of Natural History
New York, New York   10024

Gamal A. M. Singer
Max Planck Institute for Biochemistry
Munich, Germany

Michael J. Stanhope
Department of Anatomy and
Cell Biology
Wayne State University School of
Medicine
Detroit, Michigan   48201

Frederick S. Szalay
Department of Anthropology
Hunter College, City University of
New York
New York, New York   10021

J. G. M. Thewissen
Department of Biological
Anthropology and Anatomy
Duke University Medical Center
Durham, North Carolina   27710
Present address:
Department of Anatomy
NEOUCOM
Rootstown, Ohio 44272

John R. Wible
Department of Anatomical Sciences
and Neurobiology
School of Medicine
University of Louisville
Louisville, Kentucky   40292

Preface

This book has the modest aim of bringing together methodological, theoretical, and empirical studies that bear on the phylogenetic placement of primates and their relatives, and continues a tradition started by *Phylogeny of the Primates: A Multidisciplinary Approach* (edited by W. P. Luckett and F. S. Szalay; Plenum Press, 1975) and *The Comparative Biology and Evolutionary Relationships of Tree Shrews* (edited by W. P. Luckett, Plenum Press, 1980). Although there are several recent compendia of studies of primate relationships, most of these are exclusively concerned with the internal arrangement of clades within the order, not with the place of primates and their relatives on the eutherian cladogram.

Evolutionary theory predicts that primates must be more closely related to some nonprimate mammals than to others, but a continuing problem has been to find reliable procedures for recovering historical relationships among taxa. Before the 1970s, higher-level relationships among primates and eutherian mammals that might be closely related to them were rarely treated in detail. Outstanding exceptions, like Le Gros Clark's *Antecedents of Man*, were just that—exceptions. (Clark himself essentially stopped with making a case for tree shrews; he did not, for example, explore whether bats and colugos were also related to primates.) In the 1970s and 1980s, the rise of cladistic techniques and advances in molecular methods began to transform primate systematics. So did new controversies, such as the "flying primate" hypothesis, which had the salubrious effect of refocusing attention on how phylogenetic arguments ought to be justified. In April 1991, an opportunity to reflect on what to date this new knowledge has brought us came in a symposium organized for the 60th annual meeting of the American Association of Physical Anthropologists, entitled "Debating the Superordinal Relationships of Primates: Issues, Evidence, and Proposed Solutions." And debate there was, as is reflected in this volume, composed of papers that emerged from that meeting. I am grateful to all of those who participated in this undertaking.

For various forms of assistance, including help with organizing the AAPA symposium, refereeing submitted papers, discussing fine points of systematic methodology, and other matters, I thank John Wible, Rob DeSalle, Chris Beard, Ward Wheeler, Darrel Frost, Fred Szalay, Pat Luckett, Matt Cartmill, Audrone Biknevicius, and Todd Preuss. My editor at Plenum, Mary Born, encouraged as well as insisted—a useful combination—and I am grateful to her for helping this project through its many stages. But most of all I thank my wife Veronica, for her invaluable assistance throughout.

Ross D.E. MacPhee

New York

Contents

The Importance
of Methods

1

Archontan Phylogeny
and Cladistic Analysis
of Morphological Data

NANCY B. SIMMONS

Introduction

Perusal of the recent systematic literature leaves the impression that cladistic methodology has contributed little to resolving the relationships between primates and other placental mammals. Despite numerous cladistic analyses, no viable consensus has yet been reached among workers studying morphology (e.g., Novacek and Wyss, 1986; Wible and Novacek, 1988; Pettigrew *et al.*, 1989; Novacek, 1990; Dumont, 1992; Beard, 1993) or molecular data (e.g., Ammerman and Hillis, 1990, 1992; Baker *et al.*, 1991a; Mindell *et al.*, 1991; Adkins and Honeycutt, 1991; Bailey *et al.*, 1992). Most workers agree that primate origins lie somewhere among "archontan" mammals (tree shrews, bats, gliding lemurs, extinct plesiadapiformes), but there is little agreement concerning relationships among these groups and monophyly of Archonta is still an open question (Fig. 1). If cladistic analysis is the powerful tool that we

NANCY B. SIMMONS • Department of Mammalogy, American Museum of Natural History, New York, New York 10024.

Primates and Their Relatives in Phylogenetic Perspective, edited by Ross D.E. MacPhee. Plenum Press, New York, 1993.

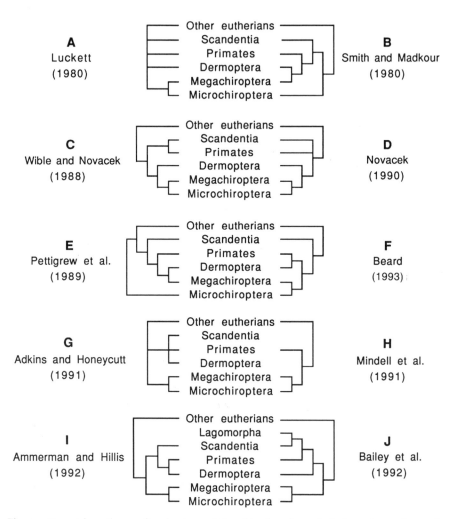

Fig. 1.   Recent hypotheses of archontan relationships based on cladistic analyses of morphological features (A–F) or molecular data (G–J). (A) Luckett's (1980) hypothesis was based on an analysis of fetal membranes and osteological characters. (B) Smith and Madkour (1980) analyzed structure of the penis, brain, teeth, and patagium. The hypotheses of (C) Wible and Novacek (1988) and (D) Novacek (1990) were based on analyses of features of the skull, postcranial skeleton, patagium, musculature, vascular system, reproductive tract, and fetal membranes. (E) Pettigrew *et al.* (1989) based their hypothesis on analysis of features of the central nervous system. (F) Beard's (1993) phylogeny is derived from analysis of osteological characters in living and fossil taxa. Three of the molecular studies (G–I) are based on analysis of mitochondrial genes. (G) Adkins and Honeycutt (1991) based their study on the cytochrome oxidase subunit II gene; (H) Mindell *et al.* (1991) examined the 12 S ribosomal RNA gene and the cytochrome oxidase subunit I gene; and (I) Ammerman and Hillis (1992) analyzed the 12 S ribosomal RNA gene. (J) Bailey *et al.* (1992) based their hypothesis on analyses of the nuclear ε-globin gene. Note that monophyly of Archonta is indicated by only five of the ten phylogenies shown here.

suppose, then why haven't the relationships of primates been resolved? Part of the answer may lie in the different approaches and methods employed by workers using cladistic methods to address archontan relationships.

Cladistic analysis, often termed phylogenetic analysis, is a method of developing hypotheses of evolutionary relationships among organisms based on shared, derived similarities (synapomorphies). Much has been written concerning the theory and practice of cladistics, including many books (e.g., Hennig, 1966; Eldredge and Cracraft, 1980; Nelson and Platnick, 1981; Wiley, 1981; Ax, 1987; Sober, 1988; Brooks and McLennan, 1991; Harvey and Pagel, 1991; Wiley et al., 1991), edited volumes (e.g., Funk and Brooks, 1981; Joysey and Friday, 1982; Platnick and Funk, 1983; Duncan and Stuessy, 1984), and journal articles (for a recent review see Sereno, 1990). While debate continues concerning some basic assumptions of cladistic theory (e.g., validity of the parsimony criterion, importance of an a priori belief in descent with modification), few modern workers challenge the value of cladistics as a tool for investigating phylogeny. Most discussions now center on more practical issues: taxonomic sampling, choice of characters and character states, recognizing and measuring homoplasy, etc. These issues play a central role in arguments concerning the relationships of archontan mammals.

Most recent workers interested in archontan phylogeny have begun their studies with the assumption that the relationships of primates to other mammals can be resolved using cladistic methods. Inherent in this assumption is the idea that archontan evolution involved a phylogenetic branching sequence of some kind, not simultaneous origination of all of the major lineages from a single ancestral species. An immediate goal of phylogenetic analysis is to correctly identify the branching sequence. By developing a new data set, or by including new fossil material, each additional study suggests that it can offer not only a specific phylogenetic hypothesis, but one that is "better" than any previously offered (i.e., one supported by more characters or more external evidence; in short, one more likely to reflect the true phylogeny). Confidence in a new hypothesis is generally argued by reference to one or more broad methodological considerations. Wible and Novacek (1988), for example, stressed the importance of considering data from a wide variety of anatomical systems, arguing that taxonomic groupings supported by diverse data deserve more confidence than grouping supported by data from only one or two character complexes. Alternatively, Pettigrew and his co-workers (e.g., Pettigrew et al., 1989; Pettigrew, 1991a,b) have argued that hypotheses of archontan relationships should be based on anatomical systems hypothesized to be relatively free of homoplasy (such as neural characters), rather than on data sets with supposedly high probabilities of homoplasy (such as postcranial characters). Taking a different approach, Beard (1993, this volume) has emphasized the value of fossils, developing a new series of osteological characters and stressing the importance of considering extinct archontans (e.g., paromomyids, plesiadapids) in analyses of primate relationships. These different approaches have produced different phylogenetic hypotheses, each of which appears strongly supported by the character systems used in the analysis (see Fig. 1).

It is tempting, in the face of data that support mutually exclusive hypotheses, to conclude that the underlying method (cladistic analysis) is simply incapable of resolving the issue of primate relationships. Perhaps all of the archontan lineages did in fact originate simultaneously from some sort of ancestral stock. Alternatively, if the origin of the archontan lineages involved a sequence of dichotomous branching, perhaps these events occurred so fast—and so long ago—that the "signal to noise" ratio in any data set will make it impossible to reconstruct the branching sequence [an explanation recently suggested by Baker *et al.* (1991a) to explain the results of their study of ribosomal DNA]. Confronting these possibilities, MacPhee (1991, p. 122) recently suggested that

> The problem now is to know how to evaluate these high-level "crossing synapomorphies": for example, is wingedness in microbats and megabats impervious proof of chiropteran monophyly, or is this grouping broken by retinotectal pathway similarities in megabats and primates? Since one or the other set of similarities must be due to convergence of an unusually detailed sort, arguably undetectable by standard morphological methods, it is difficult to see how cladistic procedures (including ad hoc parsimony manipulations) can offer any resolution. Perhaps it is time to return to scenarios about adaptational history as better devices for understanding primate evolution.

This provocative statement suggests that cladistics cannot resolve archontan relationships because (in several critical instances) detailed studies of morphology do not permit us to distinguish homoplasy from homology. However, this judgment overlooks the fact that phylogeny, not morphology, provides the crucial test of homology. No matter what sort of data are being analyzed (e.g., morphological, molecular, developmental), homology and homoplasy are both hypotheses that follow from prior hypotheses of evolutionary relationships (Rieppel, 1980; Patterson, 1982).

MacPhee (1991) implied that homoplasy is identified by finding differences between two structures, thus refuting the hypothesis of their homology. However, many of the biological processes cited to explain cases of convergence or reversal (e.g., developmental canalization, paedomorphosis) predict that morphologically identical structures may evolve more than once (Gould, 1977; Kluge, 1988, 1989; Rieppel, 1989; Wake, 1991). In practice, homoplasy is often detectable only in the context of a specific phylogeny (Rieppel, 1980; Patterson, 1982; Farris, 1983; Sanderson and Donoghue, 1989). The chief difficulty in recognizing homoplasy in archontan evolution is simply that there are too many conflicting phylogenetic hypotheses. Rather than advocating a return to "scenarios" as the best way to elucidate primate origins, it seems prudent to first examine the sources of the various conflicting results of cladistic analyses of archontan mammals. Disagreements among workers clearly exist at many levels (see below), and it is premature to reject cladistics as a tool for elucidating primate relationships before resolving such issues.

All workers seem to agree that more detailed comparative morphological studies of archontan mammals are needed. Welcome examples of recent contributions include Thewissen and Babcock's (1991, 1992, this volume) work on the occipitopollicalis muscle in bats and dermopterans, and Dumont's (1990, 1992) examination of enamel microstructure across many archontan lineages. Reexamination of some character has already suggested that several putative synapomorphies are not supported by unambiguous observational data as originally claimed (e.g., see the review of neural characters presented by Kaas, 1993). Pending detailed evaluations of all of the characters relevant to archontan relationships (an undertaking that will probably take years), some immediate insights may be realized by scrutinizing the methods used to transform observational data into complex phylogenetic hypotheses. Most disagreements about archontan relationships center around the relative usefulness of data from different anatomical systems (e.g., postcranial versus neural characters), problems of taxonomic sampling (both within and among groups), and procedures for identifying homologies and homoplasies (Baker *et al.*, 1991b; Simmons *et al.*, 1991). No less important are several problems involving the nuts and bolts of coding observational data for computer-assisted parsimony analysis. Resolution (or at least discussion) of these issues should precede the next generation of analyses.

Published Data Sets: A Summary

I and others have recently argued that archontan relationships cannot be adequately addressed without reference to all of the different types of morphological data that are thought to be relevant to relationships among the principal taxa (Greenwald, 1990, 1991a,b; Baker *et al.*, 1991b; Simmons *et al.*, 1991). A wealth of morphological characters exist in the literature, yet these have never been drawn together into a single data set. These data represent an obvious starting point for any study of archontan phylogenetics, and thus provide excellent examples for discussing issues of methodology and the interpretation of results. To this end, the data set presented in Appendixes 1 and 2 represents a compilation of the morphological characters previously used in cladistic analyses of archontan mammals. This "summary data set" was constructed by adding together various published data sets representing a wide range of anatomical systems. The data set includes cranial characters from Novacek (1980, 1985, 1986), Wible and Novacek (1988), and MacPhee *et al.* (1988), postcranial characters from Novacek (1980), Wible and Novacek (1988), and Beard (1993), characters of the reproductive tract and fetal membranes from Luckett (1977, 1980a,b, 1985) and Smith and Madkour (1980), and neural characters from Johnson *et al.* (1982a,b) and Pettigrew *et al.* (1989). Redundant characters (those described in two or more studies) have been removed from this data set, which nevertheless includes over 150 characters. The taxa scored in the summary data set (see Appendix 2) include all of the extant "archontan" lineages plus three extinct groups (Paromomyidae, Micromomyidae, and Plesiadapidae). Polarity for each character was established by including a "hypothetical ancestor" taxon (ANCESTOR) in the data matrix. The primitive condition of each character (used to construct this hypothetical taxon) was taken from the literature; polarity assessments therefore represent the opinions of the original authors of each character description (see references in Appendix 1).

It should be noted that all of the information encoded in the summary data set comes directly from the literature. No effort was made to modify the character descriptions, even though some appear to be seriously flawed in terms of character independence, definition of character states, or polarity assessments (see below). Similarly, the published distributions of character states have not been changed, even when gaps exist that might now be filled (e.g., by discoveries of new fossils). For these and other reasons discussed below, the relationships that appear in the trees presented here should be regarded with some skepticism.

The summary data set was analyzed using Swofford's (1989) PAUP (Phylogenetic Analysis Using Parsimony) program version 3.0. Exact methods were used to identify the most parsimonious trees (see discussion below under "Finding the Shortest Tree: Algorithms and Search Techniques"). The results of various analyses are discussed separately below. Unless otherwise stated, all characters were treated as unordered and all transformations were given equal weight.

Choice of Taxa for Analysis

Taxonomic sampling is an aspect of cladistic analysis that is rarely accorded much attention. Because it is impractical to sample every species in an analysis of higher-level relationships, some selection method must be adopted. Taxonomic sampling may best be considered as a two-part problem. First, one must decide which presumably monophyletic lineages will form the operational taxonomic units (OTUs) of an analysis. After the OTUs have been chosen, another problem involves breadth of sampling *within* each OTU. These two issues frequently become intertwined, especially when one seeks to address higher-level systematic problems. If primates are included in an analysis as a single OTU, the diversity of morphologies found within that group may make it difficult to score many characters. Alternatively, diving primates into multiple OTUs (e.g., Strepsirhini, Tarsiiformes, Anthropoidea) may complicate the analysis by forcing one to consider numerous additional characters (e.g., autapomorphies of primates, character relevant only to relationships among primates). The number of OTUs may also affect the type of analyses possible. When too many taxa are present, exact algorithms [i.e., methods that guarantee identification of the most parsimonious tree(s)] may not be able to operate, forcing one to use heuristic methods to find the shortest trees (see discussion under "Finding the Shortest Tree: Algorithms and Search Techniques"). It is generally most efficient to collect data for the maximum number of OTUs that may be of interest, and later lump taxa together if it seems appropriate as analyses progress. This may be particularly important when monophyly of the larger groups is uncertain.

The summary data set (Appendix 2) includes ten presumably monophyletic OTUs: Scandentia, Micromomyidae, Paromomyidae, Galeopithecidae, Plesiadapidae, Strepsirhini, Tarsiiformes, Anthropoidea, Megachiroptera, and Microchiroptera. Figures 2 and 3 illustrate the results of analysis of this data set. Interestingly, five equally parsimonious trees were found—and primates (Strepsirhini + Tarsiiformes + Anthropoidea) appears as a monophyletic group in only three of these trees. Given this result, the separation of primates into several components is clearly essential: the possible relationship between Strepsirhini and two of the fossil taxa (Plesiadapidae and Micromomyidae) would have been missed if the extant primate lineages had been lumped into one OTU. This example illustrates that even when most or all workers agree that a group is monophyletic (as in case of primates), the data used in a particular study may not support such an assumption.

Within-Group Variation

Within-group variation almost always creates problems in phylogenetic analyses involving higher-level taxa. When an OTU exhibits more than one character state among its members, it may be difficult or impossible to identify the condition that is primitive for the OTU as a whole. An example is Character 131 in the summary data set: type of chorioallantoic placenta. Strepsirhines exhibit an epitheliochorial placenta (state 0), while tarsiers and anthropoids have a hemochorial placenta (state 2). What is the primitive condition for primates? Luckett (1977, 1980a,b) argued that an epitheliochorial placenta is primitive for eutherians, and strepsirhines are often regarded as the basal (and thus most primitive) branch of the primate family tree. On this basis it seems reasonable to accept an epitheliochorial placenta as the primitive primate condition. This technique (generalized outgroup analysis combined with hypotheses of ingroup relationships) is the standard method used to identify the primitive condition for a higher-level group—the condition that will subsequently be used to assess the relationships of that group to other lineages. When one applies this procedure to all of the OTUs in an analysis, the result is a data matrix in which each matrix cell contains only a single character state (e.g., the matrices presented by Novacek, 1989, 1990).

Unfortunately, the method described above is prone to overestimating what we really know about various OTUs. No matter what the relationships among extant lineages of primates, the distribution of epitheliochorial and hemochorial placentas leaves the primitive condition for primates ambiguous (Fig. 4); outgroup comparisons are essential to resolve this issue. Because we

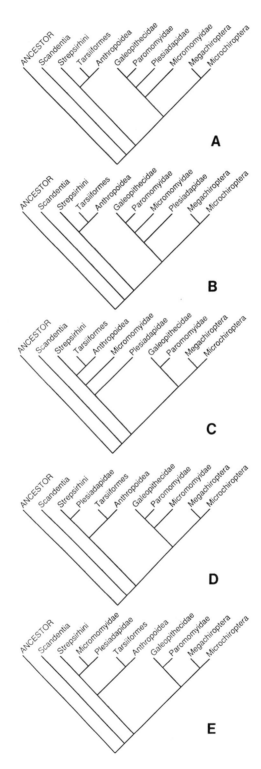

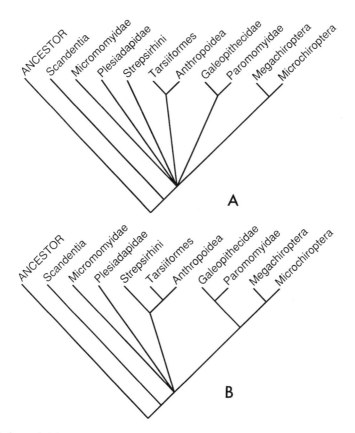

Fig. 3. Strict and Adams consensus trees constructed from the five trees presented in Fig. 2. The strict consensus tree (A) identifies only those clades that occur in all five trees; the Adams consensus tree (B) additionally recognizes other relationships that remained stable in all trees (for a review of the various types of consensus trees, see Swofford, 1991). Comparison of these trees reveals that the lack of resolution seen in the strict consensus tree (A) is the result of the variable positions of only two taxa, Micromomyidae and Plesiadapidae.

cannot agree on the relationships of primates to other mammals (resolving this problem is actually one goal of the analysis), it simply is not possible to identify the primitive condition for primates without adding an extra layer of assumptions. All we really know about primates is that an endotheliochorial placenta (state 1) does not appear in the group. The problem of polymorphism is greatest when we know little about the branching sequence of taxa within an OTU, when relationships among various outgroups are uncertain,

Fig. 2. Five equally parsimonious optimal trees identified by an exhaustive enumeration analysis of the summary data set (Appendix 2; transformations unordered and equally weighted). Each tree requires 318 steps; CI = 0.626. Note that living primates form a monophyletic group relative to the fossil taxa Plesiadapidae and Micromomyidae in only three of the five trees (A–C).

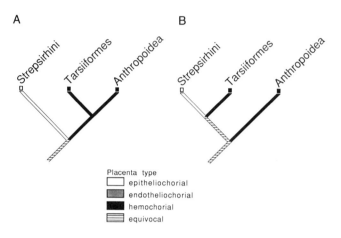

Fig. 4. Distribution of placental types among living primates (see Character 131 in Appendix 2). (A) If Tarsiiformes is the sister group of Anthropoidea, the common ancestor of this clade would be expected to have a hemochorial chorioallantoic placenta. Without reference to outgroups, it is not possible to unambiguously identify the placental type that is primitive for Primates because strepsirhines exhibit a different condition (epitheliochorial placenta). (B) An alternative phylogenetic arrangement of the three living primate lineages. If Strepsirhini and Tarsiiformes form a monophyletic group, the placental type primitive for this clade cannot be determined without reference to outgroups; similarly, the condition primitive for Primates is also unresolved. No matter which phylogeny of primates is favored (A or B), it is not possible to unambiguously determine the primitive condition for Primates prior to an analysis of their relationships to other mammalian taxa.

and when there is variation within the outgroups themselves. Unfortunately, these factors apply in almost all studies of relationships among higher-level mammalian taxa.

Before the advent of the latest generation of computer programs (e.g., PAUP 3.0), it was necessary to choose a single character state to describe each OTU (presumably the condition primitive for the lineage). The level of uncertainty produced by repeatedly forcing such assumptions, particularly when OTUs are diverse higher-level taxa, can severely flaw otherwise robust analyses. Now, however, cases of polymorphism within supraspecific OTUs can be coded as such (see Appendix 2). Global parsimony, rather than a priori assumption, can thus be used to identify the condition that is most likely primitive for any given group—but only after the phylogenetic analysis has been completed.

Nixon and Davis (1991) used the term taxonomic polymorphism to refer to cases in which a supraspecific taxon exhibits more than one character state among its member species. This type of polymorphism is distinct from "within-species polymorphism," a phenomenon of population genetics that cannot be adequately addressed by computer phylogenetics programs such as PAUP (see warning in PAUP user's manual; Swofford, 1989). PAUP assumes that any taxon coded as having multiple states represents a monophyletic collection of subtaxa, each of which is monomorphic (Swofford, 1989). In searching for the most parsimonious trees, the program assumes that the ancestor of a polymorphic OTU had only one of the observed states; there is thus no provision for polymorphic ancestral taxa (Swofford, 1989). This limitation makes it essential to clearly distinguish between taxonomic and within-species polymorphism when scoring taxa for a cladistic analysis. Only taxonomic polymorphism may be treated by scoring OTUs as having multiple character states; within-species polymorphism requires different techniques, such as definition of additional character states to describe polymorphic conditions.

Taxonomic polymorphism is relatively low in the summary data set. The highest level of polymorphism is seen in Scandentia, for which 10% of the characters were scored as polymorphic (the other OTUs were scored as polymorphic for 0 to 5% of the characters included in the data set). Given these low numbers, one might expect that eliminating polymorphism from the matrix (by choosing a single state to represent each OTU) would have little effect on the outcome of the analysis. As an experiment, this was done with the summary data set. When an OTU exhibited both the primitive and a derived condition of a character, it was assumed that the condition defined as primitive for the analysis (0) was also the ancestral condition for that OTU. When an OTU exhibited two or more derived conditions, the least derived (i.e., least changed from the primitive state) was chosen as the likely ancestral state. Analysis of the recoded data produced only two most parsimonious trees, identical to Fig. 2A and 2B. The consensus tree (Fig. 5) constructed from these trees shows much more resolution than seen previously (compare with Fig. 3); primates, for example, is now resolved as a monophyletic group. But

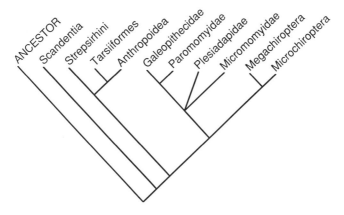

Fig. 5.   Results of an analysis exploring the effects of removing polymorphisms from the summary data set (see text for discussion). An exhaustive enumeration analysis of the modified data set resulted in identification of two equally parsimonious optimal trees requiring 323 steps (CI = 0.614). The tree shown here is a strict consensus tree constructed from these two trees; compare this topology with that of the strict consensus tree shown in Fig. 3A.

this new degree of resolution is not based on any new observational data. Rather, an additional layer of assumptions—hidden from view in the data matrix—is responsible for the apparent improvement in results. Increased resolution does *not* always reflect increased knowledge; "pseudoresolution" of the sort produced by eliminating polymorphisms may be extremely misleading. By suggesting that a given node is fully resolved (rather than unresolved), attention is drawn away from a potentially interesting phylogenetic problem (e.g., the relationship of plesiadapids to extant primates), and the level of support for some relationships may be overestimated.

Outgroups

In addition to the ingroup taxa included in a phylogenetic analysis, most workers agree that at least two outgroups should be considered in order to properly polarize characters (Maddison *et al.*, 1984; Wiley *et al.*, 1991). The outgroups can be specified and included in the analyses as OTUs, or a "hypothetical ancestor" can be created to summarize the results of the outgroup comparisons. In the case of "Archonta," which may not be a monophyletic group, choice of outgroups becomes difficult. Frequently, characters are polarized by reference to the condition found in "most eutherians" (e.g., Beard, 1993), or ". . . marsupials, monotremes . . . and . . . putatively plesiomorphic eutherians" (Pettigrew *et al.*, 1989, p. 496). This technique makes it difficult to evaluate the polarity assessments made by these authors. Identification of specific taxa as outgroups is preferable because this limits assumptions about generality of the presumed primitive condition. An organism considered a "plesiomorphic eutherian" by one worker may be considered highly derived by another; there is simply no way to evaluate these assumptions if outgroups are not explicitly designated.

So what are appropriate outgroups for an analysis of archontan relationships? This represents a real problem because of the diversity of opinions concerning higher-level mammalian relationships. If one accepts the phylogenies offered by Novacek and his co-workers (e.g., Novacek and Wyss, 1986; Novacek *et al.*, 1988; Novacek, 1990), reasonable outgroups might include any of several other epitherian (advanced eutherian) lineages, e.g., Lipotyphla (insectivores), Macroscelidea (elephant shrews), Lagomorpha (rabbits and pikas), Carnivora (carnivores). However, Pettigrew *et al.* (1989) have recently suggested that microchiropterans may be closely related to nonepitherian bradypodid edentates (tree sloths). To properly test the latter hypothesis, both edentates and at least two noneutherian lineages (such as monotremes and marsupials) should be included in an analysis. Inclusion of taxa that might be outgroups (or alternatively might fall within the smallest clade containing all five archontan lineages) can clearly contribute substantially to both the outcome and the potential significance of an analysis. Bailey *et al.* (1992) recently reported results of a molecular study indicating that lagomorphs may be the sister group to scandentians, nesting well within the archontan clade (Fig. 1). Accordingly, it is now essential that Lagomorpha be included in studies designed to test archontan monophyly.

The monophyly of outgroups, which is not often discussed in published reports of cladistic analyses, is of potential concern when ingroup monophyly has not been adequately demonstrated. Monophyly of Lipotyphla, for example, has yet to be firmly established (MacPhee and Novacek, 1993). If bats are diphyletic, perhaps microchiropterans are more closely related to some lipotyphlans than to others (similarities have been noted between the teeth of microchiropterans and erinaceoid liptyphlans; Slaughter, 1970; Koopman and MacIntyre, 1980). In this context, it may be most appropriate to choose lower-level taxa as outgroups. For example, Erinaceidae (hedgehogs) and other families that appear monophyletic (Frost *et al.*, 1991; MacPhee and Novacek, 1993) might be included as separate OTUs rather than lumping them together under Lipotyphla.

Unfortunately, there is no easy end to the process of adding potentially relevant taxa. Careful consideration of the goals of a project should precede random inclusion of "extra" outgroups or ingroup taxa. Relevance of particular taxa to any given phylogenetic problem may be tested by running a series of separate cladistic analyses in which various taxa are added to or subtracted from the data set (e.g., Donoghue *et al.*, 1989; Simmons, 1993).

Fossils

Fossils can be treated in many different ways in a phylogenetic analysis. They may be completely ignored (a common practice when the majority of characters in a study involve soft tissue structures), or fossils may be included as subunits of largely extant OTUs (e.g., consideration of Omomyidae when scoring characters for Tarsiiformes), or they may be included as separate OTUs (e.g., Paromomyidae in the summary data set). In practice, lack of comparable data has often discouraged workers from attempting to consider nonosteological characters in analyses that include fossils, or from including fossil taxa as separate OTUs in analyses that involve many characters of soft anatomy (Novacek, 1989; Beard, 1993). This is unfortunate, because fossils sometimes preserve information crucial to resolving relationships among extant taxa. For example, Gauthier *et al.* (1988) demonstrated that addition of fossil taxa to a data set comprising both osteological and soft anatomical characters resulted in substantial changes in the apparent relationships among *extant* lineages of amniotes.

Fossil taxa are significant because they often preserve critical information in the form of unique combinations of primitive and derived character states (Gauthier *et al.*, 1988; Donoghue *et al.*, 1989; Novacek, 1992). The relative importance of a taxon in terms of its effect on relationships among other taxa seems to be a function of the degree of character conflict already present in the data set, and the amount (and type) of additional character conflict introduced by adding the new taxon (Gauthier *et al.*, 1988; Donoghue *et al.*, 1989; Novacek, 1992). The completeness of a taxon (i.e., the percentage of characters for which it can be scored in a given analysis) is not necessarily relevant, nor is the age of the taxon (Gauthier *et al.*, 1988; Donoghue *et al.*, 1989; Simmons, 1993).

In addition to the potential of fossils to alter topologies of phylogenetic trees, fossils are also important because they frequently affect character optimizations (the parsimonious arrangement of character state transformations on a tree). Because the biological interpretation of morphological changes in phylogeny depends on such optimizations, fossils may affect conclusions concerning the degree of support for various monophyletic groups, taxonomic diagnoses, character independence, homoplasy, and the relative timing of various evolutionary events (see discussion below under "Character Optimization").

In the current debate concerning archontan relationships, several extinct taxa may be particularly important: Paromomyidae, Micromomyidae, Plesiadapidae, Microsyopidae, and Mixodectidae (Szalay, 1969; Szalay and Drawhorn, 1980; Szalay and Lucas, 1990; Kay *et al.*, 1990; Beard, 1990, 1991, 1993, this volume). While more data are published on these taxa every year, some are still quite poorly known. The summary data set (Appendix 2) includes three of these taxa: Plesiadapidae (53% complete in the context of the matrix), Paromomyidae (19%), and Micromomyidae (10%). When all of these taxa are removed from the matrix, analysis results in a single most parsimonious tree (Fig. 6). Comparison of this tree with those in Fig. 2 and 3 reveals that inclusion of the fossil taxa did not alter the relationships among the extant lineages. However, some character optimizations and evolutionary interpretations are substantially affected. For example, if the tree in Fig. 2A is correct,

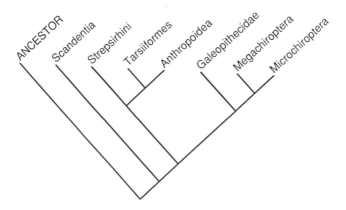

Fig. 6.   Results of an exhaustive enumeration analysis that included only the living taxa of archontan mammals (see text for discussion). Only one most parsimonious optimal tree was found (294 steps, CI = 0.658).

several skeletal modifications apparently related to flight [e.g., lengthening of the forelimbs (Character 59) and reduction of the ulna (Character 68)] may have evolved twice, once in bats and once within dermopterans (Galeopithecidae), rather than just evolving once (in the common ancestor of dermopterans + bats) as proposed by Wible and Novacek (1988).

Choice of Characters and Character States

Choice of Anatomical Systems

The choice of characters for analysis is a complex process that depends on many factors, including the phylogenetic question(s) under consideration, the types of specimen preparations available, the areas of expertise of the researcher, the amount of time available, and, in the case of costly laboratory studies, the amount of funding that can be obtained. As a result of various constraints, most workers generally concentrate on one or two anatomical systems (e.g., Pettigrew *et al.*, 1989; Beard, 1993). In the area of archontan phylogenetics, a very wide variety of data have been used in different studies (see discussion above under "Published Data Sets: A Summary"), which has led to discussions concerning the relative "usefulness" of different character systems.

If phylogenetic analysis of characters from different anatomical systems produced trees with identical topologies, choice of character systems would not be a problem: one could simply choose the system that was easiest to study and base all phylogenetic hypotheses on analyses of that system. In the case of archontan mammals, however, every anatomical system apparently preserves a *different* phylogenetic signal. Figure 7 illustrates strict consensus trees based on separate analyses of characters from six different anatomical systems: auditory system (33 characters), nonauditory cranial features (20 characters), nervous system (23 characters), anterior axial skeleton and forelimb (31 characters), hindlimb (35 characters), and reproductive tracts and fetal membranes (12 characters). The topology of every tree is different, and none of these trees is identical to the tree generated from the complete summary data set (154 characters; Fig. 6). These conflicting phylogenetic signals are probably the largest single factor responsible for the current controversy over archontan relationships.

Recently a debate has developed concerning the relative value of characters from different anatomical systems (see Pettigrew *et al.*, 1989; Greenwald, 1990, 1991a,b; Pettigrew, 1991a,b; Baker *et al.*, 1991b; Simmons *et al.*, 1991). Pettigrew and his co-workers have argued that the nervous system is apparently less subject to homoplasy—and therefore a better guide to phylogeny—than other anatomical systems, particularly the postcranial musculoskeletal system (Pettigrew *et al.*, 1989; Pettigrew, 1991a,b). Data from previous studies (of non-archontan taxa) were cited to support this idea:

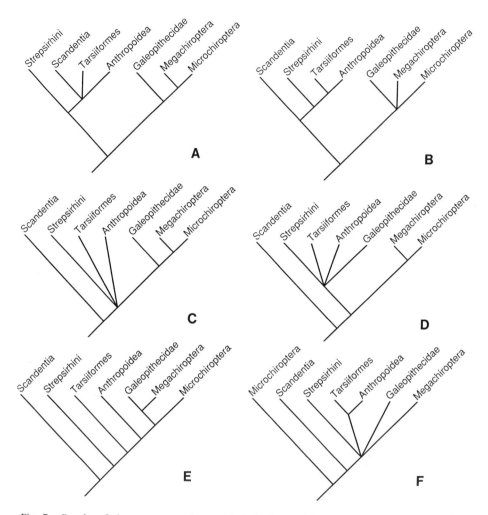

Fig. 7. Results of six separate analyses each including a different subset of characters (see Appendix 1 for contents of each data subset). Taxa known only from fossils (Plesiadapidae, Paromomyidae, and Micromomyidae) were excluded from all analyses. (A) Nonauditory cranial characters (strict consensus tree of three optimal trees of 53 steps each; CI = 0.660); (B) auditory characters (strict consensus of two optimal trees of 38 steps each; CI = 0.730); (C) characters of the anterior axial skeleton and forelimb (strict consensus of nine trees of 45 steps; CI = 0.789); (D) hindlimb characters (strict consensus of two optimal trees of 64 steps each; CI = 0.677); (E) characters of the reproductive tract and fetal membranes (one optimal tree, 28 steps; CI = 0.679); and (F) neural characters (strict consensus of four optimal trees of 38 steps each; CI = 0.895).

> The genetic programs for brain development are highly constrained across all Animalia. . . . Functional convergence is also relatively easy to spot in brain data because of the diversity of distinguishable wiring diagrams that can produce the same end result. [Pettigrew, 1991a, p. 201]

Because some cases of homoplasy are easy to identify in the nervous system (i.e., when structures fail an *a priori* test of homology because of lack of morphological similarity), we are led to believe that all cases of homoplasy in the nervous system should be detectable based on morphological differences. But why? Presumably the same developmental process (e.g., canalization, paedomorphosis) that may produce homoplastic yet identical structures in other systems may also affect the nervous system. Northcutt (1984) identified several cases of homoplasy in the vertebrate central nervous system by mapping character state distributions on cladograms derived from other types of data; these instances of convergence were not detected by simple observation of the neural structures in question. It seems clear that the only way to identify cryptic cases of convergence and reversal—where morphologies, not just functions, appear extremely similar—is to conduct a phylogenetic analysis. When derived conditions are being considered, detailed similarities that appear in closely related taxa are probably homologous; similarities that occur in distantly related taxa are probably the result of homoplasy. These methods apply to neural characters as well as all other types of morphological characters.

As argued above, *hypotheses of convergence and reversal depend on prior hypotheses of relationship.* In this context, phylogenetic studies of a variety of groups have suggested that characters that show little homoplasy in some clades may exhibit high levels of homoplasy in other clades. For example, viviparity appears to have evolved only once in mammals, but many times in reptiles (Tinkle and Gibbons, 1977). Even if a character system were a completely reliable indicator of phylogeny at one level or in one clade, the possibility of homoplasy at other levels or in other clades cannot be ruled out *a priori.* Accordingly, it is impossible to determine which character systems preserve the most information about relationships within a group (i.e., contain the least "noise") before conducting a phylogenetic analysis of that group. In the case of archontan mammals, consideration of all of the relevant morphological data is clearly warranted.

Kluge (1989) described two ways in which different subsets of data can be compared and combined to produce a single phylogenetic hypothesis; these techniques employ principles that he termed *taxonomic congruence* and *character congruence.* Using the principle of taxonomic congruence, different character data are used to produce "fundamental cladograms" (e.g., the separate cladograms shown in Fig. 7). These cladograms are then compared using consensus methods, and a consensus cladogram is constructed (see Swofford, 1991, for a review of consensus methods). In the case of the archontan taxa considered here, taxonomic congruence among the six cladograms is zero; a consensus cladogram would be completely unresolved. The alternative method, using character congruence as the guiding principle, is more productive. Under this method, all of the data subsets are added together to form a larger data set (e.g., the summary data set used in this chapter), and a cladogram is produced based on analysis of these combined data (e.g., the cladogram in Fig. 6). This cladogram can be interpreted as representing the most robust phylogenetic hypothesis possible (i.e., the most likely to withstand future testing) because it maximizes character congruence across all systems. However, a key assumption is that all of the characters used to construct this cladogram are independent.

Character Independence

Character independence is an assumption central to all phylogenetic studies. Kluge (1989, p. 11) states that:

> The assumption of independence of evidence exists in all empirical sciences. Ideally in phylogenetic systematics, each synapomorphy in the character matrix is assumed to count as a separate piece of evidence, viz., has the potential to confirm, or disconfirm, taxonomic relationships independent of all other synapomorphies considered. . . . Obviously in the context of phylogenetic systematics, character independence doesn't mean that the characters will be uncorrelated, because independent characters may be correlated due to common descent. The term separate seems to imply that two or more novelties cannot have been the result of the same biological process or mechanism. Unambiguous dependent change occurs instantaneously, the responsible mechanisms may include (1) genetic factors, such as pleiotropy, gene associations (linkage), and polyploidy, (2) ontogenetic factors, such as allometry and paedomorphosis, and (3) compensatory functional changes. . . .

While it is fairly simple to describe the sorts of factors that may result in dependent characters, it is far more difficult to distinguish independent from dependent characters in practice. For most morphological characters, consideration of possible linkage at the genetic level is next to impossible, because we do not know which genes and epigenetic phenomena affect which morphological traits. Potential cases of ontogenetic and/or functional dependence may be somewhat easier to identify. However, detection of both sorts of dependence requires a clear understanding of the ontogeny and functions of the characters in question, something that is rarely realized.

Shaffer *et al.* (1991) pointed out that conflicts involving well-supported (yet apparently mutually exclusive) hypotheses of relationship often constitute *prima facie* evidence that many of the characters are either genetically or selectively correlated. This observation has serious implications for the study of archontan phylogeny, because it suggests that many of the characters previously employed in phylogenetic analyses (e.g., the characters in Appendix 1) may not be independent. Because neural characters and characters of the postcranial musculoskeletal system strongly support conflicting hypotheses (bat diphyly and bat monophyly, respectively), these might be considered the most suspect. However, all characters should be examined as closely as possible before they are included in future phylogenetic analyses.

The easiest practical method of identifying dependent characters is to consider the distributions of states of characters that appear at the onset to share some developmental or functional relationship. If two or more such characters have identical character state distributions, then they may be considered to be dependent, and a new composite character should be described to replace them. An example are Characters 88–90 in the summary data set (Appendix I). Character 88 describes the shape of the acetabulum: circular (0) or elliptical (1). Character 89 involves the buttressing around the acetabulum: evenly developed around circumference (0) or heavily emphasized at cranial end of acetabulum (1). Character 90 describes the position of the fovea capitis femoris, the pit on the head of the femur that receives the ligamentum capitis femoris: fovea located centrally (0) or located well posterior to midline (1). All three of these characters involve related components of the hip joint, and there are obvious reasons to suspect that they may not be independent. Changes in shape of the acetabulum could easily affect the buttressing, which in fact forms the edge of the socket. Changes in the acetabulum could also affect the relative position of the fovea capitis femoris, because changes in the shape of the acetabulum would likely be accompanied by changes in the head of the femur (and changes in acetabular shape and buttressing could additionally affect the origin and course of the ligamentum capitis femoris).

Despite the fact that these characters may be dependent, we cannot assume so *a priori*. If the primitive and derived states of these characters are not correlated in all taxa (e.g., an elliptical acetabulum is sometimes found in the absence of cranially emphasized buttressing and a posteriorly located fovea), then an *a priori* assumption of independence would be justified. However, a survey of the data indicates that every taxon with the derived condition of Character 90 also has the derived condition of the other two characters, and every taxon exhibiting the primitive condition of Character 90 also has the primitive condition of the other characters. In the absence of any evidence to the contrary, these identical character state distributions indicate that the features in question may be best considered as aspects of a single transformation involving restructuring of the hip.

Beard (1993), who originally described the hip joint characters discussed above, was well aware that they may not be independent. He noted that these characters appear to be functionally correlated, and further pointed out that the derived conditions of each have apparently evolved convergently in other mammals, including *Petaurus* and *Petauroides* (Beard, 1993). Because this suite of characters appears both functionally and evolutionarily correlated, Beard (1993) indicated that he considered the phylogenetic value of each of the hip joint characters to be relatively weak in comparison with most of the other characters used in the analysis. Nevertheless, all three of these characters were included in the PAUP analysis that concluded Beard's study. This amounts to counting what may be one evolutionary change (reorganization of the hip joint) three times in the cladistic analysis. Particularly in the case of studies employing numerical parsimony techniques (e.g., PAUP), inclusion of characters that appear to be dependent may effectively bias an analysis in favor of the relationships implied by these characters. This bias can be investigated only by running a series of analyses in which potentially dependent characters are treated in different ways (i.e., left as separate characters or combined into a single character). Because Beard (1993) does not describe results of any such tests, we cannot be sure whether his phylogenetic results were significantly affected by the decision to treat the three hip features as separates characters.

Hierarchical Characters

A special type of character dependence involves what are best termed *hierarchical characters*. A hierarchical series of characters is a set of characters that are partially dependent on one another in a hierarchical fashion (e.g., Character C is partially dependent on Character B, which in turn is partially dependent on Character A). The nature of this hierarchical dependence can be construed as a series of prerequisites: in order to have the derived condition of Character C, you must first have the derived condition of Character B; and in order to have the derived condition of Character B, you must first have the derived condition of Character A. What makes these characters somewhat confusing is that the dependence is only one way (Character B is dependent on Character A, but not vice versa), and that the dependence is not complete (knowledge of the distribution of states of Character A allows inference of the condition of Character B only under certain circumstances).

An example of a hierarchical character series is the subset comprising Character 138, 139, and 141 in the summary data set. Character 138 involves presence/absence of laminar differentiation of the lateral geniculate nucleus (LGN) in the visual system: absent (0) or present (1); Character 139 refers to differentiation of magnocellular layers in the LGN: not differentiated (0) or differentiated (1), and Character 141 describes the position of the magnocellular layers relative to the optic tract: not adjacent (0) or adjacent (1). Pettigrew *et al.* (1989), the original authors of these character descriptions, scored each of these characters as if they were completely independent; that is, each OTU was scored as exhibiting either state 0 or state 1 for each character (see Appendix 2). But this scoring method overlooks the hierarchical relationship among these characters. If an organism exhibits the primitive condition for Character 138 (absence of laminar differentiation in the LGN), then it cannot have the derived conditions of the other two characters, both of which require the presence of differentiated laminae (layers). The position of the magnocellular layers relative to the optic tract (Character 141) depends on existence of differentiated magnocellular layers (derived state of Character 139), which in turn depends on existence of lamination in the LGN (derived state of Character 138). Yet these three characters are not completely linked. An organism that has a laminated LGN may or may not have differentiated magnocellular layers; there is no way to predict the state of Character 139 if an organism exhibits the *derived* condition of Character 138. Similarly, an organism with differentiated magnocellular layers (derived state of Character 139) may or may not have magnocellular layers adjacent to the optic tract (Character 141); there is no way to predict the state of Character 141 if an organism exhibits the derived condition of Character 139.

A common approach to hierarchical characters is to lump them together into a single multistate character, and then to specify that change must occur in a particular direction (see discussion of how this could be done for Characters 138–140 below under "Ordering Transformations"). Unfortunately, this process may not always be appropriate because potential drawbacks become more severe as the number of hierarchical levels increases. In the case of the LGN example discussed above, as many as 10 characters (with 27 character states) appear to be hierarchically related (i.e., Characters 137–146 in Appendix 1). Coding these features as a single multistate character is impractical because several LGN cell layers are involved in this character complex (parvocellular and koniocellular in addition to magnocellular layers), and changes in one layer cannot be assumed to be correlated with changes in others. In such instances it may be difficult or impossible to code the observational data so as to simultaneously preserve information about homology (e.g., of various cell layers) and limit the number of assumptions made about the direction of evolution.

Another way to deal with hierarchical characters—without introducing unnecessary assumptions—is to keep the characters separate, but to compensate for their hierarchical dependence during the process of scoring OTUs for analysis. This process involves recognition that some characters simply cannot be validly scored for all of the taxa under consideration (see discussion under "Dealing with Missing Data" below). Modern computer parsimony packages (e.g., PAUP) allow matrix cells to be effectively left empty; "?," "9," or "N" are variously used as place-keepers to indicate absence of a distinct character state ("0," "1," etc.) in a given cell (see discussion below under "Dealing with Missing Data"). When a series of hierarchical characters are scored, character dependence can be indicated by omitting data in clearly dependent (and thus redundant) cases of character state distribution. For example, Megachiroptera has well-differentiated magnocellular layers located adjacent to the optic tract (Pettigrew *et al.*, 1989). Because Megachiroptera exhibits the derived condition of Character 138 (lamination of the LGN), it can be validly scored for the next level character in the hierarchy (Character 139); because it exhibits the derived condition of Character 139 (presence of a differentiated magnocellular layer), it can be validly scored for the last character in the hierarchy (Character 141, relative position of the magnocellular layers). By following this hierarchy in this way, we can verify that all three characters can be scored in Megachiroptera. But the situation is different in Microchiroptera. Microchiroptera exhibits the primitive condition for the first character in the hierarchy (Character 138, lack of lamination in the LGN). Because there are no laminae in the LGN, there can be no magnocellular layers (Character 139), and we cannot logically discuss the relative position of magnocellular layers relative to the optic tract (Character 141). Accordingly, because of character dependence, Microchiroptera can validly only be scored for the first character in the hierarchy (Character 138). In practice, this can be indicated by scoring Microchiroptera "N" (no valid data) for Characters 139 and 141 in the data matrix.

The significance of treating hierarchical characters in this fashion can be appreciated by comparing the different scoring methods from the perspective of a parsimony analysis. If all three of the characters discussed above are treated as independent (*sensu* Pettigrew *et al.*, 1989), Megachiroptera would be scored as having the derived state of each character, and Microchiroptera the primitive condition of each. This would result in a record of three differences between the bat suborders, differences that would be counted as three evolutionary "steps" (transformations) in a parsimony analysis. Under the hierarchical scoring scheme, however, only one difference is noted between Megachiroptera and Microchiroptera: Megachiroptera has a laminated LGN (derived condition of Character 138), while Microchiroptera does not (primitive condition). Details of morphology within a laminated LGN (Characters 139 and 141, which are relevant for Megachiroptera and many other mammals) are scored for Megachiroptera, but not for Microchiroptera, to which they do not properly apply. The difference between one step and three steps in an evolutionary tree may not sound like much, but small numbers of transformations can have a significant effect in the outcome of a parsimony analysis. The impact of different coding schemes should not be underestimated.

Dealing with "Missing Data"

The issue of "missing data" (when a particular taxon cannot be scored for a particular character) arises in a variety of different circumstances in phylogenetic analyses. As discussed earlier, missing data may influence the choice of characters for an analysis (e.g., when a worker excludes soft structure characters from an analysis that includes fossils), and can also affect the outcome of an analysis by affecting resolution, interpretation of hierarchical characters, and so forth. Because these factors need to be taken into account when it is time to interpret the results of a phylogenetic analysis, it is often useful to differentiate among different types of missing data (Rowe, 1988; Donoghue *et al.*, 1989; Platnick *et al.*, 1991b).

A common kind of missing data results from lack of preservation. Fossil taxa are often known from incomplete specimens; when a particular element (e.g., the femur) has not yet been found, we cannot score characters involving that element. In the case of an extinct terrestrial mammal, we can hope that a femur might someday be recovered and the corresponding data matrix cells filled. Soft structures (such as the uterus) may not ever be preserved, but we can assume that features like Character 120 (degree of fusion of the uterus) could be scored if we had a live specimen (this assumption is based on the fact that all known female mammals have uteri). In other words, a character may be relevant for a particular taxon even if we can never score that character because of the incompleteness of the fossil evidence. A similar type of missing data occurs when a particular character has not yet been studied in a group.

This problem generally affects only characters that involve complex laboratory work, such as details of the nervous system and fetal membranes (e.g., Characters 125–154 in Appendix 1). Missing data resulting from nonpreservation and lack of study are similar because we have the expectation that, for the taxon and character in question, some definite character state *should* fill the matrix cell, even if we do not know what it is. Cases such as this may be scored "?" in a matrix, to indicate "character valid, but state unknown" (Rowe, 1988; Donoghue *et al.*, 1989).

As discussed earlier, matrix cells may be left empty in an analysis because some characters are not relevant for all OTUs in an analysis. Hierarchical characters provide one example, but not all cases involve direct conflict with other characters explicitly included in an analysis. Occasionally a feature in one or more OTUs has been so modified by evolution that it is simply not possible to score it within a logical sequence of character states. An example involves the muscles associated with the patagium (flight membranes) in dermopterans and bats. One of these muscles, m. occipitopollicalis (Chapter 84), is present on the leading edge of the propatagium in Megachiroptera and Microchiroptera, but a homologue is apparently absent in Dermoptera (Wible and Novacek, 1988; Thewissen and Babcock, 1991, this volume). Scoring this character in these volant mammals appears to be a relatively simple matter [muscle absent (0) or present on leading edge of propatagium (1)], but how should this character be scored in nonvolant forms? Given the structure and position of m. occipitopollicalis, it seems unlikely that it could exist in the absence of a patagium. "Presence of a patagium" is not a character in the summary data set, so this might not be considered a case of character hierarchy; nevertheless, the same philosophy should apply. If a patagium is a prerequisite for having an occipitopollicalis muscle, it is not valid to score nonvolant mammals (which lack a patagium) for Character 84. Nonvolant mammals should be scored as having "missing data" in this matrix cell. This is not the same type of missing data as described above for fossils; in this case, missing data results from intrinsic factors (evolutionary divergence), not extrinsic factors (nonpreservation, lack of study). Cases such as this may be scored "N" in a matrix, to indicate "character not applicable (invalid) in this OTU" (Rowe, 1988; Donoghue *et al.*, 1989).

The distinction between nonapplicable (N) and unknown (?) character states is made principally to aid in communication and interpretation of results. The computer programs currently used to build trees by numerical parsimony methods (e.g., PAUP, Hennig86) do not distinguish between different types of missing data. PAUP constructs branching diagrams using only "positive" data [e.g., "0," "1," "2"; character states representing distant primitive or derived morphologies (Swofford, 1989)]. With PAUP, missing data ("N" and "?," often both coded as "9" in data matrices) play a role in tree construction insofar as they generally increase uncertainty and decrease resolution in an analysis (see Platnick *et al.*, 1991b, for a discussion of these and other possible effects).

Before the advent of the most recent generation of computer programs, cases of taxonomic polymorphism were often dealt with by scoring a taxon "?" to indicate that the primitive state for that OTU was unknown. As discussed above under "Within-Group Variation," this is no longer necessary, and it is clearly not advisable. Because missing data are ignored by the algorithm during the process of tree building, one may effectively throw information away when taxonomic polymorphisms are coded as missing data. Even more significant may be the fact that scoring polymorphic characters as missing data may introduce error into the calculation of tree lengths because some character change is hidden within terminal branches (Nixon and Davis, 1991). Returning to Character 131 (type of chorioallantoic placenta) as an example, one might be tempted to score primates as "?" for this character because we cannot tell which condition [epitheliochorial (0) or hemochorial (2)] is primitive for primates (Fig. 4). However, this method overlooks the important fact an endotheliochorial placenta (1) is absent in primates. It is clearly more informative to score Primates "0,2" rather than "?" because this coding provides positive data for the algorithm to consider. If primates are scored as "?," the computer will assume that all character states ("0," "1," "2," etc.) have an equal probability of representing the primitive condition for primates.

Nixon and Davis (1991) recently demonstrated that, in some cases, the trees found when polymorphisms are scored as missing data may not include any of the most parsimonious trees found when taxa are broken up into monomorphic subunits and scored separately. While such taxonomic splitting is not always practical in higher-level systematic studies, this observation clearly argues against the use of missing data to describe polymorphic characters.

Quantitative Characters

Many of the morphological characters that have been brought to bear on archontan relationships are qualitative characters with discrete states [e.g., Character 7: palatal fenestrae absent (0) or present (1)]. Other characters represent continuous quantitative measurements (e.g., Character 141, ratio of ipsilateral to contralateral input to LGN), but such characters make up only 3% of the summary data set (Appendix 1). The majority of characters employed in studies of archontan relationships appear to fall somewhere in between. These characters are described using nonnumerical terms (e.g., "large," "narrow") and discrete character states, but the patterns described may represent continuously varying features. An example is Character 1, which describes the proportions of the premaxilla [small, restricted to anterior portion of snout (0), or large, nearly one-half surface area of snout (1), or very small (2)]. While this character is described in terms of discrete character states, it is likely that it represents a continuous character. Novacek (1985), the original author of this character, did not discuss intermediates between the states (or the absence of such intermediates), so the magnitude of differences between the states is not clear. This is a common problem that makes it difficult to interpret the published results of phylogenetic analyses.

Much has been written in recent years concerning the difficulties of coding continuous quantitative data for phylogenetic analysis (e.g., Mickevich and Johnson, 1976; Simon, 1983; Archie, 1985; Felsenstein, 1988a; Chappill, 1989; Stevens, 1991). These discussions seem to have little effect on studies of higher-level relationships of mammalian taxa, where workers continue to describe potentially continuous variation in poorly defined, semiqualitative terms. Most studies seem to employ a form of "gap-coding," in which apparent gaps in a continuum are used to distinguish between character states (e.g., the states "big" and "small" may be recognized if no animal in the study exhibits a structure that appears "medium" in size). While this method may seem straightforward, the criteria used to define these different states are rarely explicitly described. This makes it very difficult for subsequent workers to interpret character states without going back to specimens, which are rarely listed in phylogenetic publications and may be inaccessible (especially in the case of fossils). An even more difficult situation occurs when the character states themselves are not defined, but simply listed. Pettigrew *et al.* (1989) are the original authors of Character 143, ratio of ipsilateral to contralateral input to LGN. Of this character they write only "We have scored this character on a scale from 0 (less than 10% ipsilateral input) to 6 (50% ipsilateral input)" (p. 501). This is an inadequate description for a continuous quantitative character: not only do we not know what coding technique was used (e.g., gap-coding, range-coding), but we also do not know the boundaries of the character states recognized by these authors. Even if comparable data were available for an additional taxon (e.g., "35% ipsilateral input"), it would not be possible to conduct a new analysis including these data because we do not know which character state ("2," "3," "4," or "5") would appropriately describe the data in the framework developed by Pettigrew *et al.* (1989).

There are no easy answers to the problems encountered when trying to describe continuous morphological variation in discrete terms. Workers scoring quantitative characters for cladistic analysis should specify the coding technique used [e.g., "simple gap-coding" *sensu* Mickevich and Johnson (1976), "generalized gap-coding" *sensu* Archie (1985), "segment-coding" *sensu* Chappill (1989), "generalized range-coding" *sensu* Simon (1983)]. Presence or absence of conditions intermediate between recognized states should be discussed, as should cases of polymorphism. The most useful character descriptions are generally those that are most explicit. If a future researcher does not agree with the coding techniques or character state definitions one has employed, explicit character descriptions will at least communicate the nature of the original observational data. The primacy of data should never be forgotten; the enduring value of a cladistic study often lies more in the data reported than in the trees constructed from those data. Stevens (1991, pp. 576–577) recently argued that

The justification of character states should be far more extensively documented than at present. This recommendation entails a change in both reviewing and editorial practice. Because the basic data of phylogenetic analyses are character states, more space should be devoted to the justification of the delimination of these states. The reader is all too often simply presented with lists of character states, or there may be brief mention of the extent of the variation of a particular character. . . . In some cases the requisite documentation of the existence (or otherwise) of character states is quite simple. . . . In less obviously quantitative characters, attention should first be paid to analyzing the basic variation, only later seeing how that variation pattern can be expressed using conventional descriptive terms. We should not assume that these terms represent discrete variation in the real world.

Weighting Characters

Character weighting is one of the most controversial aspects of cladistic studies because of the clear potential for introducing unacceptable biases into an analysis. Many workers utilize what are called "unweighted characters" because of a professed desire to eliminate these biases, but no analysis can be considered truly unbiased because weights must always be applied at some level.

The first level of weighting is perhaps the most simple: characters that are included in an analysis are always given more weight than characters that are not included in the analysis. Thus the analysis of archontan relationships by Pettigrew *et al.* (1989) is highly weighted toward neural characters, just as Beard's (1993) analysis is heavily weighted toward postcranial features. Before any thought is given to weighting some characters more than others within an analysis, initial choice of characters reflects a judgment concerning the relative usefulness of various features.

Once characters and character states have been defined, one next must decide how to weight these constructs. The most common weighting scheme is to weight each *transformation* equally (a transformation is a change from one character state to another, e.g., $0 \rightarrow 1$). Because multistate characters involve more transformations than binary characters, the end result is that some characters may contribute more information (have more weight in determining the outcome of an analysis) than other characters. Because weights are applied at the level of transformation rather than character, workers often refer to characters as being "unweighted" in such an analysis.

The principal justification for weighting all transformations equally is that this method limits assumptions concerning the relative importance of different changes; a transformation from state "0" to state "1" in a character from the nervous system is of equal importance as a change from state "1" to state "2" in a character related to fetal membranes. This scheme is clearly appropriate when one does not wish to bias an analysis by making *a priori* decisions about the relative importance of different morphological changes.

For this reason, all of the analyses described thus far in this chapter (Figs. 2–7) were run with transformations weighted equally.

When equal weighting is applied to each transformation, the number of characters becomes unimportant. Transformations, which contain the real information, may be packaged into any number of characters without altering the outcome of the analysis. For example, changes in the shape of the snout might be described in terms of several (e.g., three) binary characters or one multistate character (e.g., with four discrete states); each scheme involves the same number of transformations (three). If all taxa can be scored for all characters, and each transformation is given equal weight, these two coding schemes would produce the same result upon analysis. However, a worker may believe that the individual characters are more important than the number of transformations, particularly when continuously varying features have been described using variables and arbitrarily defined numbers of states. In this case, one can choose to weight each character equally. It must be recognized, however, that weighting in this fashion may bias an analysis because it places undue emphasis on how one defines "a character."

Weighting characters equally has the effect of minimizing the importance of individual transformations in characters that have many states, while maximizing the importance of changes in binary characters. The difference between the results of an "unweighted" analysis (in which transformations are weighted equally) and the results of an analysis in which characters are weighted equally may therefore depend largely on the relative proportions of binary versus multistate characters. If there are numerous multistate characters in an analysis, one may find that these two weighting schemes result in completely different most parsimonious trees.

The summary data set (Appendix 1) contains 112 binary characters, 38 multistate characters with three states (i.e., two transformations), and 4 characters that have four or more states. If each *transformation* is weighted equally, five equally parsimonious trees result from a branch and bound PAUP analysis (Fig. 2). If each *character* is weighted equally, only one most parsimonious tree is found (Fig. 8), a tree that is different from any of those found earlier.

A third possible weighting scheme involves weighting each character system equally, regardless of the number of characters and transformations. This procedure has been discussed principally in the context of combined molecular and morphological studies, where information from numerous molecular characters may be expected to overwhelm that from less abundant morphological data (Miyamoto, 1985; Kluge, 1989). Kluge (1989) suggested that weighting data subsets in this manner may be warranted especially when character incompatibility is high and cladograms resulting from separate analyses of the data subsets have markedly different topologies (e.g., Fig. 7). Weighting subsets of morphological data in this way makes some sense, because different anatomical systems are often considered to represent independently evolving entities. One problem with this technique, however, is that any data set can be carved up into the anatomical systems in many different

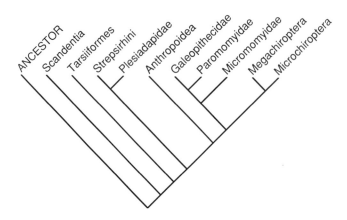

Fig. 8.   The single most parsimonious tree identified in an analysis of the summary data set (Appendix 2) in which each *character* was weighted equally (see text for discussion). This tree requires 253,164 weighted steps (CI = 0.602). Compare this tree with Fig. 2, which illustrates the five optimal trees that were identified in a similar analysis in which transformations (rather than characters) were equally weighted.

ways. For example, one might choose to give equal weight to each of the six anatomical systems previously defined (e.g., cranial features excluding auditory region, auditory region, axial skeleton and forelimb, hindlimb, reproductive tract and fetal membranes, and nervous system). Another worker might prefer to lump all of the cranial features together, and/or all of the postcranial features. Osteological characters might be separated from muscular characters; vascular characters might be considered as representing yet another system. When a system-based weighting scheme is used, how one defines an "anatomical system" may be the single most important factor governing the outcome of the analysis.

If weights are applied to the summary data set such that each of the six anatomical systems is weighted equally (with each transformation within a given system having equal weight), one parsimonious tree results from a branch and bound analysis. This tree is identical in topology to that shown in Fig. 2E. If only four systems are defined (cranial, postcranial, reproductive, and neural), two equally parsimonious trees result; these have the same topology as the trees in Fig. 2D and 2E.

Farris (1969) introduced a different kind of weighting procedure that effectively gives relatively higher weight to characters that are congruent with many other characters, while giving lower weights to characters that show lower levels of congruence. *Successive approximations* weighting is an iterative procedure that begins with generation of an initial tree (or trees) from an unweighted data set (i.e., one in which transformations are equally weighted). For each character, the consistency index (CI) is calculated, and this index (or the average of indices from several trees) is used to weight characters in a second analysis. The consistency index is a measure of homoplasy that is calculated by dividing the minimum number of transformations possible in a character (1 for a binary character, 2 for a character with three states, etc.) by the number of transformations required by a particular tree topology (see discussion below under "Measuring Homoplasy"). By using these values as character weights, those characters that exhibit a high level of homoplasy in an initial analysis are given relatively low weights, while characters with little apparent homoplasy are given higher weights. In the successive approximation weighting process, consistency indices from a second analysis are then used as weights in a third analysis, and so on until some specified stopping point is reached (usually when the tree stabilizes or only one tree is found).

Successive approximations weighting is often used by researchers who wish to extract a single parsimonious tree from a relatively noisy data set (e.g., when an "unweighted" analysis of a data set produces a large number of equally parsimonious trees). However, successive approximations weighting may be used regardless of the number of trees discovered in the initial unweighted analysis, and there is no guarantee that the weighting process will produce fewer trees than the original analysis. Platnick *et al.* (1991a) suggested that some data sets that produce a single most parsimonious tree when "unweighted" may produce several trees when successive approximations weighting is applied. As described above, an analysis of the summary data set with transformations weighted equally (characters "unweighted") resulted in five optimal trees (Fig. 2). Application of successive approximations weighting produced a single tree after one repetition; this tree was identical to that shown in Fig. 2A.

Sometimes the successive weighting procedure may produce trees that are not among the optimal trees identified in the initial "unweighted" analysis. An example is the study of haplogyne spider phylogeny presented by Platnick *et al.* (1991a). Platnick and colleagues recovered ten optimal trees of 184 steps each in an unweighted analysis; successive weighting resulted in the discovery of six different trees of 568 "weighted" steps each. When the character weights were returned to their original values, four of the six trees were found to require 187 steps; the other two required 185 steps. Of the six trees, Platnick *et al.* (1991a) chose one "preferred cladogram" that was among the shortest when measured in terms of both the weighted and unweighted data (e.g., required 185 steps) and was the best resolved (with only one trichotomy).

While the process described above permits identification of a single "preferred" tree, the large number of assumptions built into the process raise troubling questions concerning the value of this method. Several theoretical objections have been raised to successive approximations weighting, principally because it is an *a posteriori* method subject to circularity (Neff, 1986; Cannatella and de Queiroz, 1989; Swofford and Olsen, 1990). Cannatella and de Queiroz (1989, pp. 58–59) commented that

> Although this is a convenient way of reducing the number of equally parsimonious topologies resulting from an unweighted analysis, it obscures alternative hypotheses of relationships that result from incongruent characters. Furthermore, while we do

> not reject out-of-hand the use of character weighting, doing so on the basis of the fit of the characters to a tree derived from these same characters can be criticized on the grounds that this practice reduces the independence of characters.

Successive approximations weighting assumes that the overall homoplasy level of a character is an appropriate measure of the phylogenetic information contained in that character. This is an overly simplistic assumption, because some characters that exhibit a great deal of homoplasy in one part of a tree may exhibit no homoplasy in other parts of the tree (e.g., viviparity in amniotes). Successive weighting may thus downplay the value of characters that contain crucial phylogenetic information. It is not clear how this procedure will necessarily improve the accuracy of a phylogenetic analysis.

An additional problem that may be encountered with successive approximations weighting involves missing data. The presence of missing data in a matrix can bias the weighting process because missing data often masks the presence of homoplasy. Characters scored with high levels of missing data may be given greater weights than characters with no missing data, thus skewing the results of the successive weighting process (M. J. Novacek, personal communication). This artificial effect may present a major problem in analyses including fossil taxa.

It is clear that different weighting schemes involve a wide variety of different assumptions, and that they frequently result in construction of trees with very different topologies. "Unweighted" analyses (transformations weighted equally) apparently require the fewest *a priori* assumptions about how evolution works; accordingly, this method is preferred by most workers. If another weighting scheme is to be used, the most appropriate technique is to run both weighted and unweighted analyses and to compare the trees that result from each. Publication of only one weighted tree (or set of trees) is frustrating for readers who do not agree with the weighting scheme that was used. In the case of the summary data set, comparison of the diverse trees described above (illustrated in Fig. 2 and 8) reveals several relationships that remained constant no matter what weighting scheme was applied: bats always appear to form a monophyletic group; living dermopterans (Galeopithecidae) are more closely related to bats than to living primates; and tree shrews are the most distal branch within Archonta. The fact that these relationships remained stable under different weighting schemes can be interpreted as evidence of a particularly strong "phylogenetic signal" within the data.

Ordering Transformations

Prior to a parsimony analysis, it is necessary to decide how to count changes within multistate characters. This is, in effect, a kind of weighting process. In "unordered" characters all changes are treated as nonadditive; that is, a change from one state to any other state is treated as a single transformation (e.g., transformations $0 \rightarrow 1$, $1 \rightarrow 0$, $0 \rightarrow 2$, $2 \rightarrow 0$, $1 \rightarrow 2$, and $2 \rightarrow 1$ are each counted as a single evolutionary step). Alternatively, a character may be treated as an additive "ordered" character, in which passage through intermediate stages is required in some instances (e.g., $0 \rightarrow 1$ and $1 \rightarrow 2$ each count as one step, but $0 \rightarrow 2$ requires two steps because passage through state 1 is necessary to change from state 0 to state 2). It is not necessary for every character in an analysis to be treated the same way; modern phylogenetic computer programs permit a separate transformation series to be constructed for each character. When a character has multiple states, there are a large number of possible ways to configure a transformation series. Mabee (1989) described three types of ordered transformation series: linear, reticulate, and branched. Using Character 148 (proportion of the superior colliculus taken up by ipsilateral representation) as an example, Fig. 9 illustrates four possible transformation series for a five-state character. Because parts of a transformation series may be unordered and other parts ordered (e.g., Fig. 9C), the number of possible transformations series increases exponentially relative to the number of defined character states. As one would expect, the effect of using ordered rather than unordered transformation series is most keenly felt in analyses involving a large number of multistate characters.

Specifying ordered transformation series is often used as a technique for dealing with hierarchical characters. Returning to the example cited above (see "Hierarchical Characters"), Characters 138, 139, and 141 might be combined into a single multistate character with four states: LGN not laminated (0), LGN laminated, but magnocellular layer not differentiated (1), magnocellular layer present, not adjacent to optic tract (2), or magnocellular layer present, adjacent to optic tract (3). If this were treated as a linear ordered transformation series ($0 \leftrightarrow 1 \leftrightarrow 2 \leftrightarrow 3$), the homologies implicit in the original character descriptions would be preserved. However, theoretical problems intrude when we consider the possibility of reversal. For loss of the magnocellular layer to occur in a lineage that originally exhibited condition "3" (magnocellular layer adjacent to the optic tract), must the magnocellular layer first move back to a position not adjacent to the optic tract (e.g., state "2") before being lost (returned to state "0" or "1")? A complex transformation series (e.g., $0 \leftrightarrow 1 \leftrightarrow 2 \leftrightarrow 3$, plus $3 \rightarrow 1$, $3 \rightarrow 0$, $2 \rightarrow 0$) could eliminate such assumptions in this case. However, as many as ten characters may be involved in this hierarchical character system (Characters 137–146). Because these other characters involve different layers in the LGN, it would be impossible to score all of these features as a single multistate character without either sacrificing information on homology of the various layers or introducing numerous assumptions about the direction of evolutionary change. The coding scheme described in the section on "Hierarchical Characters" (using "N") is more efficient in this regard because of the ease with which it can deal with complex hierarchical character systems without requiring assumptions about the direction(s) of evolutionary change. Some authors have suggested that hypotheses of character state order are more informative than hypotheses of

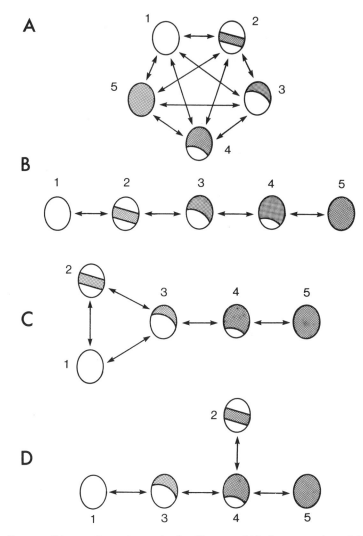

Fig. 9. Four possible transformation series for Character 148, the proportion of the superior colliculus taken up by ipsilateral representation (colliculi redrawn from Fig. 15 of Pettigrew *et al.*, 1989). The stippling represents the proportion of the dorsal surface of the right (ipsilateral) superior colliculus taken up with label transported from the right eye; anterior is to the top. Numbers refer to the character states described in Appendix 1. Each arrow indicates a potential transformation (one evolutionary "step"). (A) An unordered transformation series; (B) a linear ordered transformation series suggested by Pettigrew *et al.* (1989); (C) an alternative transformation series combining both ordered and unordered components into a reticulate pattern; (D) an ordered transformation series combining linear components to form a branching transformation series. Note that the transformation series illustrated here (A–D) represent only a few of the many alternatives possible for a five-state character.

unorder, and that use of ordered transformation series may restrict the number of equally parsimonious trees and increase phylogenetic resolution (e.g., Mabee, 1989). However, a recent study of 27 data sets by Nixon and Davis (1991) demonstrated that ordered characters can produce more, equal, or less equally parsimonious trees and can increase, decrease, or have no effect on tree resolution. As mentioned above, the summary data set contains 112 binary characters and 42 multistate characters. All of the analyses previously discussed (illustrated in Figs. 2–8) were run with all multistate characters unordered. As one might expect, applying the simplest possible ordering system to the multistate characters (linear transformation series, e.g., $0 \leftrightarrow 1 \leftrightarrow 2 \leftrightarrow 3 \leftrightarrow 4 \leftrightarrow 5 \leftrightarrow 6$) produces different phylogenetic results. An unweighted analysis using *unordered* multistate characters produced the five trees shown in Fig. 2. A similar unweighted analysis using *ordered* multistate characters resulted in identification of seven equally parsimonious trees, none of which matched the trees in Fig. 2 (a strict consensus tree constructed from the seven trees is shown in Fig. 10). Because separate transformation series (often termed *character state trees*) can be specified for every character, these is no need to use the same type of transformation series (e.g., linear) for each character. Repetition of the experiment using reticulate or branched transformation series for some characters (e.g., Character 148, shown in Fig. 9) might produce different results.

One potential problem with ordering characters is circularity. Specifying an order of transformation prior to phylogenetic analysis may be a self-fulfilling prophesy; ordering characters may influence the topology of a resulting tree in such a way that transformation series then appear to reflect the actual pattern of change within a group. The practice of ordering characters may thus introduce unacceptable biases into an analysis unless each transfor-

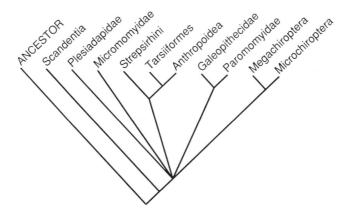

Fig. 10.   Strict consensus tree based on seven equally parsimonious optimal trees generated from an unweighted data set in which multistate characters were *ordered* rather than unordered (compare with Fig. 3A).

mation series can be justified *a priori*. Traditional hypotheses of character evolution are usually based on previous hypotheses of relationships within a group, or on perceived patterns of change in other taxa, both of which must be considered suspect in terms of their abilities to contribute real information to a phylogenetic analysis *a priori*. The most compelling evidence that may justify ordering comes from examination of ontogenetic series. Mabee (1989) discussed the use of ontogeny as a criterion for building character state trees, including an example in which the ontogeny of opercle shape in centrarchid fishes was used to unambiguously order a five-state character. However, Mabee (1989) noted that if phylogenetic change occurs through the addition or deletion of two or more states from a character ontogeny, hypotheses of character state order based on ontogenetic criteria will not reflect phylogenetic order. In other words, patterns of evolutionary change may not always mirror patterns of ontogenetic change, in which case character state trees based on ontogeny may be misleading.

Whenever *a priori* evidence supporting a particular transformation series is weak, application of ordering may prejudice an analysis through circularity. Leaving such characters unordered alleviates this problem, making it possible to draw relatively unbiased conclusions concerning patterns of transformation after the phylogenetic analysis has been completed. Some researchers run all multistate characters unordered to avoid introducing extra assumptions (and potential biases) into their analyses (e.g., Rowe, 1988).

Finding the Shortest Tree: Algorithms and Search Techniques

Once initial decisions concerning taxa, character states, weighting, and ordering of characters have been made, the next step is to run a parsimony analysis to discover the phylogenetic hypotheses best supported by the data. Two basic classes of techniques exist for searching for "optimal" (shortest) trees: exact methods and heuristic methods. Exact methods guarantee the discovery of all optimal trees; heuristic procedures are approximate methods that generally find optimal trees, but are not guaranteed to do so.

The simplest exact method is an "exhaustive search" that evaluates the length of every possible tree in order to identify all optimal trees. Because all possible trees are examined, an exhaustive search additionally allows examination of the frequency distribution of trees of different lengths. Unfortunately, exhaustive methods are limited by the number of taxa that can be practically analyzed. On most computers, exhaustive enumeration is currently feasible only for data sets including 11 or fewer taxa, which requires examination of up to 34,459,425 different tree topologies (Swofford and Olsen, 1990). Because large numbers of trees must be considered, exhaustive enumeration methods are often very time-consuming. PAUP 3.0 (running on a Macintosh IIx computer) required over 33 h to analyze the summary data set (Appendix 2), which contains 11 taxa. Five optimal trees of 318 steps (those in Fig. 2) were found by this method. The frequency distribution of trees measured in this analysis is shown in Figs. 11 and 12.

A second category of exact methods includes techniques known as "branch-and-bound" methods (Hendy and Penny, 1982; Swofford and Olsen, 1990). These techniques involve systematic evaluations of large subsets of tree topologies, allowing identification of all optimal trees without actual enumeration of all of the possible trees. Branch-and-bound methods permit exact solutions for 20 or more taxa, depending on the algorithm used, the speed of the computer, and the "messiness" of the data set (Swofford and Olsen, 1990). One of the principal advantages of branch-and-bound methods is that they are much faster than exhaustive searches; a PAUP branch-and-bound analysis of the summary data set (Appendix 2) required only 28 s to identify the five optimal trees.

In contrast to exact methods, heuristic approaches do not guarantee discovery of all most parsimonious trees. These techniques, which sacrifice the guarantee of optimality in favor of reduced computing time, are generally used only when exact methods are not practical because of large numbers of taxa, messy data, and/or limited computing facilities (Swofford and Olsen, 1990). Heuristic searches are often the only option available when large numbers of relatively incomplete fossil taxa are considered in an analysis (e.g., Simmons, 1993). Despite the absence of any guarantees, heuristic procedures

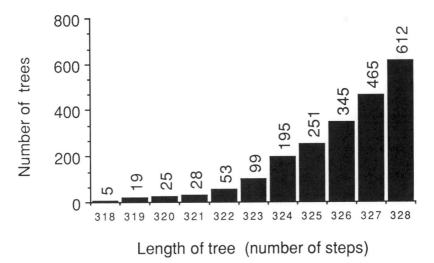

Length of tree (number of steps)

Fig. 11.   Frequency distribution of trees that are within ten steps of the length of the optimal (most parsimonious) tree length. Five optimal trees (shown in Fig. 2) were identified, but over 200 other trees exist that are within five steps of the length of the most parsimonious trees. These data were collected during an exhaustive enumeration analysis of the summary data set; Fig. 12 illustrates the complete frequency distribution for these data.

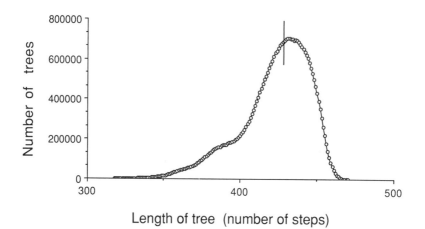

Fig. 12. Frequency distribution of trees examined during an exhaustive enumeration analysis of the summary data set (characters unordered, transformations weighted equally). All possible topologies for the 11 taxa (34,459,425 tree topologies) were examined to produce this distribution graph. The vertical line near the peak of the distribution curve indicates the mean tree length for the summary data set.

frequently yield the same results as exact methods. For example, a PAUP heuristic search of the summary data set (Appendix 2) required only 10 s to discover the same five optimal trees obtained using exact methods (Fig. 2).

PAUP (Swofford, 1989) has been used in all of the examples reported here because it is the program most often used by the author. Researchers should be aware, however, that two other widely available computer packages can also be used to identify optimal trees: Hennig86 (Farris, 1988) and PHYLIP (Felsenstein, 1990). Each of these programs has somewhat different capabilities and limitations; for example, Hennig86 is less user-friendly than PAUP but may be much faster when implementing exact methods (Sanderson, 1990). Another program, MacClade (Maddison and Maddison, 1992), is not intended as a "tree-finding" program but is very useful for inputting data, analyzing character evolution, and testing alternative phylogenetic hypotheses. MacClade and PAUP share a common file format, enabling easy transfer of information between these programs. For a detailed review of these programs, see Sanderson (1990).

Interpreting Results of a Cladistic Analysis

As discussed throughout the preceding sections, the outcome of a cladistic analysis is highly dependent on the assumptions that are built into the data set and analytical methods. Interpretation of results, in order to be valid, must take these assumptions into account. To complicate the issue even further, the interpretive process itself involves a number of assumptions and decisions concerning the levels at which various synapomorphies apply, strength of support for particular relationships, etc.

Two broad categories of information can be retrieved from a cladistic analysis: information concerning relationships (phylogeny), and information concerning patterns of character state changes (character evolution). Phylogenetic results usually come in the form of branching diagrams. If these diagrams are rooted by use of outgroups or a hypothetical ancestor, phylogenetic trees (cladograms) are produced (e.g., Fig. 2). Information concerning character evolution must be retrieved through a process called *character optimization,* in which character states are mapped onto the most parsimonious tree(s) in order to determine the branches along which various transformations occurred.

Evaluating Phylogenetic Results

The principal questions that must be asked about phylogenetic results concern reliability and significance. Reliability estimates, which address the probability of chance errors, have been discussed in detail by Felsenstein (1985, 1988b), Penny and Hendy (1986), and Sanderson (1989). The method most frequently used is the bootstrap, a powerful technique designed to provide confidence estimates for statistics with unknown sampling distributions. Bootstrapping phylogenies involves repeated resampling of the character data to produce new data matrices, which are subsequently analyzed using cladistic methods. A typical bootstrap analysis involves 100 sample matrices, each of which includes the same taxa and the same number of characters as the original data set. These matrices are created by sampling (with replacement) columns of character data from the original matrix. Each new matrix is separately analyzed to identify optimal trees, and the resulting trees are then compared to identify the groupings that occur most often. Figure 13 illustrates the results of bootstrap analyses of the summary data set. All 11 taxa were included in the first analysis (Fig. 13A). Note that only one grouping—Megachiroptera + Microchiroptera—appeared in 100% of the bootstrap trees. Many of the other groupings did not even appear in 50% of the trees. Figure 13B illustrates results of a bootstrap analysis including only the living archontan taxa. With the fossil taxa removed, Anthropoidea + Tarsiiformes + Strepsirhini appeared in 71% of the bootstrap trees, compared with appearance in only 32% of the trees in the previous analysis (i.e., Fig. 13A).

Several workers have suggested that bootstrap results may be used to place confidence limits on phylogenies (e.g., Felsenstein, 1985; Sanderson, 1989). Taken most simply, a taxonomic grouping that appears in 95% or more of the trees resulting from bootstrap replicates could be considered to be significantly supported at the 95% level (Felsenstein, 1985). However, this

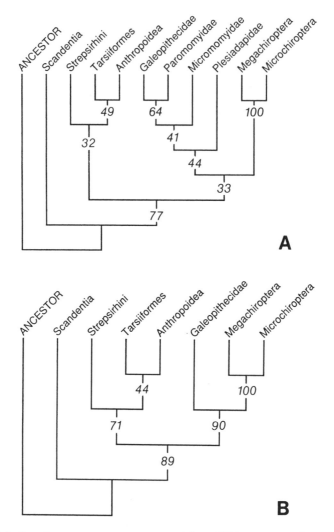

Fig. 13. Majority rule consensus trees constructed from 100 bootstrap replicate analyses. The numbers represent the percentage of bootstrap trees in which a particular clade appeared. (A) Results of an analysis including all 11 taxa in the summary data set; (B) results of an analysis in which the fossil taxa have been removed.

probabilistic interpretation requires a number of assumptions that may be difficult to support in the context of most phylogenetic analyses (see discussions in Felsenstein, 1985, Sanderson, 1989, and Swofford and Olsen, 1990). Biases may result from many factors that may or may not be avoidable. Among other things, a high character-to-taxon ratio (often unattainable) is necessary to avoid bias in bootstrap results (Sanderson, 1989). In addition, because large data sets with many taxa generally contain more character conflict than data sets with fewer taxa (Sanderson and Donoghue, 1989), boot-strap analyses of even moderately large data sets will often lead to the conclusion that little statistical support exists for the monophyletic groups identified by the analysis (Sanderson, 1989). Presence of missing data in a data matrix may also reduce the apparent support for clades that include relatively incomplete taxa. Given these problems, it may be most appropriate to view boot-strap results as a *relative* measure of the support for different groupings (when they occur together in a particular topology), rather than as an index of the statistical significance of the evidence supporting monophyly of particular clades. Bootstrap results may also be used to place constraints on the phylogenetic position of taxa by examining the relative size of the smallest monophyletic groups that contain them (see Sanderson, 1989, for a description of this technique).

The strength of the phylogenetic signal supporting the topology of the most parsimonious tree is another issue that should be considered in any phylogenetic analysis. The level of support for a particular tree topology can be estimated by examining the skewness of the frequency distribution of tree lengths (e.g., Fig. 12) generated by an exhaustive enumeration analysis (Fitch, 1979, 1984; Hillis, 1991; Huelsenbeck, 1991). Tree length distributions ("All Tree Histograms") with strong left skewness (a long "tail" on the left side of the distribution) indicate that the data contain more phylogenetic structure (characters consistent with one tree) than more symmetrical distributions. Skewness of tree length distributions may be measured using the g_1 statistic (for details of the use of this statistic see Hillis, 1991, and Huelsenbeck, 1991). By comparing tree length distributions generated from random data, Hillis (1991) calculated a series of critical values for studies including six, seven, or eight taxa. Values of g_1 generated from real tree distributions can be compared with these critical values to determine if the observed skewness is statistically significant at the 95 or 99% level. In effect, these calculations allow evaluation of the character data supporting the more parsimonious trees. If the signal produced by the data falls within the range of the critical values, one may be justified in claiming that the data strongly support those trees that fall at the end of the left tail of the distribution curve. However, this measure does not address the degree of confidence assignable to the most parsimonious tree(s) vis-à-vis those trees that are next most parsimonious.

Although bootstrap analyses and statistical evaluation of skewness may give useful estimates of the support for different groupings, one of the easiest ways to evaluate phylogenetic results is simply to look at the "next best" trees—those trees that are within only a few steps of the optimal tree length. The number, distribution, and topology of these trees can provide a perspective even when these data are not formally analyzed. For example, analysis of the complete summary data set resulted in identification of 5 optimal trees of 318 steps (i.e., Fig. 2). However, over 200 other trees were found that were within 5 steps of the optimal tree length (Fig. 11), suggesting that the optimal trees may not reflect a clear phylogenetic signal (this is a crude but effective use of skewness information). Examination of the topology of "next best" trees may be particularly enlightening in such cases. In the summary analysis, certain clades appeared in all of the 229 trees with 323 or fewer steps: living primates formed a monophyletic group relative to the other extant lineages, and bats formed a monophyletic group relative to all other taxa. These relationships thus appear to be supported despite confusion concerning other relationships (note that these results echo the conclusions suggested by the bootstrap analyses). However, care must be taken in interpreting observations concerning clade stability because both the number of characters and the number of taxa may affect the stability of groups identified in an analysis (Novacek, 1991). As a result of homoplasy, analyses including large numbers of taxa and/or a low character-to-taxation ratio may be less likely to produce stable clades than analyses including fewer taxa and/or a high character-to-taxon ratio.

Examination of trees of varying lengths is also important if one is interested in comparing two alternative hypotheses of relationship. Finding that a cladistic analysis produces an optimal tree compatible with one hypothesis is not enough to demonstrate superiority of that hypothesis; one must first examine the topology of any trees close to the optimal length. If the second hypothesis appears supported by any of these trees, the results indicate that the analysis was not capable of distinguishing between the hypotheses. Baker *et al.* (1991a) were forced to such a conclusion in their recent analysis of bat relationships using ribosomal DNA. Bat diphyly (*sensu* Pettigrew *et al.,* 1989) required a minimum of 45 steps in the context of their analysis; bat monophyly (*sensu* Wible and Novacek, 1988) required 46 steps, only one step more. The small difference in length between these two trees was interpreted as insignificant; therefore, Baker *et al.* (1991a) concluded that their ribosomal DNA cistron data were not capable of resolving the controversy over bat monophyly.

Measuring Homoplasy

Measurements of homoplasy are cited in discussions of phylogenetic results because the amount of homoplasy is often interpreted as evidence of the goodness of fit of data to a tree topology. The more homoplasy detected, the greater the amount of "noise" in the data set, and the weaker the support for a particular tree topology. To measure the amount of overall homoplasy, it is necessary to have both a data set and a specific tree topology; mapping the character transformations on the tree provides a means of estimating the number of convergence and/or reversal events.

A wide variety of indices are available for use in measuring overall homoplasy, including the "consistency index" (CI; Kluge and Farris, 1969), the "retention index" (RI; Farris, 1989), and "homoplasy excess ratios" (HER, REHER, and HERM; Archie, 1989). Unfortunately for most workers, there has been a great deal of argument concerning the relative faults and merits of these statistics (e.g., Archie, 1989, 1990; Farris, 1989, 1990, 1991; Sanderson and Donoghue, 1989) but no comprehensive review is available. The two measures most commonly used, CI and RI, are automatically calculated by most computer parsimony programs. Both indices are expressed as values between zero and one, with higher values indicating relatively less homoplasy than lower values.

The simplest measure of homoplasy is the consistency index, which is calculated as:

$$CI = \frac{\text{Minimum possible number of steps}}{\text{Observed number of steps}}$$

The "minimum possible number of steps" is determined entirely by the structure of the data matrix; it is simply the number of character states minus the number of characters. The "observed number of steps" refers to both the data and the tree topology; it is the total number of state changes (transformations) required to most parsimoniously fit all of the characters on the tree under consideration (Kluge and Farris, 1969).

Several problems exist with using CI as a direct measure of the support for a particular tree. First, CI is inflated (less overall homoplasy is detected) whenever autapomorphic characters are included in an analysis. For this reason, it is customary to use only phylogenetically informative characters (non-autapomorphies) when calculating CI. Second, CI is highly correlated with the number of taxa in analysis, with homoplasy increasing as the number of taxa increases (Archie, 1989; Sanderson and Donoghue, 1989). Accordingly, it is inappropriate to compare the CI of trees containing different numbers of taxa (e.g., the tree in Fig. 6 with those in Fig. 2) when attempting to determine which tree is best supported.

Before concluding that a particular tree is poorly supported (e.g., that there is an unusually high level of homoplasy), one must first consider the expected level of homoplasy given the number of taxa in the analysis. Sanderson and Donoghue (1989) compared the results of 60 cladistic analyses involving a wide variety of taxa and characters, and computed a formula that provides an estimate of the expected CI for analyses including 3 to 60 taxa:

$$CI = 0.90 - 0.022(\text{number of taxa}) + 0.000213(\text{number of taxa})^2$$

For 11 taxa (the number of taxa found in the trees in Fig. 2), the expected CI is 0.674; the observed CI was 0.626, only slightly less than the expected value. Even though these trees exhibit a CI well below 1.000, this does not appear to represent an unusually large amount of homoplasy given the number of taxa included in the analysis.

Sanderson and Donoghue (1989) found no significant relationship between CI and the number of characters used in an analysis, or between CI and the taxonomic rank of terminal taxa (OTUs). However, there is a clear relationship between CI and the amount of missing data in a matrix (Sanderson and Donoghue, 1989). Because missing data (specifically missing data resulting from unknown information, coded "?") may mask the presence of homoplasy, higher CI values should always be expected when there are large numbers of matrix cells scored as "?" in a matrix. Taking this into account, the CI calculated for the trees in Fig. 2 should be considered to represent only an upper bound on the homoplasy implied by these topologies.

A more complex measure of homoplasy is the retention index of Farris (1989). This index is sensitive not only to the total homoplasy (the number of times that character states have evolved more than once), but also to the potential for homoplasy given the structure of the data matrix. Some characters have more potential for homoplasy than others in any given data set. A character that exhibits a derived condition in two taxa clearly may have evolved twice. However, a derived state seen in five taxa has a greater potential for homoplasy—it may have evolved as many as five times. The retention index is calculated as:

$$\text{RI} = \frac{\text{Maximum possible number of steps} - \text{Observed number of steps}}{\text{Maximum possible number of steps} - \text{Minimum possible number of steps}}$$

The "maximum number of steps" is defined as the greatest number of steps a character may require on any tree (Farris, 1989). By taking the potential for homoplasy into account, the RI may offer a more comprehensive evaluation of the level of homoplasy than the CI. However, there seems to be little consensus concerning the properties of RI (see Archie, 1990; Farris, 1991), making interpretation of this index difficult at present.

Both consistency and retention indices may be applied to individual characters as well as to complete data sets (Kluge and Farris, 1969; Farris, 1989). These measurements are sometimes used to distinguish characters that may be particularly informative (in terms of their ability to indicate phylogenetic relationships) from other characters that may exhibit too much homoplasy to be useful indicators of relationship. Monophyletic groups diagnosed by characters with low homoplasy levels (high CI or RI) are often considered to be more strongly supported than groups diagnosed by an equal number of more homoplastic characters (with lower CI or RI).

Character Optimization

The evaluation of character data in the context of an optimal tree is a separate process from the use of that data to construct the tree. Optimization of characters (the parsimonious arrangement of transformations on a tree) is done only after a cladistic analysis has been completed and the optimal tree topology has been identified (although see discussion of default options below). It is only at this point that hypotheses of homoplasy may be generated (Farris, 1983; Sanderson and Donoghue, 1989). For any given tree topology there are generally many ways to arrange transformations so that they add up to the optimal length. The number of possible arrangements is dependent on the amount of homoplasy and missing data present.

Two types of transformations may be recognized during an optimization: "unequivocal" transformations, which have only one parsimonious placement on the optimal tree, and "equivocal" transformations, which can be parsimoniously arranged in two or more ways. Figure 14 illustrates two unequivocal character optimizations. In Fig. 14A, only one step is necessary to explain the character state distribution, and there is only one parsimonious location for this transformation on the tree. This transformation (acquisition of a patagium between the digits of the manus) can therefore be interpreted as a synapomorphy for the clade including galeopithecids and bats. The transformations shown in Fig. 14B are also unequivocal, even though homoplasy is present in the form of a reversal. Any other arrangement of transformations would require additional steps. Accordingly, both transformations can be considered synapomorphies of the clades in question.

When there is homoplasy or missing data there may be more than one way to parsimoniously arrange transformations on a tree. Figure 15 illustrates the problem of optimizing a character for which there are missing data. Only one transformation is required, but there are two possible locations on the phylogenetic tree. This is essentially a problem of level; state "1" clearly represents a synapomorphy, but we cannot be sure at which level this synapomorphy applies.

Homoplasy may introduce more complex problems during character optimization. Figure 16 illustrates an ambiguous case resulting from homoplasy. A minimum of two transformations is required to explain the character state distribution, but these transformations may be arranged in different ways. Figure 16A shows an optimization that favors a hypothesis of convergence; Fig. 16B shows an alternative hypothesis of reversal. These two hypotheses are equally parsimonious, but they have very different implications for interpretation of homologies. In the first hypothesis (Fig. 16A), state "1" (circular, well-excavated central fossa on proximal radius) is not homologous in all taxa that exhibit this state; in the alternative hypothesis (Fig. 16B), character state "1" is homologous in all taxa.

In cases where character transformations are ambiguous, parsimony criteria cannot be used as a guide to optimization. Two of the most common computer optimization algorithms [ACCTRAN and DELTRAN in Swofford's (1989) PAUP package] arbitrarily place transformations in different locations when ambiguous distributions are encountered. ACCTRAN (accelerated transformation optimization) forces transformations to the lowest possible points on a tree, and thus favors hypotheses of reversal over hypotheses of convergence. ACCTRAN would "solve" the optimization problem posed above by choosing the option shown in Fig. 16B. Alternatively, DELTRAN (delayed transformation optimization) forces transformations to the highest points on a tree, thus favoring convergence over reversal; DELTRAN optimi-

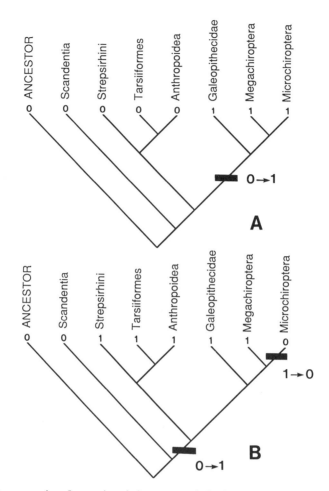

Fig. 14. Two examples of unequivocal character optimizations. (A) Optimization of Character 82: patagium not present between digits of manus (0) or continuously attached between digits of manus (1). Note that this optimization does not require any hypothesis of homoplasy. Only one transformation is necessary to explain the character state distribution, and there is only parsimonious location for this transformation on the tree. The derived condition can therefore be interpreted as a synapomorphy of galeopithecids + bats. (B) Optimization of Character 124: corpus spongiosum tissue in penis not distally expanded (0), or vascular, distally expanded (1). Although this optimization requires a hypothesis of reversal, it is unequivocal because there is only one parsimonious arrangement of transformations that is possible given the topology of the tree. Acquisition of the derived condition may be interpreted as a synapomorphy of primates + galeopithecids + bats; reversal to the primitive condition would be considered an autapomorphy of microchiropterans.

zation would place transformations as shown in Fig. 16A. Delayed transformation optimization, accelerated transformation, and other optimization techniques are discussed in detail in Swofford and Maddison (1987).

   When one is interested in evaluating the character support for a specific monophyletic group, or in tracing the pattern of evolution of a particular

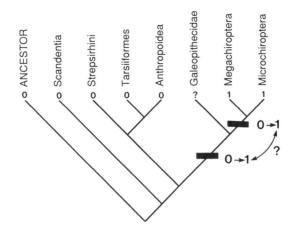

Fig. 15. An example of a character optimization that is equivocal because of missing data. The character shown here is Character 11: dorsal palatine and sphenopalatine foramina both present, not adjacent (0), or single foramen present (1). Only one transformation is required to explain the distribution of character states, but there are two possible locations in the phylogenetic tree. This is essentially a problem of level; state "1" clearly represents a synapomorphy, but we cannot be sure at which level this synapomorphy applies because we do not know the condition in galeopithecids. ACCTRAN would arbitrarily place the transformation at the lower level, implying that presence of single foramen is a synapomorphy of galeopithecids + bats. DELTRAN would choose the alternative placement, implying that the derived state is a synapomorphy of bats.

character or character complex, the ambiguity associated with equivocal transformations must be considered. The easiest way to accomplish this is to compare accelerated with delayed optimizations in order to identify all of the equally parsimonious placements for equivocal transformations. Unfortunately, some computer programs do not offer the user a choice of optimization algorithms. Nevertheless, alternative optimizations should always be investigated. Depending on the circumstances surrounding each character, different solutions may be favored—convergence in one case, reversal in another. When results of an analysis are published, it is important that these choices (and the justifications for each) are discussed in detail; it is not acceptable to simply present one optimization as if it is the only possible interpretation of character change.

It should be noted that the optimization method may affect the topology of the optimal trees output during an analysis *if* a program is instructed to collapse branches of zero length. When this option is invoked, the program automatically uses the default optimization method to measure branch lengths, and collapses branches that have a length of zero to form polytomies before presenting these trees to the researcher. Because a branch may comprise several steps under one method (e.g., ACCTRAN) but measure zero steps under a different method (e.g., DELTRAN), the trees output under this option may have polytomies in different parts of the tree depending on the optimization method designated as the default method. This effect is not significant if one is careful to check the optimization method (and compare

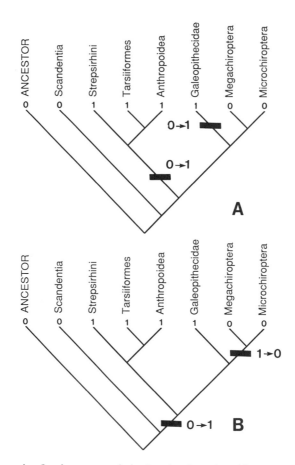

Fig. 16.   An example of a character optimization that is equivocal because of homoplasy. The character shown here is Character 69: central fossa on proximal radius ovoid and poorly excavated (0), or circular and well excavated (1). A minimum of two transformations is required to explain the character state distribution, but these transformations may be arranged in different ways. (A) An optimization that favors a hypothesis of convergence; this optimization would be favored by DELTRAN. (B) An alternative hypothesis of reversal; this optimization would be favored by ACCTRAN. These two hypotheses are equally parsimonious, but they have very different implications for interpretation of homologies. In the first hypothesis (A), state "1" is not homologous in all taxa that exhibit this state; in the alternative hypothesis (B), character state "1" is homologous in all taxa where it appears.

the results of different methods) whenever polytomies appear in trees generated by a search algorithm.

Discussion and Conclusions

The vast array of choices and assumptions embodied in every cladistic analysis significantly affects the outcome of that analysis. In studies involving archontan mammals, little attention has been paid to investigating the effects of various methodological choices. Studies of archontan relationships are generally reported in the literature in terms of results of a single analysis, with cladistic analysis treated as some sort of "black box" technique. If analyses have been repeated using alternative sets of assumptions (e.g., different sets of taxa, character coding schemes, weighting systems, and/or different optimization techniques), the results of such experiments have not been described. This has left the literature curiously bereft of any middle ground; most studies are portrayed as unambiguously supporting a single phylogenetic hypothesis, thus leaving little room for discussion.

It seems clear that much can be gained by investigating the impact that methodological choices have on the outcome of an analysis. By manipulating the assumptions in a series of replicate analyses, it is possible to identify the most stable elements in resultant phylogenies. Ambiguities are simultaneously highlighted, thus suggesting areas for future research. If adding or subtracting taxa significantly affects the outcome of an analysis, more attention might be paid to taxonomic sampling in future analyses. If ordering characters changes the results, more attention might be paid to understanding the development and function of various characters or character systems. The effects of choice of taxa, characters, coding schemes, weighting, and optimization can similarly be investigated through repeated manipulation of the assumptions made throughout the course of analyses. The extra effort involved in conducting multiple analyses is more than compensated for by the increased understanding of phylogenetic results and character evolution that may be gained from iterative analyses.

The last decade has seen numerous important contributions to our knowledge of the archontan morphology. Unfortunately, this increased knowledge has not led to resolution of many of the central questions of archontan systematics. Sophisticated anatomical techniques for identifying and describing morphological similarities and differences are not enough; equally sophisticated analytical methods for evaluating these data must be employed. Cladistics provides a powerful tool, but its effectiveness depends on clear understanding of the numerous assumptions built into every analysis. Making choices and assumptions is unavoidable during the course of cladistic analysis; the challenge is to manipulate these factors in such a way as to maximize the information gained from a phylogenetic study.

Acknowledgments

I thank D. Frost, M. Holden, R. MacPhee, M. Novacek, T. Rowe, M. Stiassny, R. Voss, J. Wahlert, and one anonymous reviewer for helpful comments on earlier versions of this manuscript. Special thanks to R. MacPhee, G. Musser, and M. Novacek for their support of my work on archontan mammals. This study was supported by NSF grant BSR-9106868 and an American Museum of Natural History Kalbfleisch-Hoffman Research Fellowship.

Appendix 1: Character Descriptions

The following character descriptions have been taken from the literature to serve as an example for this chapter. No attempt has been made to modify these descriptions, although the wording has been changed in some cases to make them easier to understand in the absence of lengthy text descriptions. As specified below, characters have been drawn from Beard (B = 1993), Johnson *et al.* (J = 1982a,b), Luckett (L1 = 1977, Table 3; L2 = 1980a, Table 6; L3 = 1980b, Table 1; L4 = 1985, Table 5), Novacek (N1 = 1980, Tables 1–5; N2 = 1985, Table 1; N3 = 1986, Table 3), MacPhee *et al.* (M = 1988, Table 2), Smith and Madkour (S & M = 1980, Figure 5), Wible and Novacek (W & N = 1988, Table 3), and Pettigrew *et al.* (P = 1989, Table 3). The primitive condition of each character [as assessed by the original author(s)] is coded "0"; derived conditions are coded "1," "2," "3," etc.

Cranial Features Excluding Auditory Region

1. **Premaxilla** small, restricted to anterior portion of snout (0), or large, nearly one-half surface area of snout (1), or very small (2). N2, N3, W & N
2. **Infraorbital canal** of large caliber (0), or narrow (1). N3
3. **Nasals** anteriorly projecting, not retracted (0), or retracted to a moderate degree (1). N3
4. **Nasals** broad posteriorly, expanded (0), or narrow (1). N1, N3
5. **Palatine contribution to palate** much larger than that of premaxilla (0), or smaller (1). N2
6. **Incisive foramina** separate (0), or coalesce with modification of premaxilla (1). N3
7. **Palatal fenestrae** absent (0), or present (1). N1
8. **Posterior margin of palate** located posterior to last molars (0), or between last molars (1), or far anterior to last molars (2). N3
9. **Maxilla orbital wing** small or absent (0), or moderately developed (1). N1, N2, N3
10. **Palatine orbital wing** with large orbital exposure (0), or orbital wing small or absent (1). N1, N2, N3
11. **Dorsal palatine and sphenopalatine foramina** both present, not adjacent (0), or single foramen present (foramina united as a single foramen or dorsal palatine foramen absent) (1). N3
12. **Lacrimal facial wing** large (0), or small or absent (1). N1, N2, N3
13. **Alisphenoid orbital wing** small (0), or large (1). N1
14. **Optic foramen** moderately large, cleftlike or ellipsoidal, opens in posterior orbitosphenoid (0), or large, circular, opens laterally (1). N3
15. **Suboptic foramen** present (0), or absent (1). N1

16. **Ethmoid foramen** located in orbital wall above and well posterior to palate (0), or more anteriorly, above posterior palate (1). N3
17. **Temporal fossa** distinct, extensive (0), or obscured by lateral expansion of alisphenoid (1). N2
18. **Sphenorbital fissure** moderate sized, crescentic (0), or very large, dorsoventrally expanded, crescentic or slitlike (1). N2
19. **Foramen rotundum** confluent with sphenorbital fissure (0), or separated from sphenorbital fissure (1). N1, N3
20. **Ramus infraorbitalis of stapedial artery** passes ventral to the alisphenoid (0), or passes through an alisphenoid canal (1), or passes through the cranial cavity dorsal to the alisphenoid (2). N1, N3, W & N
21. **Blood supply to meninges** mostly or exclusively from ramus superior of stapedial artery (0), or from other vessels (1). M
22. **Postorbital process of frontal** absent (0), small (1), or large, forming complete postorbital bar (2). N1, N3, B
23. **Supraorbital foramen** absent (0), or present (1). N1
24. **Zygomatic arch** with large jugal (0), or with reduced jugal (1). N1, W & N
25. **Occiput** not posteriorly expanded (0), or posteriorly expanded, visible in dorsal view (1). N1
26. **Jaw condyle** distinctly lower than coronoid process (0), or higher or subequal to height of coronoid process (1). N1, N2
27. **Origin of internal pterygoid muscle** confined to ectopterygoid process (0), or extends to medial wall of orbit (1). N1
28. **Origin of temporalis muscles** extends anteriorly over frontals (0), or posteriorly confined, barely reaching frontals (1). N1
29. **Groove for digastric muscle** on mastoid process deep (0), or very faint or absent (1). N3
30. **Mastoid** exposed extensively in occipital region (0), or exposure weak, narrow (1). N3
31. **Mastoid foramen** large (0), or very small (1). N3
32. **Paroccipital process** strong (0), or weak or absent (1). N3
33. **Ventral basioccipital and basisphenoid** with median ridge and a pair of shallow depressions for rectus capiti muscles (0), or with sculpturing very faint or absent (1). N3

Auditory Region

34. **Pneumatization of lateral part of middle ear** absent (0), or true mastoid cavity present (1). M
35. **Epitympanic recess** with bony lateral wall formed primarily by squamosal (0), or formed by part of ectotympanic meatal tube (1). N3

36. **Tegmen tympani** moderately developed (0), or expanded ante-rolaterally to roof epitympanic recess (1), or tapers to a slender process that projects ventrally into the middle-ear cavity medial to the epitympanic recess (2), or is greatly reduced and the roof of the recess is completed by the squamosal (3). W & N

37. **Meatal arch of squamosal** slightly below or level with tympanic roof (0), or well ventral to tympanic roof (1). N3

38. **Rostral entotympanic** absent (0), or present (1). N1, W & N

39. **Caudal entotympanic** absent (0), or present (1). N1, W & N

40. **Caudal entotympanic** not associated with internal carotid artery (0), or grooved by (or forming a canal around) the internal carotid artery (1). W & N

41. **Entotympanic elements** contact only the petrosal medially (0), or contact both the basioccipital and the petrosal medially (1). B

42. **Ectotympanic** annular (0), or moderately expanded ventrally (1), or markedly expanded into external auditory tube or medial auditory bulla (2). N1

43. **Ectotympanic** inclined at acute angle to horizontal plane of crani-um (0), or relatively vertical (1). N1

44. **Internal carotid artery** with transpromontorial course in adult (0), or with perbullar course (1), or involuted (2). B

45. **Anterior carotid foramen in basisphenoid** not converted into a long tube (0), or converted into a long tube (1). W & N

46. **Ossified tubes for intrabullar vessels** absent (0), present for some vessels (1), or most vessels (2). N1, B

47. **Medial promontorium crest** weak or absent (0), or present, elon-gate, often fused with medial wall or ossified bulla (1). N3

48. **Ramus inferior or stapedial artery** arises from proximal stapedial artery within middle ear space and travels forward ventral to the tegmen tympani (0), or arises within cranial cavity, runs forward dorsal to tegmen tympani (1), or ramus inferior absent (2). N1, N3, W & N

49. **Fenestra rotundum** not shielded ventrally (0), or shielded by en-larged carotid artery or bony tubes or flanges (1). N1, N3

50. **Fenestra rotundum** faces posterolaterally (0), or directly posteriorly (1). N3, W & N

51. **Subarcuate fossa** moderately deep (0), or very deep, but dorsal semicircular not separated from the endocranial wall of the squa-mosal (1), or very deep, dorsal semicircular clearly separated from the endocranial wall (2). N3, W & N

52. **Tympanohyal** process small (0), or large, isolates stylomastoid fora-men from tympanic chamber (1). N1

53. **Tympanic process of petrosal** lacking in auditory bulla (0), or par-tial contribution to bulla (1), or primary bullar element (2). N1, N3

Anterior Axial Skeleton and Forelimb

54. **Neural spines on cervical vertebrae 3–7** not reduced (0), or weak or absent (1). W & N
55. **Ribs** not flattened (0), or flattened, especially near their vertebral ends (1). W & N
56. **Manubrium** of sternum not enlarged (0), or enlarged, often with ventral keel (1). N1
57. **Vertebral border of scapula** short and curved (0), or longer, less curved (1). N1
58. **Metacromion process** pronounced (0), or weak or absent (1). N1
59. **Forelimbs** not markedly elongated (0), or markedly elongated (1). W & N
60. **Lesser tuberosity of humerus** gracile, does not show strong medial protrusion (0), or robust, protrudes more medially (1), or projects prominently beyond level of humeral head (2). B
61. **Deltopectoral crest** on humerus strong (0), or weak (1), or very strong (2). N1
62. **Deltopectoral crest** situated on anterior aspect of humeral shaft (0), or on lateral aspect (1). B
63. **Entepicondylar foramen** of humerus present (0), or absent (1). N1, W & N
64. **Supracapitular fossa** shallow (0), or deep (1). N1
65. **Supratrochlear depression and supinator ridge** present on distal humerus (0), or absent (1). W & N
66. **Olecranon fossa** shallow (0), or deep (1). N1
67. **Capitulum** spindle shaped (0), or spherical (1). N1, B
68. **Ulna** robust, with sigmoid lateral curvature (0), or slender, only slightly curved, not proximally or distally reduced (1), or reduced both proximally and distally (2). N, W & N
69. **Central fossa on proximal radius** ovoid and poorly excavated (0), or circular and well excavated (1). B
70. **Lateral lip about perimeter of proximal radius** forms broad articular surface that is limited to lateral side (0), or forms narrow, ribbonlike articular surface that extends halfway around perimeter of proximal radius (1). B
71. **Ulnocarpal articulation** mediolaterally and dorsopalmarly extensive, lies in transverse plane (0), or limited to the radial and palmar aspects of the distal ulna, lies in proximodistal plane (1), or articulation between ulna and triquetrum absent (2). B
72. **Scaphoid and lunate** unfused (0), or fused (1). N1, B
73. **Lunate** located ulnar to scaphoid (0), or distal to scaphoid (1). B
74. **Triquetrum** (cuneiform) contacts only lunate (0), or both lunate and scaphoid (1). B

75. **Triquetrum shape** in dorsal view roughly quadrate (0), or triangular (1). B
76. **Pisiform** moderately large (0), reduced (1), or absent (2). B
77. **Manus** not rotated (0), or rotated 90° from position typical of quadrupedal mammals (1). W & N
78. **Digits 2–5 of manus** not elongated (0), or greatly elongated (1). W & N
79. **Proximal phalanges** longer than intermediate phalanges (0), or shorter (1). B
80. **Distal phalanges** moderately laterally compressed (0), or highly compressed mediolaterally and extremely tall dorsoventrally (1), or dorsoventrally flattened and mediolaterally wide (2). B
81. **Claws or nails** present on terminal phalanges of digits 3 and 4 of forelimb (0), or absent (1). W & N, B
82. **Patagium** not present between digits of manus (0), or continuously attached between digits of manus (1). W & N
83. **Humeropatagialis muscle** absent (0), or present, inserts into plagiopatagium (1). W & N
84. **Occipitopollicalis muscle** absent (0), or present on leading edge of propatagium (1). W & N

Hindlimb

85. **Iliosacral angle** obtuse, greater than 30° (0), or acute, less than 30° (1). N1
86. **Pelvic–sacral fusion** limited (0), or extensive (1). N1
87. **Anterior inferior iliac spine weak** (0), or pronounced (1). N1
88. **Acetabulum** roughly circular in lateral view (0), or strongly elliptical in outline (1). B
89. **Buttresses around acetabulum** evenly developed around circumference (0), or heavily emphasized on the cranial side of acetabulum (1). B
90. **Fovea capitis femoris** centrally placed on femoral head (0), or located well posterior to midline (1), or absent (2). B
91. **Hindlimbs** not rotated (0), or rotated from typical quadrupedal orientation (1). W & N
92. **Greater trochanter** prominent, exceeding height of femoral head (0), or shorter than femoral head (1), or weak or absent (2). N1
93. **Lesser trochanter** large, lamelliform (0), or reduced or absent (1). N1
94. **Gluteal tuberosity** weak or absent (0), or prominent (1). N1
95. **Femoral trochlea** broad and shallow (0), or elongate, deeply excavated (1). N1
96. **Patellar groove** long and narrow (0), or short and wide (1), or very deep, elevated anteriorly (2). B

97. **Insertion of quadratus femoris muscle** limited to posterior aspect of proximal femoral shaft (0), or enlarged to fill triangular area between trochanters (1). B

98. **Sartorius muscle** present (0), or absent (1). W & N

99. **Distal tibiofibular joint** = syndesmosis (0), synovial (1), or synostosis (2). B

100. **Metatarsals** not greatly elongated (0), some elongated (1), or all markedly elongated (2). N1

101. **Calcaneal fibular facet** pronounced (0), or weak or absent (1). N1

102. **Calcaneal peroneal tubercle** strong (0), or weak or absent (1). N1

103. **Calcaneal astragalocalcaneal facet** oriented obliquely to long axis of calcaneum (0), or facet acutely angled, nearly parallel or nearly transverse to long axis of calcaneum (1). N1

104. **Calcaneal astragalocalcaneal facet** convex (0), or concave depression or trough (1). W & N

105. **Secondary articulation** between posterior side of sustentaculum tali and astragalus absent (0), or present (1), or sustentaculum greatly reduced or absent (2). B

106. **Calcaneal cuboid facet** of calcaneum moderately concave or flat, articulates with evenly convex oval facet on cuboid (0), or characterized by a plantar concavity on distal calcaneum, which articulates with proximally projecting process on cuboid (1). B

107. **Calcaneal cuboid facet** oriented obliquely to long axis of calcaneum (0), or oriented transversely to long axis of calcaneum (1). N1

108. **Calcar and depressor osseous styliformes muscle** absent (0), or present (1). W & N

109. **Astragalar tibial trochlea** broad and short (0), or moderately long and narrow (1), or extremely long and narrow (2). N1

110. **Lateral astragalar trochlear border** low and rounded (0), or high and often crestiform (1). N1

111. **Medial astragalar trochlear border** low and rounded (0), or high and often crestiform (1). N1

112. **Astragalar trochlea** limited to body (0), or extending to neck (1). N1

113. **Superior astragalar foramen** present (0), or absent (1). N1

114. **Astragalar fibular shelf** prominent (0), or weak or absent (1). N1

115. **Astragalar sustentacular facet** separate from distal astragalar facets (0), or confluent with distal astragalar facets (1). N1

116. **Calcaneoastragalar facet on astragalus** slightly concave (0), or strongly concave (1). N1

117. **Naviculoastragalar and spring ligament facets** moderately convex (0), or strongly convex or flattened (1). N1

118. **Groove for m. flexor digitorum fibularis** on astragalus lies in midline position with respect to posterior part of astragalar trochlea (0), or occurs lateral to posterior part of astragalar trochlea (1), or groove is absent (2). B

119. **Entocuneiform distal facet** very narrow distally with strong plantodistal process (0), or wider distally with reduced or absent plantodistal process (1), or proximodistally short with distinct facet for first metatarsal (2). B

Reproductive Tract and Fetal Membranes

120. **Uterus** duplex (0), or partially fused (1), or simplex (2). L2, L3
121. **Sublingula** present (0), or intermediate condition [unspecified] (1), or absent (2). L2
122. **Os penis** present (0), or absent (1). L2, S & M
123. **Accessory cavernous tissue** present in penis (0), or absent (1). S & M
124. **Corpus spongiosum** tissue in penis not distally expanded (0), or vascular, distally expanded (1). S & M
125. **Orientation of embryonic disk** at implantation antimesometrial (0), or orthomesometrial (1), or embryonic disk oriented relative to tubo-uterine junction rather than mesometrium (2). L2, L3, L4
126. **Blastocyst attachment** at paraembryonic pole (0), or embryonic pole (1), or abembryonic pole of blastocyst (2). L2, L4
127. **Amniogenesis** by folding (0), or cavitation followed by folding (1), or by cavitation only (2). L1, L2, L4
128. **Preplacenta** diffuse (0), horseshoe-shaped (1), or discoidal (2). L3
129. **Allantoic vesicle** large (0), moderate size (1), or reduced or vestigial (2). L1, L2
130. **Mesodermal body stalk** absent (0), or temporary (1), or permanent (2). L2
131. **Chorioallantoic placenta** epitheliochorial (0), endotheliochorial (1), or hemochorial (2). L1, L2, L3

Nervous System

132. **Olfactory bulbs** large (0), moderately reduced (1), or small (2). L2
133. **Neocortex** lissencephalic (0), or gyrencephalic (1). J
134. **Triradiate calcarine sulcus complex** absent (0), or present (1). L2
135. **Cortical somatosensory representation of forelimb** with distal elements of forelimb located rostal relative to more proximal elements in somatotopic map (0), or forelimb representation reversed (1). W & N
136. **Horizontal streak** in retina inferior to optic nerve head (0), or superior (1). P
137. **Demarcation between ipsilateral and contralateral inputs to lateral geniculate nucleus (LGN)** not lacunar (0), or lacunar (1). P
138. **Laminar differentiation of LGN** absent (0), or present (1). P

139. **Magnocellular layer(s)** not differentiated in LGN (0), or differentiated (1). P
140. **Ipsilateral and contralateral magnocellular layers** not separate (0), or separate (1). P
141. **Magnocellular layers** not adjacent to optic tract (0), or adjacent (1). P
142. **Ipsilateral magnocellular layer** external to contralateral magnocellular layer (0), or internal to contralateral layer (1). P
143. **Ratio of ipsilateral to contralateral input to LGN** < 10% (0), or higher (1), (2), (3), (4), (5), up to 50% (6). P
144. **Paired ipsilateral and contralateral koniocellular and parvocellular layers** not present in LGN (0), one pair present (1), or two pairs present (2). P
145. **Parvocellular laminae** with input from each eye not segregated (0), or segregated (1). P
146. **Parvocellular–magnocellular differentiation** absent (0), or visible (1), or concealed (2). P
147. **Ratio of ipsilateral to contralateral input to superior colliculus (SC):** ratio = 0 (0), or higher (1), (2), (3), up to ratio = 1 (4). P
148. **Proportion of SC taken up by ipsilateral representation** negligible (0), or higher (1), (2), (3), up to entire rostrocaudal SC (4). P
149. **Ratio of retinogeniculate (RG) to retinotectal (RT) ganglion cells:** ganglion cells mostly RT (0), or increasing RG (1), (2), (3). P
150. **Decussation of retinotectal ganglion cell population not sharp** (0), or sharp (1). P
151. **Medial terminal nucleus** prominent (0), or reduced (1). P
152. **Ratio of inferior colliculus (IC) to superior colliculus:** IC > SC (0), or SC > IC (1). P
153. **Middle temporal visual area** absent (0), or present (1). P
154. **Premotor area C** absent (0), or present (1). P

Appendix 2: Data Matrix

The following taxon–character matrix was constructed from the literature as described in Appendix 1. Distributions of character states were obtained from the references associated with each character in Appendix 1; no new observations are reported here. The taxon listed as ANCESTOR represents a hypothetical ancestor used to root the tree; this construct is coded as exhibiting the primitive condition of every character [as assessed by the original author(s) of the characters]. Character states shown in brackets (e.g., {01}) indicate cases of taxonomic polymorphism.

ANCESTOR

000000000000000000000000000000000000000000000000000000000000
000000000000000000000000000000000000000000000000000000000000
000000000000000000000000000000000000

Scandentia

010100{01}0000{01}111001{01}192{01}01011001109111110900001210100
110000{01}00100000{01}1000{01}00000000000000{01}0000000{01}000001
{01}{01}0001011{01}{01}1{01}11{01}0010110000{12}00110001111100311
111000100

Micromomyidae

99999999999999999999999999999999999999999999090999999999999919
199991911099999999999999991119999911999999991199999999991999
99999999999999999999999999999999

Paromomyidae

999999999999999999999919999999999990999999199290929999999991
9199991911191111991109999991119999911919999911999999999999119
99999999999999999999999999999999

Galeopithecidae

011011020090110111109110010119100913111012120012012101101011
100101121111111100110110010111021001101200101110211111101110
10110122012019911111111321122111119

Plesiadapidae

199119091190191910111000001199999090990099212909299902990019 10
10190101100110099019999900111900101091210091119210101111 11199
99999999999999999999999999999999

Strepsirhini

01000001010000001111{01}200009910010{01}011009901{02}1102101
12009110101000010110000000002000000 1{01}{01}{01}01110{02}101
0101001102000001111110011100000021201111111105 21133311111

Tarsiiformes

01010001100000001100{01}210119910010{01}01100992111202101120
0901091100001111000000002000090111100111290291110000011
010011101101111009222202011111111521133210119

Anthropoidea

0101000100010000110112101010100101011009921112021011200011
091100010111000000000200001009990110099011111001102001101110
122011112922221201111111062 1{12}{34}{34}311111

Megachiroptera

201101100011010110029111000011901902011010100010120011111122 0
101012002900021 10{01}1111110112120000010001 11201 10000111 00220
20112121101019111111110421122211111

Microchiroptera

2011011000119111110029011009911901902011101100001012001111112 20
10101200090002110{01}1111110111120000010011112011 11100110002219
00029{12}1{12}9{12}9{01}9100000000000000000 00000

References

Adkins, R. M., and Honeycutt, R. L. 1991. Molecular phylogeny of the superorder Archonta. *Proc. Natl. Acad. Sci. USA* **88**:10317–10321.

Ammerman, L. K., and Hillis, D. M. 1990. Relationships within archontan mammals based on 12S rRNA gene sequence. *Am. Zool.* **30**:50A.

Ammerman, L. K., and Hillis, D. M. 1992. A molecular test of bat relationships: Monophyly or diphyly? *Syst. Biol.* **41**:222–232.

Archie, J. W. 1985. Methods for coding variable morphological features for numerical taxonomic analysis. *Syst. Zool.* **34**:326–345.

Archie, J. W. 1989. Homoplasy excess ratios: New indices for measuring levels of homoplasy in phylogenetic systematics and a critique of the consistency index. *Syst. Zool.* **38**:253–269.

Archie, J. W. 1990. Homoplasy excess statistics and retention indices: A reply to Farris. *Syst. Zool.* **39**:169–174.

Ax, P. 1987. *The Phylogenetic System: The Systematization of Organisms on the Basis of Their Phylogenesis.* Wiley, New York.

Bailey, W. J., Slightom, J. L., and Goodman, M. 1992. Rejection of the "flying primate" hypothesis by phylogenetic evidence from the ε-globin gene. *Science* **256**:86–89.

Baker, R. J., Honeycutt, R. L., and Van Den Bussche, R. A. 1991a. Examination of monophyly of bats: Restriction map of the ribosomal DNA cistron, in: T. A. Griffiths and D. Klingener (eds.), *Contributions to Mammalogy in Honor of Karl F. Koopman. Bull. Am. Mus. Nat. Hist.* **206**:42–53.

Baker, R. J. Novacek, M. J., and Simmons, N. B. 1991b. On the monophyly of bats. *Syst. Zool.* **40**:216–231.

Beard, K. C. 1990. Gliding behavior and palaeoecology of the alleged primate family Paromomyidae (Mammalia, Dermoptera). *Nature* **345**:340–341.

Beard, K. C. 1991. Postcranial fossils of the archaic primate family Microsyopidae. *Am. J. Phys. Anthropol.* (Suppl.) **12**:48–49.

Beard, K. C. 1993. Phylogenetic systematics of the Primatomorpha, with special reference to Dermoptera, in: F. S. Szalay, M. J. Novacek, and M. C. McKenna (eds.), *Mammal Phylogeny,* Volume 2, *Placentals,* pp. 129–150. Springer-Verlag, New York.

Brooks, D. R., and McLennan, D. A. 1991. *Phylogeny, Ecology, and Behavior: A Research Program in Comparative Biology.* University of Chicago Press, Chicago.

Cannatella, D. C., and de Queiroz, K. 1989. Phylogenetic systematics of the anoles: Is a new taxonomy warranted? *Syst. Zool* **38**:57–69.

Chappill, J. A. 1989. Quantitative characters in phylogenetic analysis. *Cladistics* **5**:217–234.

Donoghue, M. J., Doyle, J. A., Gauthier, J. Kluge, A. G., and Rowe, T. 1989. The importance of fossils in phylogeny reconstruction. *Annu. Rev. Ecol. Syst.* **20**:431–446.

Dumont, E. R. 1990. Enamel microstructural evidence for the affinities of proposed early Tertiary archontan taxa. *J. Vertebr. Paleontol.* **10:**21A.

Dumont, E. R. 1992. Primate higher-level relationships: Evidence from tooth enamel ultrastructure. *Am. J. Phys. Anthropol.* (Suppl.) **14:**72.

Duncan, T., and Stuessy, T. F. (eds.) 1984. *Cladistics: Perspectives on the Reconstruction of Evolutionary History.* Columbia University Press, New York.

Eldredge, N., and Cracraft, J. 1980. *Phylogenetic Patterns and the Evolutionary Process.* Columbia University Press, New York.

Farris, J. S. 1969. A successive approximations approach to character weighting. *Syst. Zool.* **18:**374–385.

Farris, J. S. 1983. The logical basis of phylogenetic analysis, in: N. Platnick and V. Funk (eds.), *Advances in Cladistics*, Vol. 2, pp. 7–36. Columbia University Press, New York.

Farris, J. S. 1988. Hennig86 (computer program package distributed on floppy disk; write to James S. Farris, 41 Admiral Street, Port Jefferson Station, New York, NY 11776).

Farris, J. S. 1989. The retention index and the rescaled consistency index. *Cladistics* **5:**417–419.

Farris, J. S. 1990. The retention index and homoplasy excess. *Syst. Zool.* **38:**406–407.

Farris, J. S. 1991. Excess homoplasy ratios. *Cladistics* **7:**81–91.

Felsenstein, J. 1985. Confidence limits on phylogenies: An approach using the bootstrap. *Evolution* **39:**783–791.

Felsenstein, J. 1988a. Phylogenies and quantitative characters. *Annu. Rev. Ecol. Syst.* **19:**445–471.

Felsenstein, J. 1988b. Phylogenies from molecular sequences: Inference and reliability. *Annu. Rev. Genet.* **22:**521–565.

Felsenstein, J. 1990. Phylogeny Inference Package (PHYLIP; computer program package distributed on floppy disk; write to Joseph Felsenstein, Department of Genetics SK-50, University of Washington, Seattle, WA 98195).

Fitch, W. M. 1979. Cautionary remarks on using gene expression events in parsimony procedures. *Syst. Zool.* **28:**375–379.

Fitch, W. M. 1984. Cladistics and other methods: Problems, pitfalls, and potentials, in: T. Duncan and T. F. Stuessy (eds.), *Cladistics: Perspectives on the Reconstruction of Evolutionary History*, pp. 221–252. Columbia University Press, New York.

Frost, D. R., Wozencraft, W. C., and Hoffman, R. S. 1991. Phylogenetic relationships of hedgehogs and gymnures (Mammalia: Insectivora: Erinaceidae). *Smithson. Contrib. Zool.* **518:**1–69.

Funk, V. A., and Brooks, D. R. (eds.) 1981. *Advances in Cladistics: Proceedings of the First Meeting of the Willi Hennig Society.* New York Botanic Garden, New York.

Gauthier, J., Kluge, A. G., and Rowe, T. 1988. Amniote phylogeny and the importance of fossils. *Cladistics* **4:**105–209.

Gould, S. J. 1977. *Ontogeny and Phylogeny.* Harvard University Press, Cambridge, Mass.

Greenwald, N. S. 1990. Are bats monophyletic? Phylogenetic interpretation of neuroanatomical and other morphological data. *J. Vertebr. Paleontol.* **10:**25A–26A.

Greenwald, N. S. 1991a. Primate–bat relationships and cladistic analysis of morphological data. *Am. J. Phys. Anthropol.* (Suppl.) **12:**82.

Greenwald, N. S. 1991b. Cladistic analysis of chiropteran relationships: Why megachiropterans are not primates. *Bat Res. News* **31:**79–80.

Harvey, P., and Pagel, M. 1991. *The Comparative Method in Evolutionary Biology.* Oxford University Press, London.

Hendy, M. D., and Penny, D. 1982. Branch and bound algorithms to determine minimal evolutionary trees. *Math. Biosci.* **59:**277–290.

Hennig, W. 1966. *Phylogenetic Systematics.* University of Illinois Press, Urbana.

Hillis, D. M. 1991. Discriminating between phylogenetic signal and random noise in DNA sequences, in: M. M. Miyamoto and J. Cracraft (eds.), *Phylogenetic Analysis of DNA Sequences*, pp. 278–294. Oxford University Press, London.

Huelsenbeck, J. P. 1991. Tree-length distribution skewness: An indicator of phylogenetic information. *Syst. Zool.* **40:**257–270.

Johnson, J. I., Kirsch, J.A.W., and Switzer, R. C. 1982a. Phylogeny through brain traits: Fifteen characters adumbrate mammalian genealogy. *Brain Behav. Evol.* **20:**72–83.

Johnson, J. I., Switzer, R. C., and Kirsch, J. A. 1982b. Phylogeny through brain traits: The distribution of categorizing characters in contemporary mammals. *Brain Behav. Evol.* **20:**97–117.

Joysey, K. A., and Friday, E. A. (eds.) 1982. *Problems in Phylogenetic Reconstructions.* Academic Press, New York.

Kaas, J. H. 1993. Archontan affinities as reflected in the visual system, in: F. S. Szalay, M. J. Novacek, and M. C. McKenna (eds.), *Mammal Phylogeny:* Placentals. Springer-Verlag, Berlin (in press).

Kay, R. F., Thorington, R. W., and Houde, P. 1990. Eocene plesiadapiform shows affinities with flying lemurs not primates. *Nature* **345:**342–344.

Kluge, A. G. 1988. The characterization of ontogeny, in: J. Humphries (ed.), *Ontogeny and Systematics,* pp. 57–81. Columbia University Press, New York.

Kluge, A. G. 1989. A concern for evidence and a phylogenetic hypothesis of relationships among *Epicrates* (Boidae, Serpentes). *Syst. Zool.* **38:**7–25.

Kluge, A. G., and Farris, J. S. 1969. Quantitative phyletics and the evolution of anurans. *Syst. Zool.* **18:**1–32.

Koopman, K. F., and MacIntyre, G. T. 1980. Phylogenetic analysis of chiropteran dentition, in: D. E. Wilson and A. L. Gardner (eds.), *Proceedings of the 5th International Bat Research Conference,* pp. 279–288. Texas Tech Press, Lubbock.

Luckett, W. P. 1977. Ontogeny of amniote fetal membranes and their application to phylogeny in: M. K. Hecht, P. C. Goody, and B. M. Hecht (eds.), *Major Paterns in Vertebrate Evolution,* pp. 439–516. Plenum Press, New York.

Luckett, W. P. 1980a. The suggested evolutionary relationships and classification of tree shrews, in: W. P. Luckett (ed.), *Comparative Biology and Evolutionary Relationships of Tree Shrews,* pp. 3–31. Plenum Press, New York.

Luckett, W. P. 1980b. The use of fetal membrane data in assessing chiropteran phylogeny, in: D. E. Wilson and A. L. Gardner (eds.), *Proceedings of the 5th International Bat Research Conference,* pp. 245–265. Texas Tech Press, Lubbock.

Luckett, W. P. 1985. Superordinal and intraordinal affinities of rodents: Developmental evidence from the dentition and placentation, in: W. P. Luckett and J.-L. Hartenberger (eds.), *Evolutionary Relationships among Rodents,* pp. 227–276. Plenum Press, New York.

Mabee, P. M. 1989. Assumptions underlying the use of ontogenetic sequences for determining character state order. *Trans. Am. Fish. Soc.* **118:**151–158.

MacPhee, R.D.E. 1991. The supraordinal relationships of Primates: A prospect. *Am. J. Phys. Anthropol.* (Suppl.) **12:**121–122.

MacPhee, R.D.E., and Novacek, M. J. 1993. Definition and relationships of Lipotyphla, in: F. S. Szalay, M. J. Novacek, and M. C. McKenna (eds.), *Mammal Phylogeny: Placentals.* Springer-Verlag, Berlin (in press).

MacPhee, R.D.E., Novacek, M. J., and Storch, G. 1988. Basicranial morphology of Early Tertiary erinaceomorphs and the origin of Primates. *Am. Mus. Novit.* **2921:**1–42.

Maddison, W. P., and Maddison, D. R. 1992. MacClade, version 3.0 (computer program distributed on floppy disk; distributed by Sineauer Associates, Sunderland, MA 01375).

Maddison, W. P., Donoghue, M. J., and Maddison, D. R. 1984. Outgroup analysis and parsimony. *Syst. Zool.* **33:**83–103.

Mickevich, M. F., and Johnson, M. F. 1976. Congruence between morphological and allozyme data in evolutionary inference and character evolution. *Syst. Zool.* **25:**260–270.

Mindell, D. P., Dick, C. W., and Baker, R. J. 1991. Phylogenetic relationships among megabats, microbats, and primates. *Proc. Natl. Acad. Sci. USA* **88:**10322–10326.

Miyamoto, M. M. 1985. Consensus cladograms and general classifications. *Cladistics* **1:**186–189.

Neff, N. A. 1986. A rational basis for a priori character weighting. *Syst. Zool.* **35:**110–123.

Nelson, G., and Platnick, N. 1981. *Systematics and Biogeography: Cladistics and Vicariance.* Columbia University Press, New York.

Nixon, K. C., and Davis, J. I. 1991. Polymorphic taxa, missing values and cladistic analysis. *Cladistics* **7**:233–241.

Northcutt, R. G. 1984. Evolution of the vertebrate central nervous system: Patterns and processes. *Am. Zool.* **24**:701–716.

Novacek, M. J. 1980. Cranioskeletal features in tupaiids and selected Eutheria as phylogenetic evidence, in: W. P. Luckett (ed.), *Comparative Biology and Evolutionary Relationships of Tree Shrews*, pp. 35–93. Plenum Press, New York.

Novacek, M. J. 1985. Cranial evidence for rodent affinities, in: W. P. Luckett and J.-L. Hartenberger (eds.), *Evolutionary Relationships among Rodents*, pp. 59–82. Plenum Press, New York.

Novacek, M. J. 1986. The skull of leptictid insectivorans and the higher-level classification of eutherian mammals. *Bull. Am. Mus. Nat. Hist.* **183**:1–112.

Novacek, M. J. 1989. Higher mammal phylogeny: The morphological–molecular synthesis, in: B. Fernholm, K. Bremer, and H. Jörnvall (eds.), *The Hierarchy of Life*, pp. 421–435. Elsevier, Amsterdam.

Novacek, M. J. 1990. Morphology, paleontology, and the higher clades of mammals, in: H. Genoways (ed.), *Current Mammalogy*, Vol. 2, pp. 507–543. Plenum Press, New York.

Novacek, M. J. 1991. "All Tree Histograms" and the evaluation of cladistic evidence: Some ambiguities. *Cladistics* **7**:345–349.

Novacek, M. J. 1992. Fossils as critical data for phylogeny, in: M. J. Novacek and Q. D. Wheeler (eds.), *Extinction and Phylogeny*, pp. 46–88. Columbia University Press, New York.

Novacek, M. J., and Wyss, A. R. 1986. Higher-level relationships of Recent eutherian orders: Morphological evidence. *Cladistics* **2**:257–287.

Novacek, M. J., Wyss, A. R., and McKenna, M. C. 1988. The major groups of eutherian mammals, in: M. Benton (ed.), *The Phylogeny and Classification of the Tetrapods*, Vol. 2, pp. 31–71. Systematics Association Special Volume No. 35B. Oxford University Press (Clarendon), London.

Patterson, C. 1982. Morphological characters and homology, in: K. A. Joysey and E. A. Friday (eds.), *Problems in Phylogenetic Reconstructions*, pp. 21–74. Systematics Association Special Volume No. 21. Academic Press, New York.

Penny, D., and Hendy, M. 1986. Estimating the reliability of evolutionary trees. *Mol. Biol. Evol.* **3**:4.

Pettigrew, J. D. 1991a. Wings or brain? Convergent evolution in the origins of bats. *Syst. Zool.* **40**:199–216.

Pettigrew, J. D. 1991b. A fruitful, wrong hypothesis? Response to Baker, Novacek, and Simmons. *Syst. Zool.* **40**:231–239.

Pettigrew, J. D., Jamieson, B.G.M., Robson, S. K., Hall, L. S., McAnally, K. I., and Cooper, H. M. 1989. Phylogenetic relations between microbats, megabats and primates (Mammalia: Chiroptera and Primates). *Philos. Trans. R. Soc. London Ser. B* **325**:489–559.

Platnick, N. I., and Funk, V. A. 1983. *Advances in Cladistics*, Vol. 2. Columbia University Press, New York.

Platnick, N. I., Coddington, J. A., Forster, R. R., and Griswold, G. E. 1991a. Spinneret morphology and the phylogeny of haplogyne spiders (Araneae, Araneomorphae). *Am. Mus. Novit.* **3016**:1–73.

Platnick, N. I., Griswold, C. E., and Coddington, J. A. 1991b. On missing entries in cladistic analysis. *Cladistics* **7**:337–343.

Rieppel, O. 1980. Homology, a deductive concept? *Z. Zool. Syst. Evolutionsforsch.* **18**:315–319.

Rieppel, O. 1989. Character incongruence: Noise or data? *Abh. Naturwiss. Ver. Hamburg* **28**:53–62.

Rowe, T. 1988. Definition, diagnosis, and origin of Mammalia. *J. Vertebr. Paleontol.* **8**(3):241–264.

Sanderson, M. J. 1989. Confidence limits on phylogenies: The bootstrap revisited. *Cladistics* **5**:113–129.

Sanderson, M. J. 1990. Flexible phylogeny reconstructions: A review of phylogenetic packages using parsimony. *Syst. Zool.* **39**:414–420.

Sanderson, M. J., and Donoghue, M. J. 1989. Patterns of variation in levels of homoplasy. *Evolution* **48**:1781–1795.

Sereno, P. C. 1990. Clades and grades in dinosaur systematics, in: K. Carpenter and P. J. Currie (eds.), *Dinosaur Systematics: Perspectives and Approaches*, pp. 9–20. Cambridge University Press, London.

Shaffer, H. B., Clark, J. M., and Kraus, F. 1991. When molecules and morphology clash: A phylogenetic analysis of the North American ambystomatid salamanders (Caudata: Ambystomatidae). *Syst. Zool.* **40**:284–303.

Simmons, N. B. 1993. Phylogeny of Multituberculata, in: F. S. Szalay, M. J. Novacek, and M. C. McKenna (eds.), *Mammal Phylogeny: Mesozoic Differentiation, Multituberculates, Monotremes, Early Therians, and Marsupials.* Springer-Verlag, Berlin (in press).

Simmons, N. B., Novacek, M. J., and Baker, R. J. 1991. Approaches, methods, and the future of the chiropteran monophyly controversy: A reply to J. D. Pettigrew. *Syst. Zool.* **40**:239–243.

Simon, C. M. 1983. A new coding procedure for morphometric data with an example from periodical cicada wings, in: J. Felsenstein (ed.), *Numerical Taxonomy,* pp. 378–382. Springer-Verlag, Berlin.

Slaughter, B. H. 1970. Evolutionary trends of chiropteran dentitions, in: B. H. Slaughter and D. W. Walton (eds.), *About Bats: A Chiropteran Biology Symposium,* pp. 51–83. Southern Methodist University Press, Dallas.

Smith, J. D., and Madkour, G. 1980. Penial morphology and the question of chiropteran phylogeny, in: D. E. Wilson and A. L. Gardner (eds.), *Proceedings of the 5th International Bat Research Conference,* pp. 347–365. Texas Tech Press, Lubbock.

Sober, E. 1988. *Reconstructing the Past: Parsimony, Evolution, and Inference.* MIT Press, Cambridge, Mass.

Stevens, P. F. 1991. Character states, morphological variation, and phylogenetic analysis: A review. *Syst. Bot.* **16**:553–583.

Swofford, D. L. 1989. Phylogenetic Analysis Using Parsimony (PAUP; computer program distributed on floppy disk; write to Illinois Natural History Survey, 607 E. Peabody Dr., Champaign, IL 61820).

Swofford, D. L. 1991. When are phylogeny estimates from molecular and morphological data incongruent? in: M. M. Miyamoto and J. Cracraft (eds.), *Phylogenetic Analysis of DNA Sequences,* pp. 295–333. Oxford University Press, London.

Swofford, D. L., and Maddison, W. P. 1987. Reconstructing ancestral character states under Wagner parsimony. *Math. Biosci.* **87**:199–229.

Swofford, D. L., and Olsen, G. J. 1990. Phylogeny reconstruction, in: D. M. Hillis and C. Moritz (eds.), *Molecular Systematics,* pp. 411–501. Sinauer Associates, Sunderland, Mass.

Szalay, F. S. 1969. Mixodectidae, Microsyopidae, and the insectivore–primate transition. *Bull. Am. Mus. Nat. Hist.* **140**:193–330.

Szalay, F. S., and Drawhorn, G. 1980. Evolution and diversification of the Archonta in an arboreal milieu, in: W. P. Luckett (ed.), *Comparative Biology and Evolutionary Relationships of Tree Shrews,* pp. 133–169. Plenum Press, New York.

Szalay, F. S., and Lucas, S. G. 1990. Postcranial skeleton of *Mixodectes* and a rediagnosis of the Primates. *J. Vertebr. Paleontol.* **10**:45A.

Thewissen, J.G.M., and Babcock, S. K. 1991. Distinctive cranial and cervical innervation of wing muscles: New evidence for bat monophyly. *Science* **251**:934–936.

Thewissen, J.G.M., and Babcock, S. K. 1992. The origin of flight in bats: To go where no mammal has gone before. *BioScience* **42**:340–345.

Tinkle, D. W., and Gibbons, J. W. 1977. The distribution and evolution of viviparity in reptiles. *Misc. Publ. Mus. Zool. Univ. Mich.* **154**:1–55.

Wake, D. B. 1991. Homoplasy: The result of natural selection, or evidence of design limitations? *Am. Nat.* **138**:543–567.

Wible, J. R., and Novacek, M. J. 1988. Cranial evidence for the monophyletic origin of bats. *Am. Mus. Novit.* **2911**:1–19.

Wiley, E. O. 1981. *Phylogenetics: The Theory and Practice of Phylogenetic Analysis.* Wiley, New York.

Wiley, E. O., Siegel-Causey, D., Brooks, D. R., and Funk, V. A. 1991. The compleat cladist: A primer of phylogenetic procedures. *Univ. Kans. Mus. Nat. Hist. Spec. Publ.* **19**:1–158.

Origin and Evolution of Gliding in Early Cenozoic Dermoptera (Mammalia, Primatomorpha)

2

K. CHRISTOPHER BEARD

Introduction

Aside from the acquisition of powered flight in bats and the aquatic habits of whales, few mammalian life-styles have led to more drastic alterations of the postcranial skeleton than the evolution of gliding in the living Dermoptera (commonly known as colugos, flying lemurs, or taguans). For obvious reasons, fundamental macroevolutionary innovations such as those made by early bats, whales, and colugos are the subject of keen interest and debate among mammalian paleontologists and systematists (e.g., Jepsen, 1970; Gingerich *et al.*, 1983, 1990; Novacek, 1985; Beard, 1990a). For bats and whales these innovations were probably causal prerequisites for the adaptive radiations of these taxa. In contrast, it is much less obvious that the evolution of gliding in dermopterans led to an adaptive radiation at all, given the low diversity characteristic of the order today. Regardless of these differences in extant diver-

K. CHRISTOPHER BEARD • Section of Vertebrate Paleontology, Carnegie Museum of Natural History, Pittsburgh, Pennsylvania 15213.

Primates and Their Relatives in Phylogenetic Perspective, edited by Ross D.E. MacPhee. Plenum Press, New York, 1993.

sity, bats, whales, and colugos are distinctive components of the Earth's biota primarily because of the suites of diagnostic attributes they possess as a result of the key innovations that occurred earlier in their respective histories.

For those who are interested in phylogeny reconstruction, fundamental innovations such as powered flight in bats, aquatic adaptations in whales and gliding in colugos have both advantages and disadvantages. The advantages are self-evident: the suites of diagnostic attributes that relate to flight in bats, swimming in whales, and gliding in colugos serve as rich sets of characters, on the basis of which hypotheses of monophyly seem nearly invincible. Hence, it is not surprising that the monophyly of Cetacea and Dermoptera has never been seriously questioned (see Van Valen, 1968), and that hypotheses favoring the monophyly of Chiroptera rest heavily on characters that are functionally related to flight (Pettigrew, 1986, 1991a, b; Pettigrew *et al.*, 1989; Thewissen and Babcock, 1991, this volume; Baker *et al.*, 1991; Simmons *et al.*, 1991; Simmons, this volume). On the down side, these same suites of characters can reflect such pervasive evolutionary transformations that a phenetic chasm effectively separates the taxon in question from all other mammals. As a result, attempts to reconstruct the broader affinities of taxa such as Chiroptera and Cetacea have been severely hampered, and it is no accident that no consensus exists regarding these relationships.

One obvious means of addressing this dilemma would be to find and study fossil bats, whales, and colugos that had yet to evolve the full complement of autapomorphies otherwise characteristic of these groups. Clearly, for example, discovery of an early bat that had yet to evolve fully functional wings could yield great insight into the broader relationships of Chiroptera as a whole. This possibility of discovering primitive fossil representatives of higher taxa has long been recognized as one of the most important contributions that paleontology has to offer to phylogeny reconstruction (e.g., Gauthier *et al.*, 1988). Although early bats and whales sufficiently primitive to yield great insight into the broader relationships of these taxa have so far largely eluded paleontologists, this is not the case for Dermoptera.

Recent studies (Beard, 1989, 1990a, 1991b, 1993; Kay *et al.*, 1990) of newly recovered fossils of extinct, early Cenozoic dermopteran taxa such as Plesiadapidae, Paromomyidae, and Micromomyidae have provided vital insight into both the broader affinities of the order and the acquisition of gliding adaptations among more derived members of the group. Here, I will review the current evidence pertaining to this remarkable evolutionary transformation, whereby gliding colugos evolved from quadrupedal, nonvolant ancestors. New fossils that were not included in my earlier analyses have added significant additional information on the taxonomic distribution of gliding among early Cenozoic Dermoptera. Because the acquisition of gliding in Dermoptera also has important consequences for phylogenetic relationships among so-called "archontan" mammals, these issues will also be explored (see also Szalay and Lucas, this volume).

Sixty-Five Million Years of Gliding in Dermoptera

The term *Dermoptera* derives from Greek roots meaning "skin wing," which is an accurate verbal description of the patagium or gliding membrane of modern colugos. As this etymology implies, the dermopteran patagium is a soft anatomical structure that would rarely if ever become fossilized, barring exceptional depositional settings such as occur at the middle Eocene site of Messel in Germany (Schaal and Ziegler, 1988). Therefore, inferences concerning the presence of a patagium in extinct dermopteran taxa such as Paromomyidae must ultimately be based on osteological traits that are directly related (in a functional sense) to this structure. In spite of the drastic alterations that are found throughout the postcranial skeleton of living colugos (Leche, 1886; Shufeldt, 1911; Pocock, 1926), only the highly derived phalangeal osteology of these animals bears such a direct functional relationship to their patagium and associated gliding habits (Beard, 1989, 1990a).

However, the inference that paromomyids were gliders also rests on anatomical evidence that bears no obvious functional relationship to their inferred gliding habits (Beard, 1989, 1990a). Such a relationship is clearly necessary, because without independent evidence that paromomyids are cladistic dermopterans, morphologic correspondence between paromomyids and Recent colugos in characters thought to be related to the presence of a colugolike patagium would be virtually meaningless. In other words, if the preponderance of anatomical evidence known for paromomyids suggested close cladistic affinities with some mammalian group other than Dermoptera (e.g., primates or rodents), evidence for gliding in paromomyids would be completely lacking, even if the phalanges of paromomyids and Recent Dermoptera were absolutely identical. Therefore, the functional hypothesis (based on phalangeal anatomy) that paromomyids were gliding mammals is contingent on the phylogenetic hypothesis (based on the entire paromomyid skeleton) that paromomyids are cladistic dermopterans (Beard, 1989, 1990a, 1993). Conversely, evidence that paromomyids are cladistic dermopterans lends credence to the hypothesis that paromomyids were gliders in the same way that evidence on the phylogenetic position of any fossil animal yields insight into its ecology, behavior, or locomotion.

It must be emphasized here that this interplay between functional and phylogenetic hypotheses is not circular. As Krause (1991, p. 187) has emphasized, "Even if the hypothesis is confirmed that paromomyids are cladistic dermopterans . . . it does not ineluctably follow that paromomyids were gliders." However, if one accepts that paromomyids are cladistic dermopterans, and if paromomyids possess the same derived features of phalangeal anatomy that are directly related to the presence of a patagium in living dermopterans, it is certainly logical to hypothesize that paromomyids possessed patagia like those of their close living relatives. Indeed, there is no rational basis on which to assert otherwise.

Anatomy of Gliding in Modern Colugos

It has long been recognized that the patagium of extant colugos is more extensive than that of any other group of mammalian gliders (e.g., Leche, 1886; Chapman, 1902; Pocock, 1926). Pocock (1926), among others, subdivided the gliding membrane of colugos into the *propatagium,* located between the neck and the forelimbs, the *plagiopatagium,* located between forelimbs and hindlimbs, the *uropatagium,* located between the hindlimbs and the tail, and the *interdigital patagia,* located between the separate digital rays on both the manus and pes. The presence of interdigital patagia in colugos is unique among modern gliding mammals. This fact has led many systematists to posit a special relationship between colugos and bats, for which the supraordinal nomen Volitantia is available (e.g., Novacek, 1986). Although the monophyly of Volitantia is not supported here (see below), the interdigital patagia of modern colugos are perhaps the most conspicuous elements of their highly specialized cheiridia (Fig. 1), as the following passage from Pocock (1926, p. 436) illustrates:

> A peculiarity of the digits [of the manus] is the great length of the middle, or second phalange, as compared with the proximal or first, and the joint between them seems to be permanently bent upwards at an angle, the two phalanges being incapable of being set in a straight line. The digital pads, which are hard, horny and quite smooth, are remarkably elongated proximally. They are broadest in front and gradually taper to a point where they terminate near the middle of the lower surface of the second phalange. . . . Digits 2 to 5 are only narrowly separable proximally but distally they are wide apart, and the whole space between them is occupied by a sheet of skin, or web, which reaches to the summit of the base of the claw and is as extensive as the swimming-webs of the hind foot of a beaver or otter. A similar web connects the 1st and 2nd digits, but a portion of the attached edge of this rises from the inner margin of the sole between the pollex and the base of the 2nd digit.

As one might expect, this very unusual (if not bizarre) external anatomy of the manus in modern colugos is reflected by equally peculiar aspects of phalangeal osteology. For example, Pocock's observation that the proximal interphalangeal joints of modern colugos are permanently flexed is dictated by the anatomy of the distal articular surfaces of the proximal phalanges and especially by that of the proximal articular surfaces of the intermediate phalanges. The distal surfaces of the proximal phalanges of colugos are generally similar to those of a variety of mammals, with the exception that this surface is more trochleiform in colugos than in most mammals. In colugos and many other mammals the radius of curvature of the distal surface of the proximal phalanx varies along the arc that is formed by this surface in lateral view (Fig. 2). In contrast, the proximal surfaces of the intermediate phalanges of colugos differ from those of most mammals in being higher (dorsopalmarly) than wide (Fig. 3). As a result of these features, congruence between the conarticular surfaces that comprise the proximal interphalangeal joint varies considerably with flexion and extension in colugos. In flexed postures the proximal surface of the intermediate phalanx articulates with the palmar part of the

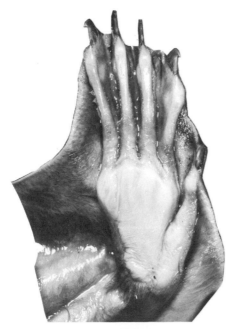

Fig. 1.   Right manus of juvenile *Cynocephalus volans* (Delaware Museum of Natural History 6244), palmar view. Note flexion at the proximal interphalangeal joints, elongated digital volar pads, and interdigital patagia.

distal surface of the proximal phalanx. This part of the distal surface of the proximal phalanx is characterized by a relatively long radius of curvature that conforms well with the long radius of curvature of the proximal surface of the intermediate phalanx. Hence, proximal interphalangeal joint congruence is maximized in flexion in colugos. In more extended postures the proximal surface of the intermediate phalanx articulates with the dorsal part of the proximal phalanx's distal surface. The much shorter radius of curvature on the latter does not match the curvature of the former, so that joint congruence is minimal in extended postures. Manipulation of isolated proximal and intermediate phalanges of colugos suggests that incongruence between their respective articular surfaces prohibits extension beyond about 140° (180° represents full extension, with the long axes of both phalanges in alignment). The possible role of soft structures, such as the collateral ligaments of the proximal interphalangeal joint, is not considered here, but these structures can only serve to further restrict the range of extension at this joint in colugos.

Among claw-bearing mammals in which full extension of the proximal interphalangeal joint is possible, such as sciurids and tupaiids, a high degree of congruence exists between the distal surface of the proximal phalanx and the proximal surface of the intermediate phalanx throughout the range of flexion–extension that occurs. Primarily, this high degree of joint congruence is maintained because the proximal articular surface of the intermediate pha-

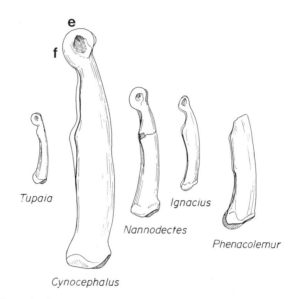

Fig. 2. Manual proximal phalanges of selected eutherians in lateral view. Note difference in radius of curvature of distal articular surface at points *f* and *e*. During flexion at the proximal interphalangeal joint, the intermediate phalanx articulates with the proximal phalanx near point *f*, which shows a relatively long radius of curvature. During more extended postures, the intermediate phalanx articulates with the proximal phalanx near point *e*, which shows a much shorter radius of curvature. Incongruence between the articular surfaces of the proximal and intermediate phalanges prevents full extension at the proximal interphalangeal joint in colugos and, apparently, in paromomyids.

lanx is relatively much lower (dorsopalmarly) in these mammals than is the case in colugos (Fig. 3). As a result, the proximal surface of the intermediate phalanx forms a much shorter arc (along a parasagittal axis) that more readily accommodates the changes in curvature of the distal surface of the proximal phalanx encountered along the flexion–extension excursion. Thus, the specialized proximal articular surfaces of the intermediate phalanges in colugos

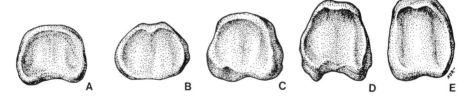

Fig. 3. Intermediate phalanges of selected eutherians in proximal view. (A) *Sciurus carolinensis*, (B) *Tupaia glis*, (C) *Nannodectes intermedius* (USNM 442229), (D) *Ignacius graybullianus* (USNM 442253), (E) *Cynocephalus variegatus*. Note relatively great height of the proximal articular surface in the paromomyid *Ignacius* and in the colugo, in contrast to other taxa.

are primarily responsible for the permanent flexion at the proximal inter-phalangeal joints that is characteristic of these animals. This and other post-cranial specializations in colugos prohibit them from engaging in generalized quadrupedal or scansorial locomotion like that of squirrels and tree shrews (Wharton, 1950; Beard, 1989; Wischusen, 1990).

Other specializations of the manus of colugos noted by Pocock, especially the presence of interdigital patagia, are of more immediate interest for the purposes of this chapter. The most obvious of these is the great elongation of the intermediate phalanges of colugos, which are highly atypical in being longer than their corresponding proximal phalanges. However, the inter-mediate phalanges of colugos are also unusual in having shafts that are ex-tremely straight, rather than being curved so that they are concave palmarly (as in many primates) or being recurved dorsally near their distal ends (as in tree shrews, squirrels, and many other mammals; Fig. 4). In tree shrews and squirrels, this dorsally recurved part of the intermediate phalangeal shafts is located above the digital volar pads, which are oval in outline and occur immediately proximal to the claws (Pocock, 1922; Haines, 1955; Cartmill, 1974). The remarkably straight and elongated intermediate phalangeal shafts of colugos are associated with digital volar pads that are also greatly elongated (Fig. 1; Pocock, 1926). The digital volar pads of colugos differ radically in this respect from those of tree shrews, squirrels, and most other mammals.

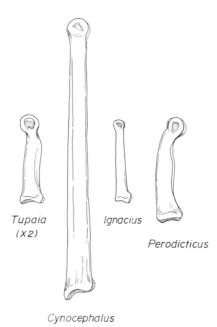

Tupaia
(*X2*)

Ignacius

Perodicticus

Cynocephalus

Fig. 4.   Intermediate phalanges in lateral view. Note similarity between *Cynocephalus* and *Ignacius* in the degree of elongation and straightness of phalangeal shaft, in contrast to other taxa.

The precise relationship between the elongated intermediate phalanges of colugos and the function of the interdigital patagia remains to be clarified by the appropriate biomechanical analyses and observations of gliding colugos in the wild. However, it is possible to model this relationship qualitatively on the basis of known principles of aerodynamics. This approach has already been applied, in a more quantitative fashion than is possible here, to flying squirrels by Thorington and Heaney (1981).

The possibility that the elongated intermediate phalanges of colugos merely occur in association with their interdigital patagia, without any causal or functional relationship, seems most unlikely. At the least, the derived intermediate phalanges of colugos have altered certain aerodynamic properties of the dermopteran patagium. This can be demonstrated by answering the following question: "Given the presence of interdigital patagia, what are the effects of elongation of the intermediate phalanges in a gliding mammal?" There are at least four ways that elongation of this sort has an impact on the aerodynamics of gliding, but only one of these appears to be of substantial importance in colugos.

First, elongation of the intermediate phalanges results in an increased surface area for the patagium as a whole by increasing the contribution made by the interdigital patagia. This increased patagial surface area in turn results in decreased patagial loading, assuming a negligible increase in body weight associated with phalangeal elongation. Thorington and Heaney (1981), on the basis of an analogy with hang gliders, suggest that variation in patagial loading has no effect on the optimal glide ratio (greatest horizontal distance/ vertical drop) in flying squirrels, although it does affect the speed at which it is necessary to glide for glide ratios to be maximized. Specifically, "Because the speed is proportional to the square root of wing loading, a flying squirrel with a patagium loading four times that of another presumably must glide twice as fast to maximize the distance of its glide" (Thorington and Heaney, 1981, p. 111). Because of the relatively small differences in patagial surface area, patagial loading, and especially optimal gliding speeds that would result from elongation of the intermediate phalanges, it seems unlikely that these factors were important determinants of this elongation in colugos.

A second effect of elongation of the intermediate phalanges on the aerodynamics of gliding occurs through modification of the aspect ratio (wing span^2/patagial surface area) of the patagium. The optimal glide ratio is dependent on the aspect ratio in hang gliders, and presumably in gliding mammals. For example, Thorington and Heaney (1981, p. 111) note that, "For a Rogallo conical wing, changing the aspect ratio from 1 to 2 changes the lift/drag ratio from 3 to 4, approximately, and therefore changes the optimal glide ratio from 3 to 4." However, elongation of the intermediate phalanges to the extent that occurs in colugos obviously has a very small effect on the aspect ratio, because the slight increase in wing span would be largely offset by a similar increase in patagial surface area. Consequently, these factors are also unlikely to have been important influences on elongation of the intermediate phalanges in colugos.

A third effect of elongation of the intermediate phalanges concerns the influence of tip shape on the aerodynamic properties of gliding structures having low aspect ratios (Zimmerman, 1932; Thorington and Heaney, 1981). Because the patagia of all gliding mammals have relatively low aspect ratios, this factor is potentially important. The experiments of Zimmerman (1932) demonstrate, for example, that maximum lift coefficients and maximum resultant force coefficients are increased in low-aspect-ratio airfoils having semicircular tips compared with those having rectangular tips. These properties in turn affect variables such as landing speed and variation in the lift/drag ratio with angle of attack. In general, Zimmerman (1932) concluded that semicircular tips are aerodynamically superior to rectangular tips in airfoils having low aspect ratios. Elongation of the intermediate phalanges in colugos acts to "round off" the tip shape of the leading edge of the patagium when the separate digits are maximally abducted (see photograph in MacKinnon, 1984, p. 447). Therefore, patagial tip shape is not static in colugos, but rather is under voluntary control to some extent. Thorington and Heaney (1981) reached a similar conclusion in the case of flying squirrels.

By far the most important effect of elongation of the intermediate phalanges on the aerodynamics of a mammalian glider having interdigital patagia relates to the phenomenon of billowing (Beard, 1989, 1990a). In inanimate gliding structures such as hang gliders, the degree of billowing is relatively static. According to Thorington and Heaney (1981, p. 112):

> Hang gliders are generally designed with 3° to 4° of billow. The less billow, the better the glide ratio can be, and hang gliders with 1° of billow have been flown. However, they are dangerous because they sideslip easily. As billowing is increased, the lift and drag coefficients are increased and air speed can be decreased, but at the cost of a decreasing glide ratio.

In at least some mammalian gliders, including flying squirrels (Thorington and Heaney, 1981) and colugos (Beard, 1989, 1990a), the degree of patagial billowing is both dynamic and under voluntary control. Flexion of the proximal interphalangeal joints in colugos results in a "parachute" effect—that is, a marked increase in patagial billowing—because of the presence of the interdigital patagia, which extend distally to the level of the distal interphalangeal joints. Elongation of the intermediate phalanges amplifies the degree to which flexion at the proximal interphalangeal joints leads to patagial billowing, for the obvious reason that a proportionately greater part of the edge of the patagium is downturned to act as a parachute. In this manner colugos are able to control the amount of billowing to which their patagia are subjected, simply by flexing and extending their proximal interphalangeal joints. As was noted above, this degree of billowing in turn affects almost all other aerodynamically significant properties of the patagium, including the glide ratio, the lift and drag coefficients, and air speed. Moreover, by flexing the proximal interphalangeal joints of one manus more than the other, colugos should be able to control patagial billowing on a differential, bilateral basis. By this means colugos should also be able to control the direction of their glides while· they are airborne.

In light of the acute influence of billowing on the aerodynamic properties of the patagium, it is interesting to note the myriad anatomical specializations involving the attachment of the patagium to the forelimb among gliding mammals. Briefly, they are as follows (see Thorington, 1984): (1) In extant Petauristinae (Rodentia, Sciuridae) the patagium attaches to a cartilaginous rod that flares ulnarly from the wrist. (2) In extant Anomaluridae (aside from *Zenkerella*, which lacks a patagium) the patagium also attaches to a cartilaginous rod, although in anomalurids this structure flares laterally from the elbow rather than the wrist. (3) In extant *Petaurus* (Marsupialia, Petauridae) the patagium attaches to the ulnar side of the fifth manual digit, which is elongated compared with the other manual digits of this genus. (This is clearly an autapomorphous condition, because in other closely related marsupials the fourth manual digit is longer than the fifth.) (4) In colugos the patagium attaches near the distal interphalangeal joints of all five manual digits and the manual intermediate phalanges are elongated. (5) In *Acrobates* (Marsupialia, Burramyidae) the patagium attaches to the ulnar side of the wrist. (6) In *Petauroides* (Marsupialia, Pseudocheiridae) the patagium attaches to the elbow.

The general pattern that emerges from the preceding overview is one in which the site for attachment of the patagium to the forelimb shows some sort of morphologic specialization that is peculiar to the specific group involved. That is, no two groups of extant gliding mammals have a similar type of patagial attachment to the forelimb (Thorington, 1984). Furthermore, in four of the six mammalian taxa that have evolved a capacity for gliding, there is some degree of elongation of the site of attachment of the patagium to the forelimb: Petauristinae and Anomaluridae have each evolved neomorphic cartilaginous rods that flare laterally from the wrist and elbow, respectively; *Petaurus* has elongated the fifth manual digit; colugos have elongated all of their manual intermediate phalanges.

Why has this elongation of the site of attachment of the patagium to the forelimb occurred in four of the six mammalian groups with a capacity for gliding? Following the example of colugos discussed above, it seems most likely that these morphologic specializations allow dynamic regulation over the degree of patagial billowing, and thus provide voluntary control over numerous aerodynamic properties of the patagium during gliding. Indeed, Thorington and Heaney (1981) have previously suggested that patagial billowing is dynamic and under voluntary control in extant Petauristinae. Patagial billowing also appears to be under dynamic control in extant Anomaluridae and *Petaurus*, judging from the anatomical specializations associated with the mode of patagial attachment to the forelimb they exhibit (see above).

Indeed, field observations of the gliding behavior of *Petaurus* attest to the refined degree of control this genus has over the aerodynamic properties of its patagium. For example, Wakefield (1970, p. 235) reports that "The aerobatic ability of the *Petaurus* was demonstrated in the Shining Gum forest near Cumberland Junction on the night of 10/9/1966, when a spotlight beam was directed at a glider which was approaching through the air. Its response was to bank steeply, execute a tight half-circle, and then land in a tree back in the direction from which it had come." In contrast, in extant *Petauroides* the patagium attaches in an anatomically simple fashion to the elbow, and *Petauroides* therefore plausibly lacks a high degree of control over its patagial billowing. Thus, it may be significant that *Petauroides* is reported to exhibit relatively poor control over its gliding locomotion. For example, according to Smith (1984, p. 857), "The heavier Greater glider [*Petauroides volans*] . . . descends steeply with limited control [while gliding], but the smaller gliders [*Petaurus* spp.] are accomplished acrobats that weave and maneuver gracefully between trees, landing with precision by swooping upwards."

To summarize, the greatly elongated intermediate phalanges of colugos appear to perform the same function—that of allowing voluntary, dynamic control over patagial billowing during gliding—that is met by such diverse structures as the neomorphic, cartilaginous rods of petauristine and anomalurid rodents, and the elongated fifth manual digits of *Petaurus*. In each of these taxa, the derived anatomical complexes relating to control over patagial billowing serve as potential characters for reconstructing both phylogenetic relationships and paleoecological attributes. Indeed, Thorington (1984) has forcefully and explicitly argued for the monophyly of the Petauristinae solely on the basis of this anatomical complex in flying squirrels. Were one to discover extinct sciurids with these same derived characters, it would be logical to conclude that they were members of the Petauristinae that were capable of gliding. Similarly, the unique phalangeal features of modern colugos discussed above have the potential to indicate the presence of a colugolike patagium in fossil dermopterans, if it can be demonstrated that the fossil form in question possesses homologously derived aspects of phalangeal anatomy.

Anatomical Evidence for Gliding in Paromomyids

Two partial skeletons of paromomyids are known, both of which preserve proximal and intermediate phalanges in variable states of preservation (Beard, 1989, 1990a). The older specimen (UM 66440 and UM 86352) represents *Phenacolemur praecox* and is early Wasatchian in age (either Wa1 or Wa2 in the zonation scheme of Gingerich, 1982, 1983). The younger specimen (USGS 17847) represents *Phenacolemur* sp., cf. *P. jepseni*, and is late Wasatchian (Lostcabinian) in age. Both specimens were recovered from the Willwood Formation in the Clark's Fork and Bighorn Basins of northwestern Wyoming. Further information on the provenance, history of discovery, and anatomy of these two specimens is provided by Beard (1989). Like the vast majority of vertebrate fossils recovered from fluvial depositional settings, the individual elements, including the phalanges, that comprise these two specimens were not recovered in articulation with one another and are only variably complete. Nevertheless, these specimens are critical in providing the only unambiguous association between paromomyid phalanges and dental remains.

More nearly complete paromomyid phalanges, including both proximal and intermediate phalanges that are very well preserved, are known from a highly fossiliferous calcareous nodule collected by Dr. Peter Houde from within or near UM locality SC-4 in the Clark's Fork Basin. This nodule, which is early Wasatchian (Wa1) in age, has yielded abundant dentitions and post-cranial fossils of paromomyid and micromomyid dermopterans, many of which have already been described (Beard and Houde, 1989; Beard, 1989, 1990a). Paromomyid phalanges from this nodule are attributed to *Phenacolemur simonsi* and *Ignacius graybullianus* on the basis of their size and provenance (Beard, 1989, 1990a).

There is a close, overall phenetic similarity between the proximal and intermediate phalanges of paromomyids and those of Recent colugos (Beard, 1989, 1990a; Fig. 4). Because the intermediate phalanges of colugos have been shown to be highly derived in ways that are directly related to the presence and function of the dermopteran patagium, I will concentrate on the anatomy of the intermediate phalanges of paromomyids here.

It is pertinent to note that Krause (1991) objects to my interpretations of phalangeal anatomy in paromomyids. He insists that without directly articulated proximal and intermediate phalanges of a single paromomyid individual, nothing can be surmised about elongation of the intermediate phalanges in these animals. Below, I demonstrate that Krause is wrong on this point. Krause (1991) also incorrectly asserts that my attribution of relatively elongated intermediate phalanges to the manus (rather than the pes) of paromomyid species is based on circular reasoning. In fact, the paromomyid intermediate phalanges I described (Beard, 1989, 1990a) included specimens showing a high degree of elongation, such as USNM 442254, and specimens that are much less elongated, such as USNM 442252 (Fig. 5). I suggested that the more elongated intermediate phalanges represent the manus of paromomyids while the less elongated specimens represent the pes, based on an analogy with extant colugos in which this is the case. This analogy does not employ circular reasoning because of the independent evidence from other aspects of the postcranial and cranial anatomy of paromomyids indicating that paromomyids are cladistic dermopterans (Beard, 1989, 1990a, 1993; Kay *et al.*, 1990). Given the phylogenetic evidence that paromomyids and living colugos are closely related, it would obviously be illogical to assume that paromomyids show the opposite pattern of elongation of their intermediate phalanges of that found in living colugos. New evidence on phalangeal anatomy in micromomyids corroborates my original attribution of relatively elongated paromomyid intermediate phalanges to the manus (see below).

One way in which the intermediate phalanges of paromomyids closely resemble those of living colugos is the anatomy of the proximal articular

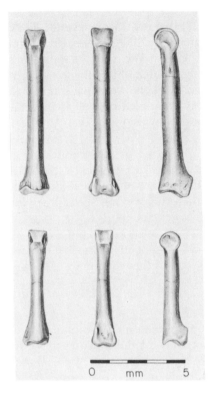

Fig. 5.   Isolated paromomyid intermediate phalanges attributed to *Phenacolemur simonsi*. Views are (from left to right): dorsal, palmar, and lateral. Top: USNM 442254, attributed to the manus. Bottom: USNM 442252, attributed to the pes.

surfaces of these bones. In both paromomyids and colugos, this surface is higher than it is wide (Fig. 3). In most other claw-bearing mammals, including sciurids and tupaiids, the proximal articular surfaces of the intermediate phalanges are wider than they are high. The anatomy of the proximal articular surface of the intermediate phalanges prevents fully extended postures at the proximal interphalangeal joints in living colugos (see above). The similar anatomy of this articular surface in paromomyids suggests that fully extended postures at the proximal interphalangeal joints were also precluded in these animals. Manipulation of isolated proximal and intermediate phalanges of paromomyids confirms this conclusion.

Paromomyid intermediate phalanges further resemble those of colugos in having extremely straight shafts that lack the distal recurvature found in tupaiids, sciurids, and many other mammals (Figs. 4 and 5). In this respect, paromomyid intermediate phalanges also differ strongly from those of primates, which are curved along the entire length of their shafts so that they are concave palmarly (Fig. 4). The extremely straight intermediate phalangeal shafts of living colugos are associated with remarkably elongated digital volar pads (Fig. 1; Pocock, 1926) that contrast markedly with the more rounded or oval digital volar pads of most other claw-bearing mammals (Pocock, 1922; Haines, 1955; Cartmill, 1974). The very straight intermediate phalangeal shafts of paromomyids suggest (but obviously do not prove) that these animals may have possessed elongated digital volar pads like those of modern colugos.

More significant in a functional context is the elongation of the intermediate phalanges in paromomyids, which is an additional way in which these animals resemble living colugos. Previously, I compared the lengths of the elongated intermediate phalanges of paromomyids with those of proximal phalanges attributed to the same species (Beard, 1989, 1990a). These comparisons suggested that the (inferred manual) intermediate phalanges of paromomyids were longer than the proximal phalanges with which they articulated, a specific similarity to extant colugos.

In light of Krause's (1991) objections to these comparisons, I have devised an alternative means of quantifying elongation of the intermediate phalanges in paromomyids, colugos, and other mammals. The new index is as follows:

$$L/(W \times H)^{1/2}$$

where L is the proximodistal length of the phalanx, and W and H are the width and height, respectively, of the phalangeal shaft at its midpoint. Because this index quantifies elongation of the intermediate phalanges without reference to the proximal phalanges, it is appropriate for use in interspecific comparisons involving fossil taxa like paromomyids, in which the proximal and intermediate phalanges are rarely preserved in articulation with one another. Values for this index, obtained from measurements of manual intermediate phalanges in a selected group of mammals, are presented in Table I. Examination of these data confirms that the isolated intermediate phalanges of paromomyids attributed to the manus (Beard, 1989, 1990a) differ from those of most other mammals, including the plesiadapids *Nannodectes intermedius* and *Plesiadapis tricuspidens,* in being more elongated. Importantly, the degree of phalangeal elongation shown by paromomyids appears to be intermediate between that of most other mammals (here assumed to be the primitive condition by virtue of its widespread taxonomic distribution) and the more extreme phalangeal elongation characteristic of modern colugos.

The strong phenetic similarity between the phalanges of paromomyids and those of modern colugos and the independent phylogenetic evidence from the entire paromomyid skeleton that paromomyids and extant colugos are closely related (Beard, 1993) suggest that the elongated intermediate phalanges of paromomyids are homologous with those of living colugos. The lesser degree of elongation of the intermediate phalanges in paromomyids may imply that the paromomyid condition is more primitive than that of colugos, but there is no reason to deny that these two character states lie on the same morphocline, each being homologously derived with respect to the

**Table I. Representative Values for Index of Elongation
of Manual Intermediate Phalanges in Selected Living and Fossil Mammals**

Species	Specimen No.	Digit(s)[a]	Index value
Didelphis virginiana	CM 55556	2	2.86
Petauroides volans	CM 50876	4	3.82
Procyon lotor	CM 5350	4	5.08
Rattus rattus	CM 48619	2	3.67
Sciurus carolinensis	USNM 528045	—	5.21
Elephantulus edwardi	CM 40807	2	3.31
Tupaia glis	USNM 397983	—	3.93
Perodicticus potto	CM 57886	3,4	3.92–4.98
Nycticebus coucang	CM 59519	3–5	4.47–5.04
Galago senegalensis	CM 6963	3,4	6.21–6.93
Nannodectes intermedius	USNM 442229	—	5.04–6.11
Ignacius graybullianus	USNM 442253	—	8.98
Phenacolemur simonsi	USNM 442254	—	10.12
Tinimomys graybulliensis	USNM 461201	—	9.79
Cynocephalus variegatus	USNM 317118	—	16.62

[a]Generally unknown in fossil taxa, occasionally unknown in extant sample available for this study.

primitive condition seen in plesiadapids and many other mammals. Alternatively, the greater elongation of the intermediate phalanges in colugos may be merely an allometric phenomenon, reflecting the increased requirements of dermopteran-style patagial billowing in animals having much greater body mass than did paromomyids.

In either case, the only coherent functional explanation for the elongated intermediate phalanges of paromomyids is that they served a purpose homologous with that of the elongated intermediate phalanges of modern colugos—that being to provide voluntary control over patagial billowing by flexion–extension at the proximal interphalangeal joints. This inferred function obviously implies the presence of a colugolike patagium, replete with interdigital patagia. Detractors of the hypothesis that paromomyids were gliding mammals have yet to offer an alternative explanation for the function of the elongated intermediate phalanges of these animals. While there is little doubt that alternative explanations could be concocted, they would be *ad hoc* by nature and less plausible as a result. Thus, available evidence indicates that paromomyid dermopterans had already evolved the capacity for gliding by the time of their first appearance in the fossil record during the middle Paleocene (Torrejonian Land Mammal Age), some 64 Ma (Archibald *et al.*, 1987). This constitutes by far the oldest evidence for the evolution of gliding behavior in Mammalia (Beard, 1990a), and must rank with the acquisition of flight by bats by at least the early Eocene (Jepsen, 1970) as one of the most extraordinary examples of precocious locomotor adaptations in the mammalian fossil record.

Anatomical Evidence for Gliding in Micromomyids

Prior to this study, only isolated postcranial elements had been attributed to the Micromomyidae (Beard, 1989). Two new partial skeletons of *Tinimomys graybulliensis*, USNM 461201 and USNM 461202, confirm that these earlier attributions were correct and provide the first information on phalangeal anatomy in Micromomyidae. Neither of the two new partial skeletons has yet been fully prepared or studied, so only a few preliminary observations on their anatomy are made here.

USNM 461201 is a partial skeleton of *T. graybulliensis* that includes an exceptionally complete dentition; semiarticulated left scapula, humerus, radius, and ulna; right scapula, humerus, and ulna (not articulated); articulated partial vertebral column with ribs; and a few phalanges and metapodials. No elements of the hindlimbs are preserved in USNM 461201. Included among the phalangeal elements is a complete intermediate phalanx (Fig. 6) that is not preserved in articulation with any other element. However, because most of both forelimbs and no elements of the hindlimbs are included in this specimen, it seems likely that the intermediate phalanx represents the manus rather than the pes. It is preserved in closest proximity to the right scapula and right ulna, so it plausibly represents the right manus rather than the left.

Given the spatial information suggesting that the USNM 461201 intermediate phalanx belongs to the manus rather than the pes, it is interesting that it shows a high degree of elongation (Fig. 6; Table I), similar to that of isolated paromomyid intermediate phalanges that were also attributed to the manus (Beard, 1989, 1990a). Along with the arguments based on homology between the intermediate phalanges of paromomyids and colugos discussed above, this new information on phalangeal anatomy in micromomyids corroborates my original attribution of relatively elongated paromomyid intermediate phalanges to the manus (Beard, 1989, 1990a). Because the USNM 461201 intermediate phalanx is not preserved in articulation with the rest of the osseous elements of the manus, it clearly does not prove that my allocation of similarly elongated paromomyid intermediate phalanges to the manus is correct. However, it stands as one more empirical observation that supports this view.

In all respects except absolute size, the USNM 461201 intermediate phalanx closely resembles those of paromomyids. Available evidence from the remainder of the micromomyid skeleton suggests that these animals are closely related to paromomyids and living colugos (Beard, 1989, 1993). Thus, micromomyids also appear to have possessed colugolike gliding membranes with interdigital patagia, for the same reasons discussed above with respect to paromomyids. Based on correlations between molar size and body mass in living primates, body mass in micromomyids is thought to have ranged between 20 and 35 g (Fleagle, 1988, p. 275). If this estimate proves accurate, micromomyids were among the smallest gliding mammals ever to have evolved. The range in body mass for micromomyids probably overlapped

Fig. 6.   Intermediate phalanx (i) of *Tinimomys graybulliensis* (USNM 461201), dorsal view. The specimen is included in a partial skeleton and associated dentition of the species (other elements are out of focus in this view, but some forelimb elements remain articulated).

those of the smallest living flying squirrels (*Petinomys* and *Petaurillus*), but may have been slightly greater than that of the gliding burramyid marsupial *Acrobates pygmaeus*, for which body masses of 12–14 g have been reported (Nowak and Paradiso, 1983, p. 72).

Relevant Phylogenetic Evidence

Independent phylogenetic evidence on the relationships of the extinct Paromomyidae and Micromomyidae plays a pivotal role in my interpretation of the functional significance of their derived phalanges. Plainly, my inference that paromomyids and micromomyids were gliding mammals is contingent on my conclusion that these animals are closely related to extant colugos. If it can be demonstrated that this phylogenetic reconstruction is incorrect, evidence for gliding in either paromomyids or micromomyids would be lacking. However, all available evidence suggests that paromomyids and micromomyids *are* closely related to colugos. Spatial constraints prohibit me from going into further detail on this topic here, but expanded discussion has been presented elsewhere (Beard, 1989, 1990a, 1991b, 1993; Kay *et al.*, 1990).

Subsidiary Evidence

One measure of the value of a hypothesis is its ability to explain seemingly unrelated empirical observations that otherwise lack a coherent explanation.

Prior to formulation of the hypothesis that paromomyids were gliding mammals closely related to extant colugos, several empirical observations about these animals were widely acknowledged but remained enigmatic, defying any unified explanation. For many years a general consensus held that paromomyids were closely related to other extinct taxa, such as plesiadapids, carpolestids, saxonellids, and micromomyids (micromomyids have often been classified as members of the Paromomyidae). Usually, all of the preceding taxa were grouped together as the Plesiadapiformes—widely considered to be the geologically earliest and anatomically most primitive members of the Primates (e.g., Szalay and Delson, 1979). Given this consensus on the relationships of the Paromomyidae, certain empirical observations seemed anomalous and required explanation.

Most prominent among these was the differential pattern of extinction shown by paromomyids and their closest extinct relatives. Plesiadapids, carpolestids, and saxonellids all became extinct in North America by the advent of the Wasatchian Land Mammal Age, or very shortly thereafter (Rose and Bown, 1982; Maas *et al.*, 1988). [It must be noted that the North American record of Saxonellidae is limited to the middle Tiffanian of Alberta, Canada (Fox, 1991), which contrasts with the much better stratigraphic and geographic representation available for Plesiadapidae and Carpolestidae.] In contrast, paromomyids survived into the Duchesnean Land Mammal Age (e.g., Krishtalka, 1978), some 15 million years after the extinction of plesiadapids and carpolestids. Maas *et al.* (1988) suggested that competitive exclusion by rodents may have led to the extinction of plesiadapids and carpolestids in North America, but these authors offered no explanation for the differential pattern of extinction shown by paromomyids on the one hand and carpolestids and plesiadapids on the other.

A second empirical observation that defied explanation was the different pattern of abundance shown by paromomyids and, especially, plesiadapids during the interval in which these taxa coexisted. Plesiadapids commonly comprised dominant components of the mammalian paleofaunas to which they belonged (Russell, 1964; Gingerich, 1976; Rose, 1981), while paromomyids were always less abundant animals, despite their much longer temporal persistence.

Finally, the dentition of paromomyids, particularly that of derived taxa such as *Phenacolemur* and *Ignacius*, was known to be unique among plesiadapiforms in being strongly convergent on that of the living marsupial sugar-glider *Petaurus* (Gingerich, 1974; Kay and Cartmill, 1977). While most workers assumed that this convergence in dental anatomy implied similar diets in paromomyids and *Petaurus*, it was not apparent how this might relate to the low abundance and delayed extinction of paromomyids compared with plesiadapids and carpolestids.

The hypothesis that paromomyids were gliding mammals closely related to living colugos provides, for the first time, a unified, coherent, and internally consistent explanation for all of these otherwise anomalous observations about paromomyid paleobiology (Beard, 1989, 1990a). This hypothesis implies that paromomyids were very close ecological analogues of modern *Petaurus*—both taxa comprise gliding mammalian gum-feeders of similar body size. As such, the hypothesis that paromomyids were gliders is fully consistent with the observation that their dentition is strongly convergent on that of modern *Petaurus*. Moreover, if the ecological analogy between paromomyids and *Petaurus* is accurate, it also provides the first plausible explanation for the persistence of paromomyids for some 15 million years beyond the extinction of plesiadapids, carpolestids, and saxonellids. Regardless of whether competitive exclusion by early rodents was involved in the extinction of plesiadapids, carpolestids, and saxonellids (Maas *et al.*, 1988), it is clear that competition between paromomyids and early rodents played no part in paromomyid extinction—the two taxa coexisted in North America for some 18 million years (Beard, 1990a). If paromomyids were close ecological analogues of modern *Petaurus,* this apparent lack of competition can readily be envisioned. Similarly, the close ecological analogy between paromomyids and *Petaurus* provides an explanation for the differences in relative abundance observed between paromomyids and, especially, plesiadapids. Available evidence suggests that plesiadapids were primitive dermopterans that lacked patagia (see below); they seem to have been ecologically more generalized, eurytopic animals than were paromomyids.

The lines of evidence that have just been discussed do not directly corroborate the hypothesis that paromomyids were gliding mammals. They are merely ancillary observations that are consistent with it, and for which other coherent explanations are lacking. In particular, workers who question the hypothesis that paromomyids were gliders have yet to offer an alternative interpretation of these phenomena.

Antecedent Conditions in Primitive Dermoptera

Archaic plesiadapiform dermopterans lacked the phalangeal specializations that are directly related to the function of the patagium in living and fossil eudermopterans (Beard, 1989, 1990a, 1993). Major limb elements of *Plesiadapis tricuspidens* (Szalay *et al.*, 1975) and the North American plesiadapid *Nannodectes intermedius* (Beard, 1989) are shorter and much more robust than those of paromomyid, micromomyid, and galeopithecid eudermopterans (Beard, 1989, 1991b, 1993), and are unlikely to have belonged to gliding mammals. More conclusive evidence that archaic dermopterans such as plesiadapids lacked a patagium is provided by the partial skeletons of *Plesiadapis insignis* known from the Paleocene of Menat, France (Russell, 1967; Gingerich, 1976, 1986). One of these partial skeletons preserves impressions of hair around its articulated caudal vertebrae, demonstrating that *P. insignis* possessed a bushy tail reminiscent of that of the extant aye-aye *Daubentonia*. It follows that *P. insignis* certainly lacked a uropatagium, in contrast to bats and living colugos, and there is little doubt that *P. insignis* lacked a patagium altogether. Direct evidence is lacking with respect to the presence or absence of a patagium in other plesiadapiform dermopterans, such as Carpolestidae and Saxonellidae. However, from available dental evidence, it appears that Plesiadapidae, Carpolestidae, and Saxonellidae are closely related (Russell, 1964; Rose, 1975; Szalay and Delson, 1979; Fox, 1991). Thus, it seems likely that archaic dermopteran taxa such as Carpolestidae and Saxonellidae lacked a patagium as well.

The absence of a patagium in archaic fossil dermopterans requires that its presence in paromomyids, micromomyids, and galeopithecids be interpreted as a key synapomorphy for the suborder Eudermoptera (Beard, 1989, 1990a, 1993). That is, available paleontological data demonstrate that the dermopteran patagium evolved *within* the order, after ancestral dermopterans had diverged from all other mammals (but see Szalay and Lucas, this volume).

Implications for Phylogeny of "Archontan" Mammals

Revised Higher-Level Systematics of Dermoptera

New fossils of the micromomyid dermopteran *Tinimomys graybulliensis* include the USNM 461201 partial skeleton and a partial skull and associated forelimb elements of a second individual, USNM 461202. Although neither of these specimens has yet been fully prepared or studied, it is already obvious that the higher-level systematics of Dermoptera must be revised in light of these new fossils.

Previously, I divided the order Dermoptera into the suborders Micromomyiformes, including Micromomyidae only, and Plesiadapiformes, including Plesiadapidae, Carpolestidae, Saxonellidae, Paromomyidae, and Galeopithecidae (Beard, 1989, 1993). Similarly, the suborder Plesiadapiformes was divided into the infraorders Plesiadapoidea, including Plesiadapidae, Carpolestidae, and Saxonellidae, and Eudermoptera, including Paromomyidae and Galeopithecidae. This systematic arrangement was based on a cladistic analysis (Beard, 1993) that has since proven to be partly deficient because of missing and erroneous information on the anatomy of the Micromomyidae. Prior to this study, nothing was known about phalangeal anatomy in micromomyids and the sole source of information on the cranial anatomy of the group was an isolated petrosal fragment that was tentatively associated with dental remains of *Tinimomys graybulliensis* by Gunnell (1989, pp. 85–88). This petrosal fragment, catalogued as part of UM 85176, no longer appears to pertain to *T. graybulliensis* because its anatomy does not match that of USNM 461202. Together, the lack of information on phalangeal anatomy in micromomyids and the erroneous information on micromomyid cranial anatomy caused these animals to appear to be the sister group of all other Dermoptera in earlier cladistic analyses (Beard, 1989, 1993). Consequently, they were elevated to a taxonomic rank consistent with their presumed basal phylogenetic position (Beard, 1989, 1993).

In contrast to these earlier results, micromomyids now appear to be nested well within the Dermoptera, being the sister group of Paromomyidae. In light of the new anatomical information on Micromomyidae, a revised classification of Dermoptera is as follows (asterisks denote extinct taxa):

Order Dermoptera Illiger, 1811
  *Suborder Plesiadapiformes Simons and Tattersall, 1972
    *Family Plesiadapidae Trouessart, 1897
    *Family Carpolestidae Simpson, 1935
    *Family Saxonellidae Russell, 1964
  Suborder Eudermoptera Beard, 1993
    *Infraorder Paromomyiformes Szalay, 1973
      *Family Paromomyidae Simpson, 1940
      *Family Micromomyidae Szalay, 1974
    Infraorder Colugodontiformes, new
      Family Galeopithecidae Gray, 1821

Volitantia Is an Unnatural Assemblage

Faced with the phenetic chasm separating living and fossil bats from all other eutherians, numerous systematists have seized on the superficial similarities between the patagia of extant colugos and bats to argue for a sister group relationship between Dermoptera and Chiroptera. This association of bats and colugos, for which the supraordinal taxon Volitantia was proposed, dates at least to the pioneering anatomical work of Leche (1886). During the early part of this century, workers such as Chapman (1902) and Pocock (1926) defended the association of bats and colugos, although their conclusions were founded on the relatively unsophisticated phylogenetic methodology prevalent at that time. For various reasons, Winge (1941), Van Valen (1967), and Jepsen (1970), among others, remained unconvinced of any special relationship between colugos and bats, and it is fair to say that few workers regarded Volitantia with any degree of seriousness until the concept was revived by Novacek and his colleagues (Novacek, 1986; Novacek and Wyss, 1986; Wible and Novacek, 1988; see also Thewissen and Babcock, 1991; Szalay and Lucas, this volume). More recently, a modified concept of Volitantia (including Dermoptera and Megachiroptera, but excluding Microchiroptera) has been advocated by Pettigrew and his co-workers, primarily on the basis of similarities shared by fruit bats, colugos, and primates in neural pathways from the eye to the midbrain (e.g., Pettigrew *et al.*, 1989).

Central to all arguments that bats (or a subset of bats such as Megachiroptera) and colugos are sister taxa is the alleged homology between the patagia of colugos and bats. It is undeniable that there are dramatic morphologic differences between the gliding membranes of dermopterans and the wings of bats. Hence, there is ample reason, on the basis of the comparative anatomy of living taxa alone, to doubt the alleged homology between these structures in extant Dermoptera and Chiroptera (e.g., Winge, 1941). Those who would argue in favor of such homology are forced to assert that the dermopteran patagium and the chiropteran wing lie on the same morphocline, despite the anatomical differences found in colugo patagia and bat wings today. In other words, if the hypothesis of homology between the patagium of colugos and the wings of bats is valid, the last common ancestor of Dermoptera and Chiroptera must also have possessed a patagium, perhaps one not unlike those of living colugos (Szalay and Lucas, this volume).

Positive evidence that the last common ancestor of Dermoptera and Chiroptera lacked a patagium is provided by primitive plesiadapiform dermopterans such as plesiadapids (see above). Accordingly, the alleged homology between the dermopteran patagium and the wings of bats is refuted, and we can be confident that the last common ancestor of bats and colugos was a nonvolant, quadrupedal mammal (as was the last common ancestor of living and fossil dermopterans).

Although the association between Dermoptera and Chiroptera was first proposed on the claimed homology between the patagia of the former and the wings of the latter, it is theoretically possible that Dermoptera and Chiroptera are sister taxa that have evolved from a nonvolant ancestor they share in common to the exclusion of other mammals. However, even this highly contrived possibility is unlikely, given new evidence from paleontology (Beard, 1989, 1991b, 1993) and molecular biology (Ammerman and Hillis, 1990; Adkins and Honeycutt, 1991; Bailey *et al.*, 1992) that Dermoptera and Primates form a clade, for which the new mirorder Primatomorpha has been proposed. Consequently, there is neither paleontological nor molecular support for the monophyly of Volitantia, and this taxonomic concept certainly represents an unnatural assemblage of taxa.

The hypothesis that Dermoptera and Megachiroptera are sister taxa, advanced in the context of chiropteran diphyly by Pettigrew and his colleagues (Pettigrew *et al.*, 1989; Pettigrew, 1991a,b), is not supported here for the same reasons cited above with respect to the lack of homology between the dermopteran and chiropteran patagium. That is, even in the unlikely event that Chiroptera proves to be diphyletic, the dermopteran fossil record requires that the patagium of colugos be considered a superficial convergence on the wings of fruit bats.

Plesiadapid and paromomyid dermopterans first appear in the fossil record during the late Puercan and early Torrejonian, respectively (early to middle Paleocene; ca. 64–65 Ma), well before the earliest appearance of primitive microchiropterans in the basal Eocene (e.g., Sigé, 1991) and much earlier than the first appearance of undoubted megachiropterans in the early Miocene (e.g., Sigé and Aguilar, 1987). These dates of first appearance can only be considered crude minimum estimates for the actual dates of establishment of the taxa in question. Nevertheless, it is worthwhile to note that these minimum ages contradict those cited by Pettigrew (1991a,b), who erroneously claimed that the earliest fossil dermopterans were intermediate in age between the oldest Microchiroptera and Megachiroptera, respectively. Using an extreme version of stratophenetics, Pettigrew (1991a,b) then listed this alleged intermediacy in the age of first appearance of Dermoptera as support for a closer relationship between Dermoptera and Megachiroptera than between fruit bats and Microchiroptera. However, such gross minimum estimates of the antiquity of higher taxa have little if any relevance to phylogeny reconstruction, and inferences concerning dermopteran relationships with primates, bats, and other mammals must be founded on a firmer biological basis if they are to have any great chance at accuracy.

Monophyly of the Primatomorpha

Based on an analysis of a large sample of postcranial fossils representing early Cenozoic Plesiadapidae, Paromomyidae and Micromomyidae, it was concluded that all of these extinct taxa were cladistic dermopterans and that Dermoptera and Primates are sister taxa within Eutheria (Beard, 1989, 1991b, 1993). The name Primatomorpha was proposed for this Primates + Dermoptera clade, and the new taxon was ranked at the level of mirorder. Importantly, at the time that Primatomorpha was originally designated, nucleotide sequence information was unavailable for Dermoptera, so that the monophyly of Primatomorpha could not be tested using discrete molecular data. However, a major prediction of this hypothesis was that when such data became available, they would support the monophyly of Primatomorpha (Beard, 1989, 1993). This prediction has subsequently been upheld (Ammerman and Hillis, 1990; Adkins and Honeycutt, 1991; Bailey *et al.*, 1992).

The monophyly of Primatomorpha is further corroborated by the immunodiffusion studies of Cronin and Sarich (1980) and by studies of soft anatomy and comparative parasitology reviewed by Beard (1993). Primatomorph monophyly is also consistent with the neurologic characters cited by Pettigrew *et al.* (1989), although the presence of similarly derived character states in Megachiroptera must be explained either by convergence between Megachiroptera and Primatomorpha in these features or by reversal to more primitive character states in Microchiroptera.

A Phylogenetic Definition of "Archaic Primates"

Given that Primates and Dermoptera form a clade, and that many of the taxa previously thought to be "archaic primates" are actually "archaic dermopterans," can any new insight be gained concerning the earliest phases of primate evolution?

No serious modern worker questions the monophyly of euprimates, a clade supported by a plethora of dental, cranial, and postcranial synapomorphies (e.g., Szalay *et al.*, 1987; Dagosto, 1988; Martin, 1990; Rose and Bown,

1991). Because of this wealth of evidence for monophyly, some workers (e.g., Kay *et al.*, 1990) have advocated restricting the order Primates to euprimates, effectively making the latter concept redundant. Gradistic, unnatural concepts such as "Proprimates" (Gingerich, 1989) would also restrict the order Primates to euprimates, recognizing euprimate monophyly at the expense of creating a new taxonomic wastebasket of mammals at the ordinal level (Beard, 1990b). While restricting the order Primates to euprimates may appear to have merit given our present knowledge of the fossil record, one can easily predict that this will not always be true.

The continuity of evolutionary descent requires that animals sharing more recent common ancestry with primates than with dermopterans existed prior to the diversification of the first euprimates. Such animals should be considered cladistic primates because of their more recent common ancestry with primates than with any other living or extinct order of mammals, despite the argument made by some systematists that taxa (especially orders of mammals) be restricted to "crown groups" rather than "closed descent communities" in the jargon of Ax (1985). To the extent that these animals retain primitive features not found in euprimates, they might readily be described as "archaic primates." Indeed, it is entirely possible that fossils representing archaic primates are already known, but are not recognized as such because of their fragmentary nature. Taxa that must be considered in this context include Purgatoriidae, Palaechthonidae, Microsyopidae (Beard, 1991a). *Altiatlasius koulchii* (Sigé *et al.*, 1990), and *Altanius orlovi* (Gingerich *et al.*, 1991), among others.

By the phylogenetic definition advocated here, "archaic primates" must lack at least some of the synapomorphies characteristic of all euprimates. Because of this very fact, some workers will no doubt raise the standard call that such animals be banished from the company of the righteous and evicted from the order Primates. However, if our goal is to gain greater understanding of the earliest phases of primate evolution, it is precisely with animals such as these that we must cast our lot.

ACKNOWLEDGMENTS

For access to fossils and to Recent comparative collections, I am grateful to Drs. T. M. Bown, P. D. Gingerich, G. K. Hess, K. D. Rose, and R. W. Thorington, Jr. I owe special thanks to Dr. P. Houde, who collected the vast majority of the fossil specimens on which this study is based. I am also indebted to E. Kasmer and A. D. Redline for producing the drawings, and to Dr. L. Krishtalka for aid with photography. I thank Dr. R.D.E. MacPhee for inviting me to participate in the symposium, and for his patience in editing and compiling this book. The financial support of NSF BSR 8801037 and NSF BSR 9020276 is gratefully acknowledged.

References

Adkins, R. M., and Honeycutt, R. L. 1991. Molecular phylogeny of the superorder Archonta. *Proc. Natl. Acad. Sci. USA* **88**:10317–10321.

Ammerman, L. K., and Hillis, D. M. 1990. Relationships within archontan mammals based on 12S rRNA gene sequence. *Am. Zool.* **30**:50A.

Archibald, J. D., Clemens, W. A., Gingerich, P. D., Krause, D. W., Lindsay, E. H., and Rose, K. D. 1987. First North American Land Mammal Ages of the Cenozoic Era, in: M. O. Woodburne (ed.), *Cenozoic Mammals of North America: Geochronology and Biostratigraphy*, pp. 24–76. University of California Press, Berkeley.

Ax, P. 1985. Stem species and the stem lineage concept. *Cladistics* **1**:279–287.

Bailey, W. J., Slightom, J. L., and Goodman, M. 1992. Rejection of the "flying primate" hypothesis by phylogenetic evidence from the ε-globin gene. *Science* **256**:86–89.

Baker, R. J., Novacek, M. J., and Simmons, N. B. 1991. On the monophyly of bats. *Syst. Zool.* **40**:216–231.

Beard, K. C. 1989. Postcranial anatomy, locomotor adaptations and paleoecology of early Cenozoic Plesiadapidae, Paromomyidae and Micromomyidae (Eutheria, Dermoptera). Ph.D. Dissertation, The Johns Hopkins University, Baltimore.

Beard, K. C. 1990a. Gliding behaviour and palaeoecology of the alleged primate family Paromomyidae (Mammalia, Dermoptera). *Nature* **345**:340–341.

Beard, K. C. 1990b. Do we need the newly proposed order Proprimates? *J. Hum. Evol.* **19**:817–820.

Beard, K. C. 1991a. Postcranial fossils of the archaic primate family Microsyopidae. *Am. J. Phys. Anthropol. Suppl.* **12**:48–49.

Beard, K. C. 1991b. Vertical postures and climbing in the morphotype of Primatomorpha: Implications for locomotor evolution in primate history, in: Y. Coppens and B. Senut (eds.), *Origines de la Bipédie chez les Hominidés*, pp. 79–87. Editions du CNRS (*Cahiers de Paléoanthropologie*), Paris.

Beard, K. C. 1993. Phylogenetic systematics of the Primatomorpha, with special reference to Dermoptera, in: F. S. Szalay, M. J. Novacek, and M. C. McKenna (eds.), *Mammal Phylogeny, Volume 2. Placentals*, pp. 129–150. Springer-Verlag, New York.

Beard, K. C., and Houde, P. 1989. An unusual assemblage of diminutive plesiadapiforms (Mammalia, ?Primates) from the early Eocene of the Clark's Fork Basin, Wyoming. *J. Vertebr. Paleontol.* **9**:388–399.

Cartmill, M. 1974. Pads and claws in arboreal locomotion, in: F. A. Jenkins, Jr. (ed.), *Primate Locomotion*, pp. 45–83. Academic Press, New York.

Chapman, H. C. 1902. Observations upon *Galeopithecus volans. Proc. Acad. Nat. Sci. Philadelphia* **54**:241–254.

Cronin, J. E., and Sarich, V. M. 1980. Tupaiid and Archonta phylogeny: The macromolecular evidence, in: W. P. Luckett (ed.), *Comparative Biology and Evolutionary Relationships of Tree Shrews*, pp. 293–312. Plenum Press, New York.

Dagosto, M. 1988. Implications of postcranial evidence for the origin of euprimates. *J. Hum. Evol.* **17**:35–56.

Fleagle, J. G. 1988. *Primate Adaptation and Evolution.* Academic Press, New York.

Fox, R. C. 1991. *Saxonella* (Plesiadapiformes: ?Primates) in North America: *S. naylori*, sp. nov., from the late Paleocene of Alberta, Canada. *J. Vertebr. Paleontol.* **11**:334–349.

Gauthier, J., Kluge, A. G., and Rowe, T. 1988. Amniote phylogeny and the importance of fossils. *Cladistics* **4**:105–209.

Gingerich, P.D . 1974. Function of pointed premolars in *Phenacolemur* and other mammals. *J. Dent. Res.* **53**:497.

Gingerich, P. D. 1976. Cranial anatomy and evolution of early Tertiary Plesiadapidae (Mammalia, Primates). *Univ. Mich. Pap. Paleontol.* **15**:1–141.

Gingerich, P. D. 1982. Time resolution in mammalian evolution: Sampling, lineages, and faunal turnover. *Third North American Paleontological Convention, Proceedings* **1**:205–210.

Gingerich, P. D. 1983. Paleocene–Eocene faunal zones and a preliminary analysis of Laramide structural deformation in the Clark's Fork Basin, Wyoming. *Wyoming Geological Association, 34th Annual Field Conference Guidebook* pp. 185–195.

Gingerich, P. D. 1986. *Plesiadapis* and the delineation of the order Primates, in: B. Wood, L. Martin, and P. Andrews (eds.), *Major Topics in Primate and Human Evolution*, pp. 32–46. Cambridge University Press, London.

Gingerich, P. D. 1989. New earliest Wasatchian mammalian fauna from the Eocene of northwestern Wyoming: Composition and diversity in a rarely sampled high-floodplain assemblage. *Univ. Mich. Pap. Paleontol.* **28**:1–97.

Gingerich, P. D., Wells, N. A., Russell, D. E., and Shah, S.M.I. 1983. Origin of whales in epicontinental remnant seas: New evidence from the early Eocene of Pakistan. *Science* **220**:403–406.

Gingerich, P. D., Smith, B. H., and Simons, E. L. 1990. Hind limbs of Eocene *Basilosaurus:* Evidence of feet in whales. *Science* **249**:154–157.

Gingerich, P. D., Dashzeveg, D., and Russell, D. E. 1991. Dentition and systematic relationships of *Altanius orlovi* (Mammalia, Primates) from the early Eocene of Mongolia. *Geobios* **24**:637–646.

Gunnell, G. F. 1989. Evolutionary history of Microsyopoidea (Mammalia, ?Primates) and the relationship between Plesiadapiformes and Primates. *Univ. Mich. Pap. Paleontol.* **27**:1–157.

Haines, R. W. 1955. The anatomy of the hand of certain insectivores. *Proc. Zool. Soc. London* **125**:761–777.

Jepsen, G. L. 1970. Bat origins and evolution, in: W. A. Wimsatt (ed.), *Biology of Bats*, Vol. I, pp. 1–64. Academic Press, New York.

Kay, R. F., and Cartmill, M. 1977. Cranial morphology and adaptations of *Palaechthon nacimienti* and other Paromomyidae (Plesiadapoidea, ?Primates), with a description of a new genus and species. *J. Hum. Evol.* **6**:19–53.

Kay, R. F., Thorington, R. W., Jr., and Houde, P. 1990. Eocene plesiadapiform shows affinities with flying lemurs not primates. *Nature* **345**:342–344.

Krause, D. W. 1991. Were paromomyids gliders? Maybe, maybe not. *J. Hum. Evol.* **21**:177–188.

Krishtalka, L. 1978. Paleontology and geology of the Badwater Creek area, central Wyoming. Part 15. Review of the late Eocene primates from Wyoming and Utah, and the Plesitarsiiformes. *Ann. Carnegie Mus.* **47**:335–360.

Leche, W. 1886. Über die Säugethiergattung *Galeopithecus:* Eine morphologische Untersuchung. *K. Svenska Vet. Akad. Handl.* **21**:1–92.

Maas, M. C., Krause, D. W., and Strait, S. G. 1988. The decline and extinction of Plesiadapiformes (Mammalia: ?Primates) in North America: Displacement or replacement? *Paleobiology* **14**:410–431.

MacKinnon, K. 1984. Flying lemurs: Order Dermoptera, in: D. MacDonald (ed.), *The Encyclopedia of Mammals*, pp. 446–447. Facts on File, Inc., New York.

Martin, R. D. 1990. *Primate Origins and Evolution: A Phylogenetic Reconstruction*. Princeton University Press, Princeton, N.J.

Novacek, M. J. 1985. Evidence for echolocation in the oldest known bats. *Nature* **315**:140–141.

Novacek, M. J. 1986. The skull of leptictid insectivorans and the higher-level classification of eutherian mammals. *Bull. Am. Mus. Nat. Hist.* **183**:1–112.

Novacek, M. J., and Wyss, A. R. 1986. Higher-level relationships of the Recent eutherian orders: Morphological evidence. *Cladistics* **2**:257–287.

Nowak, R. M., and Paradiso, J. L. 1983. *Walker's Mammals of the World*, Vol. 1, 4th ed. The Johns Hopkins University Press, Baltimore.

Pettigrew, J. D. 1986. Flying primates? Megabats have the advanced pathway from eye to midbrain. *Science* **231**:1304–1306.

Pettigrew, J. D. 1991a. Wings or brain? Convergent evolution in the origins of bats. *Syst. Zool.* **40**:199–216.

Pettigrew, J. D. 1991b. A fruitful, wrong hypothesis? Response to Baker, Novacek, and Simmons. *Syst. Zool.* **40**:231–239.

Pettigrew, J. D., Jamieson, B.G.M., Robson, S. K., Hall, L. S., McNally, K. I., and Cooper, H. M. 1989. Phylogenetic relations between microbats, megabats, and primates (Mammalia: Chiroptera and Primates). *Philos. Trans. R. Soc. London Ser. B* **325**:489–559.

Pocock, R. I. 1922. On the external characters of the beaver (Castoridae) and of some squirrels (Sciuridae). *Proc. Zool. Soc. London* **1922**:1171–1212.

Pocock, R. I. 1926. The external characters of the flying lemur (*Galeopterus temminckii*). *Proc. Zool. Soc. London* **1926**:429–444.

Rose, K. D. 1975. The Carpolestidae: Early Tertiary primates from North America. *Bull. Mus. Comp. Zool.* **147**:1–74.

Rose, K. D. 1981. Composition and species diversity in Paleocene and Eocene mammal assemblages: An empirical study. *J. Vertebr. Paleontol.* **1**:367–388.

Rose, K. D., and Bown, T. M. 1982. New plesiadapiform primates from the Eocene of Wyoming and Montana. *J. Vertebr. Paleontol.* **2**:63–69.

Rose, K. D., and Bown, T. M. 1991. Additional fossil evidence on the differentiation of the earliest euprimates. *Proc. Natl. Acad. Sci. USA* **88**:98–101.

Russell, D. E. 1964. Les mammifères Paléocènes d'Europe. *Mem. Mus. Natl. Hist. Nat. Paris Ser. C* **13**:1–324.

Russell, D. E. 1967. Sur *Menatotherium* et l'âge Paléocène du gisement de Menat (Puy-de-Dôme), in: *Problèmes Actuels de Paléontologie (Évolution des Vertébrés)*, pp. 483–490. Colloques Internationaux du CNRS No. 163. Éditions du CNRS, Paris.

Schaal, S., and Ziegler, W. (eds.). 1988. *Messel—Ein Schaufenster in die Geschichte der Erde und des Lebens*. Verlag Waldemar Kramer, Frankfurt am Main.

Shufeldt, R. W. 1911. The skeleton in the flying lemurs, Galeopteridae [Part 2]. *Philippine J. Sci.* **6D**:185–211.

Sigé, B. 1991. Rhinolophoidea et Vespertilionoidea (Chiroptera) du Chambi (Eocène inférieur de Tunisie): Aspects biostratigraphique, biogéographique et paléoécologique de l'origine des chiroptères modernes. *Neues Jahrb. Geol. Palaeontol. Abh.* **182**:355–376.

Sigé, B., and Aguilar, J.-P. 1987. L'extension stratigraphique des mégachiroptères dans le Miocène d'Europe mériodionale. *C. R. Acad. Sci.* (Ser. II) **304**:469–474.

Sigé, B., Jaeger, J.-J., Sudre, J., and Vianey-Liaud, M. 1990. *Altiatlasius koulchii* n. gen. et sp., primate omomyidé du Paléocène supérieur du Maroc, et les origines des euprimates. *Palaeontogr. Abt. A* **214**:31–56.

Simmons, N. B., Novacek, M. J., and Baker, R. J. 1991. Approaches, methods, and the future of the chiropteran monophyly controversy: A reply to J. D. Pettigrew. *Syst. Zool.* **40**:239–243.

Smith, A. P. 1984. Ringtails, pygmy possums, gliders, in: D. MacDonald (ed.). *The Encyclopedia of Mammals*, pp. 856–861. Facts on File, Inc., New York.

Szalay, F. S., and Delson, E. 1979. *Evolutionary History of the Primates*. Academic Press, New York.

Szalay, F. S., Tattersall, I., and Decker, R. L. 1975. Phylogenetic relationships of *Plesiadapis*—Postcranial evidence. *Contrib. Primatol.* **5**:136–166.

Szalay, F. S., Rosenberger, A. L., and Dagosto, M. 1987. Diagnosis and differentiation of the order Primates. *Yearb. Phys. Anthropol.* **30**:75–105.

Thewissen, J.G.M., and Babcock, S. K. 1991. Distinctive cranial and cervical innervation of wing muscles: New evidence for bat monophyly. *Science* **251**:934–936.

Thorington, R. W., Jr. 1984. Flying squirrels are monophyletic. *Science* **225**:1048–1050.

Thorington, R. W., Jr., and Heaney, L. R. 1981. Body proportions and gliding adaptations of flying squirrels (Petauristinae). *J. Mammal.* **62**:101–114.

Van Valen, L. 1967. New Paleocene insectivores and insectivore classification. *Bull. Am. Mus. Nat. Hist.* **135**:217–284.

Van Valen, L. 1968. Monophyly or diphyly in the origin of whales. *Evolution* **22**:37–41.

Wakefield, N. A. 1970. Notes on the glider-possum, *Petaurus australis* (Phalangeridae, Marsupialia). *Victorian Nat.* **87**:221–236.

Wharton, C. H. 1950. Notes on the life history of the flying lemur. *J. Mammal.* **31**:269–273.

Wible, J. R., and Novacek, M. J. 1988. Cranial evidence for the monophyletic origin of bats. *Am. Mus. Novit.* **2911**:1–19.

Winge, H. 1941. *The Interrelationships of the Mammalian Genera*, Vol. 1. C.A. Reitzels Forlag, Copenhagen.

Wischusen, E. W. 1990. The foraging ecology and natural history of the Philippine flying lemur (*Cynocephalus volans*). Ph.D. dissertation, Cornell University, Ithaca, N.Y.

Zimmerman, C. H. 1932. Characteristics of Clark Y airfoils of small aspect ratios. *Nat. Advis. Comm. Aeronaut. 18th Annu. Rep.* **431**:581–590.

The Implications of the Propatagial Muscles of Flying and Gliding Mammals for Archontan Systematics

3

J.G.M. THEWISSEN and S.K. BABCOCK

Introduction

Gregory (1910) proposed the term *Archonta* for four Recent mammalian orders: Menotyphla, Dermoptera, Chiroptera, and Primates. Archonta is a clade for which few synapomorphies have been identified (Novacek and Wyss, 1986), even when Macroscelidea (included in Gregory's Menotyphla) is excluded. There is little consensus concerning internal relationships among the archontan taxa, and the affinities of Chiroptera are especially controversial. Of special importance is whether the two suborders of Chiroptera (Megachiroptera and Microchiroptera) constitute a monophyletic group. Several investigators (Smith and Madkour, 1980; Pettigrew *et al.*, 1989; Pettigrew, 1991a,b) have proposed that Megachiroptera is the sister group of Primates, and that Microchiroptera is not closely related to either group. In this hypoth-

J.G.M. THEWISSEN and S. K. BABCOCK • Department of Biological Anthropology and Anatomy, Duke University Medical Center, Durham, North Carolina 27710. *Current address for J.G.M.T:* Department of Anatomy, NEOUCOM, Rootstown, Ohio 44272.
Primates and Their Relatives in Phylogenetic Perspective, edited by Ross D.E. MacPhee. Plenum Press, New York, 1993.

esis, Dermoptera is closely related to the megachiropteran–primate clade. If this is correct and if primates did not have flying ancestors, then active flight evolved twice in mammals (Pettigrew *et al.*, 1989).

Other authors continue to uphold the traditional view (e.g., Wible and Novacek, 1988; Baker *et al.*, 1991; Simmons *et al.*, 1991): Mega- and Microchiroptera are sister groups and active flight evolved only once in a common ancestor of all bats. Many of these authors regard Dermoptera as the sister group to bats (Leche, 1886; Wible and Novacek, 1988).

At the root of the controversy are the causes for the similarities between Megachiroptera and Primates on the one hand and between Mega- and Microchiroptera on the other hand. Pettigrew *et al.* (1989) suggested that many of the resemblances between Mega- and Microchiroptera are homoplastic, selected for independently in both groups during the acquisition of active aerial locomotion. But not all similarities between Mega- and Microchiroptera are related to flight (Wible and Novacek, 1988), and some resemblances in the nervous system between Megachiroptera and Primates proposed by Pettigrew may be related to selection for enhanced vision. Insight into the evolution of characters is critical to the controversy: did rampant homoplasy occur in the visual system of Megachiroptera and Primates, in the locomotor system of Mega- and Microchiroptera, or did reversals occur in the evolution of the visual system of Microchiroptera (Simmons *et al.*, 1991)?

Of particular interest in this controversy is a muscle complex that occurs in the leading edge of the airfoil of all bats (Fig. 1) and gliding mammals. We refer to this structure as the propatagial muscle complex to avoid implying untested homologies. Its gross structure varies within these groups and it has no obvious homologue to the musculature in mammals without an airfoil. Names applied to the propatagial muscle complex reflect uncertainty about homology. American authors working on bats usually refer to it as M. occipitopollicalis (e.g., Vaughan, 1970; Strickler, 1978; Altenbach, 1979), but the most detailed description of the complex is by Schumacher (1931), who proposed individual names for each constituent muscle belly. Propatagial muscles also occur in mammalian gliders. Leche (1886) described the propatagial muscles of *Cynocephalus* (see also Thewissen and Babcock, 1991). He described a dorsal muscle sheet formed in part by platysma myoides, and in part by a muscle called "M. jugalis propatagii" (Leche, 1886, p. 14). Leche (1886, p. 14) did not name a second, ventral muscle layer of the propatagium, but referred to it as "ventralen Schicht" (= ventral layer). Johnson-Murray described the propatagial muscles of flying squirrels (Sciuridae) as part of platysma II (Johnson-Murray, 1977) and named those of gliding marsupials "sphincter colli profundus pars protagialis" (Johnson-Murray, 1987).

The propatagial muscle complex has been used as a character supporting bat monophyly (Wible and Novacek, 1988; Baker *et al.*, 1991). Conversely, Pettigrew *et al.* (1989, p. 542) make the following claim for the propatagial muscle complex: "This is really a group of muscles of uncertain origin with a number of clear and consistent differences between Megachiroptera and Mi-

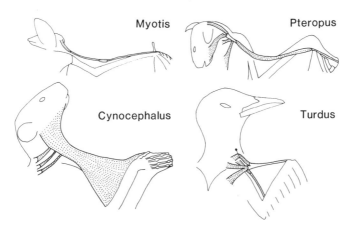

Fig. 1.   The propatagial muscle complex in a microchiropteran (*Myotis*), a megachiropteran (*Pteropus*), a flying lemur (*Cynocephalus*), and a bird (*Turdus*). *Cynocephalus* is in dorsal, the others in ventral view. Muscle bellies are stippled. The propatagial complex of all bats has an occipital origin, and is innervated by cranial nerve VII and cervical spinal nerves; it is homologous between Mega- and Microchiroptera. Two separate muscles occur in *Cynocephalus*. These are innervated by the same nerves as in bats. The propatagial complex of birds is homoplasious to that of bats and *Cynocephalus*. *Myotis*, *Pteropus*, and *Cynocephalus* modified after Thewissen and Babcock (1991; *copyright AAAS 1991*), *Turdus* after Thewissen and Babcock (1992; *copyright American Institute of Biological Sciences*, 1992).

crochiroptera. the presence of the muscle is essential to maintain a taut leading edge of the wing, despite changes in the separation of the thumb from the body. It may therefore be an obligatory functional convergence in the two kinds of bats." In order to investigate the possible homology between the propatagial muscle complex of Mega- and Microchiroptera, we have studied overall structure and innervation of the propatagial complex. Of special importance in determining homologies of muscles is their innervation, as muscle innervations are evolutonarily conservative.

Few authors have attempted to determine the innervation of propatagial muscles. The propatagial complex of bats usually consists of several muscle bellies in series (Fig. 1), the most proximal of which originates from the occiput. MacAlister (1872) proposed that the occipital belly was innervated by cranial nerve XI, partly on the assumption that this belly is the homologue of the cranial part of trapezius in quadrupeds (absent in bats). Jazuta (1937) stated that cranial nerve VII innervated the occipital belly in *Pteropus*, and held the muscle to be homologous to sphincter colli profundus. Schumacher (1931) determined that cervical spinal nerve 5, and "possibly 4" (p. 670) innervated a muscle belly in the propatagium of *Pteropus*, although Leche (1886, p. 19) thought that cranial nerve VII innervated this belly in the same genus. Mori (1960) stated that the brachial plexus (and hence cervical spinal nerves) innervated all bellies of the propatagial complex of *Pteropus* without giving details. Kyou-Jouffroy (1971) suggested that cervical spinal nerves 4 and 5 innervated the distal bellies of *Versperugo* (*Pipistrellus* of modern authors). Leche (1886; see also Ura, 1937, Figs. 28 and 29) reported that cranial nerve VII innervated the propatagial muscles in the dermopteran *Cynocephalus*.

We believe that the inconsistencies in the supposed innervations of propatagial muscles reported by these authors are the result of studies having been limited to gross dissections. Because of the small size of most bats, we have studied histological sections in addition to gross dissections to determine the innervation of the propatagial muscles in Mega- and Microchiroptera.

Materials and Methods

We studied commercially obtained specimens, uncataloged specimens in the collection of the Duke University Medical School, and cataloged material from the University of Michigan, Museum of Zoology (UMMZ), the Mammal Division of the United States National Museum, Washington, D.C. (USNM), the mammal and paleontological collections, Senckenberg Museum, Frankfurt (SMF), and the Landessammlungen für Naturkunde Karlsruhe (LNK-NAOM).

We studied three Megachiroptera: *Pteropus* sp., *Haplonycteris fisheri* (UMMZ 159842), and *Macroglossus lagochilus* (SMF 50890); four Microchiroptera: *Myotis lucifugus, Eptesicus fuscus, Tadarida brasiliensis, Rhinopoma hardwickei* (USNM 312188); the flying lemur *Cynocephalus volans* (Dermoptera), the flying squirrel *Glaucomys volans* (Petauristinae, Sciuridae, Rodentia), the sugar glider *Petaurus breviceps* (Petauridae, Marsupialia, USNM 240270), and the American robin (Aves, Muscicapidae, *Turdus migratorius*). We also studied 137 fossil impressions of the Eocene microchiropteran *Palaeochiropteryx tupaiodon* (SMF and LNK-LAOM), found in the Messel Quarry near Darmstadt, Germany.

We prepared histological sections of the full extent of the propatagial muscles in *Haplonycteris, Rhinopoma, Myotis,* and *Tadarida* and of parts of these muscles in *Pteropus*. We commonly cut sections of 10 μ and stained these with Milligan's trichrome, Wiegert's iron hematoxylin, and/or Verhoeff's elastin stains.

Our observations on gross morphology and innervation are based on gross dissections and serial sections. This method does not permit the identification of the motor nuclei that control the propatagial muscles in the central nervous system, nor does it allow tracing individual axons as they are relayed through a plexus. Therefore, in the taxa where the propatagial muscles are innervated by a plexus, we can only indicate a range of possible innervations for each of the bellies of the propatagial muscles.

Description

Megachiroptera

In the megachiropteran *Pteropus sp.* (Fig. 1), the propatagial complex originates as four muscular slips from the occiput (occipital belly), mandibular ramus (facial belly), ventral midline of the neck (cervical belly), and midline of the chest (pectoral belly). The four bellies merge near the shoulder and form a short tendon. A fifth muscle belly occurs immediately distal to the shoulder and extends to the distal part of the forearm (distal belly). The terminal tendon of the distal belly inserts on the dorsal side of the distal phalanx of the first digit. This arrangement of the distal belly coincides with that described by Mori (1960). Schumacher (1931) and Strickler (1978) described *Pteropus* specimens in which the distal bellies were small and restricted to the ante-brachial segment of the propatagium.

The facial slip is continuous at its origin with a wide sheet of muscle that extends toward the dorsal side of the neck. The cephalic vein extends parallel and adjacent to the propatagial complex for its entire extent between shoulder and wrist.

All muscle bellies except the pectoral belly receive innervation from a plexus of nerves surrounding the external jugular vein. This plexus consists of branches from cranial nerve VII and cervical spinal nerves III through V. Cranial nerve VII apparently sends branches to all five bellies of the pro-patagial complex, and the pectoral belly receives additional innervation from the pectoral nerves.

Occipital, facial, cervical, and pectoral bellies are also present in *Haplon-ycteris fisheri* (UMMZ 159842). As in *Pteropus,* the bellies join and become tendinous at the shoulder. The tendon extends to the forearm, where a small distal belly occurs proximal to the wrist. The terminal tendon of the complex inserts on the adaxial side of the second phalanx of the thumb.

Innervation to occipital, facial, and cervical bellies is supplied by a nerve plexus associated with the external jugular vein in *Haplonycteris*. A nerve feeding into this plexus can be traced rostrally to a position just lateral to the stylomastoid foramen. The actual connection of this nerve to a trunk coming out of the skull is severed, but its position suggests that the nerve is a branch of cranial nerve VII coming out of the stylomastoid foramen.

Caudally, nerves contributing to the plexus of the external jugular vein extend medially and reach beyond the point of innervation of the propatagial muscles. Judging by their position, these branches are connected to cervical spinal roots. The actual connection was not sufficiently well preserved for histological sectioning. A nerve from the external jugular plexus also inner-vates the distal belly of the propatagial complex, but may not be the only nerve to this belly.

The propatagial complex of the megachiropteran *Macroglossus lagochilus* also includes occipital, facial, cervical, and pectoral bellies. An additional slip that joins the facial belly originates from the root of the pinna. We did not determine the innervation of the propatagial muscle complex in *Macroglossus*.

Microchiroptera

The propatagial muscle complex of the vespertilionid microchiropteran *Myotis lucifugus* consists of an occipital belly extending from the occiput to the shoulder region (Fig. 1). A tendon connects this muscle belly to a second belly located distal to the elbow (antebrachial belly), and a second tendon extends along the forearm and inserts on metacarpal I and the second digit. Strickler (1978) identified a pectoral contribution to the propatagial complex in *Myotis*, but this slip was absent in the two specimens examined. This is probably the result of intraspecific variation (see discussion of *Eptesicus* below).

Both muscle bellies of *Myotis* are innervated by a plexus around the external jugular vein that gives off several nerve branches to the propatagial muscle complex (Figs. 2A and 3A). Cranial nerve VII and cervical spinal nerves II though V contribute branches to this plexus. The branching pattern of the plexus suggests that cervical spinal nerves II and III innervate the occipital belly of the propatagial muscle complex, and that cervical spinal nerves IV and V innervate the distal belly, while the facial nerve may contribute to both bellies.

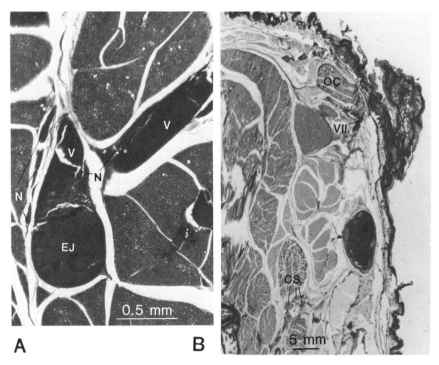

Fig. 2.   (A) Detail of a cross section through the external jugular vein (EJ) of *Myotis lucifugus* (section 6–16:2), showing some of the nerves (N) that form the plexus that innervates the propatagial muscle complex. Branches of this plexus extend with two large veins (V) dorsally and laterally. (B) Cross section through the neck of *Rhinopoma* (section A-130:1), showing the occipital belly (OC) of the propatagial complex in a dorsolateral position. Cervical spinal nerves (CS) and branches of cranial nerve VII (VII) innervate the propatagial complex. Top is dorsal, right is lateral in A and B.

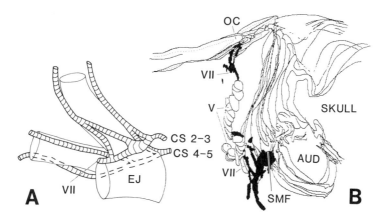

Fig. 3.   (A) The nerve plexus surrounding the external jugular vein (EJ) that innervates the propatagial muscle complex in *Myotis*. Reconstuction is based on histological serial sections (such as Fig. 2). Nerves making up the plexus are derived from cranial nerve VII (VII) and cervical spinal nerves (CS) 2 through 5. Four branches of the plexus innervate the propatagial muscle complex. (B) Three-dimensional computer reconstruction of cross sections of the head in *Tadarida*. Cranial nerve VII leaves the stylomastoid foramen (SMF) lateral to the auditory bulla (AUD). It gives off a small branch (VII) that passes dorsally along a small vein (V) to the occipital belly (OC) of the propatagial muscle complex.

The propatagial muscle complex in the molossid microchiropteran *Tadarida* contains four muscle bellies (Figs. 3B and 4). In addition to occipital and antebrachial bellies, *Tadarida* has a muscle belly just distal to the shoulder (brachial belly), as well as a pectoral slip. Unlike *Pteropus*, the latter slip is embedded in the pectoral muscles and its separation is only clear distally in histological sections (Fig. 1c of Thewissen and Babcock, 1991). The tissue linking brachial and antebrachial bellies consists mainly of elastic fibers. The propatagial muscle complex gives rise to three tendons distal to the antebrachial belly. Two of these tendons consist of collagenous tissue, and the third consists mainly of elastic fibers. The last extends dorsal to the wrist and inserts on the adaxial surface of digit 2. One of the collagenous tendons inserts on the adaxial surface of the base of metacarpal 1, while the second collagenous tendon extends palmar to the wrist, and also inserts on digit 2. The pectoral contribution to the propatagial muscle complex is innervated by the pectoral nerves, while the remaining three muscle bellies in *Tadarida* receive their innervation from cranial nerve VII.

The propatagial muscle complex of the rhinopomatid microchiropteran *Rhinopoma* consists of a single muscle belly, the occipital belly (Figs. 2B and 4). Its terminal tendon extends in the leading edge of the wing and lacks elastic tissue. The occipital muscle belly is innervated by branches from cranial nerve VII and cervical spinal nerves V and VI. Contrary to the situation in *Pteropus* and *Myotis,* branches from cranial and cervical roots do not form a plexus and are not associated with the external jugular vein.

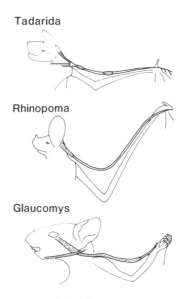

Fig. 4.   The propatagial muscle complex of the Microchiroptera *Tadarida* and *Rhinopoma* and the flying squirrel *Glaucomys*. Stippling indicates muscle tissue.

In order to evaluate intraspecific variability in the propatagial complex we dissected six specimens of the vespertilionid microchiropteran *Eptesicus fuscus*. The propatagial complex of all six *Eptesicus* specimens contains occipital, brachial, and antebrachial bellies. Origin of the occipital belly is always from the occipital side of the skull, but the actual point of origin is highly variable. In one of the six specimens, a slip from platysma joins the distal part of the occipital belly. The propatagial complex is joined by the cephalic vein at the shoulder, and the vein extends caudal to the muscle complex for its propatagial extent. The propatagial complex is also connected to the surface of the pectoral muscles by weak fascial slips in some *Eptesicus,* but in one specimen these fascial slips were condensed to form a tendon. The brachial belly is wrapped around the cephalic vein in one individual, and the antebrachial belly enfolds the vein in five of the six specimens. The tendon of insertion divides at the dorsal surface of wrist, and its two tendons insert on the thumb in all six specimens. One of these tendons inserts on the dorsal aspect of the proximal phalanx or on the common extensor tendon at phalanx I.1 in the five specimens where they could be followed. The second tendon inserts on the abaxial side of the base of phalanx I.1.

We also established the presence of propatagial muscles in the Eocene microchiropteran *Palaeochiropteryx tupaiodon* (Fig. 5). Details of the propatagium were preserved in 13 of 137 *Palaeochiropteryx* specimens recovered from the Messel Quarry (SMF-Me 74/5061, 491, 909, 957, 974, 1032, 1033, 1037, 1206, 1248, 1472, 1475 and LNK-NAOM 67–86). Irregular dark bands

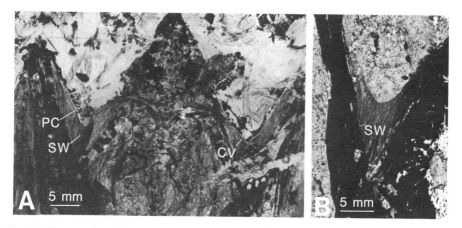

Fig. 5.   Eocene microchiropteran *Palaeochiropteryx tupaiodon*, showing details of soft anatomy (A, SMF-Me 74/5061; B, right propatagium of SMF-Me 909-B), such as skin wrinkles (SW), cephalic vein (CV), and tendon of propatagial complex (PC).

are present in the entire exposed surface of the propatagium of these specimens. These bands extend more or less parallel to the midsagittal plane from the leading edge of the propatagium to arm or forearm. In most specimens, they show an abrupt bend approximately 1 mm posterior to the leading edge, and curve gradually toward the elbow. They are irregular in width and length, and match in direction and structure skin wrinkles of Recent bat wings (SW of Fig. 5). These skin wrinkles are fixed to the elastic bands of the propatagium.

The sudden bend of the skin wrinkles posterior to the leading edge occurs at the approximate position of the cephalic vein in Recent bats. A translucent, but dark band of tissue is present parallel and caudal to the leading edge of the propatagium in SMF-Me 74/5061 (CV of Fig. 5A). Near the elbow this structure ends abruptly and is recurved on itself, suggesting postmortem deformation. The position of this structure matches that of the cephalic vein in Recent bats, and it may represent blood that filled portions of the vein. In Recent bats blood tends to pool in sections of the vein after death.

Another band that is approximately 0.25 mm wide extends immediately caudal and parallel to the leading edge of the propatagium in one bat (left side of SMF-Me 74/5061, PC of Fig. 5A). It is clearly visible in the proximal segment of the forearm, and splits proximally into two tapering parts. Structure and position suggest that this band is the tendon of the propatagial muscle. In Recent bats, this tendon may give off elastic fibers near the elbow that spread out into the propatagium. Parts of this propatagial tendon are preserved in three other Messel bats as narrow linear features that are parallel to the leading edge of the wing.

Muscle tissue in the Messel bats is generally preserved as an opaque black mass. No such structures occur in the propatagium of any Messel bats in spite of muscle preservation in other parts of the body. This suggests that the propatagial muscle of *Palaeochiropteryx* lacked muscle bellies distal to the shoulder. In this respect, *Palaeochiropteryx* is similar to the extant microchiropteran *Rhinopoma*.

Dermoptera

The propatagial muscle complex of *Cynocephalus* was described by Leche (1886). Our observations coincide with Leche's description in most respects. A dorsal sheet of muscle has a continuous origin from the side of the face between the rostral side of the mandible and external ear (Fig. 1). The muscle extends over neck and shoulder into the propatagium and inserts on the extensor aponeurosis of the forearm and first digit. Fibers of a second muscle sheet extend perpendicular and ventral to the dorsal sheet. The ventral sheet is diffuse and consists of a few, narrow, isolated strands. These strands originate in the dorsal midline of the neck and dwindle near the leading edge of the propatagium.

In our specimen of *Cynocephalus*, the rostral strand of the ventral layer sheet was larger than more caudal strands and was innervated by a branch from cervical spinal nerve I. It did not receive a branch from cranial nerve VII as reported by Leche (1886). Muscle strands caudal to it are probably also innervated by cervical spinal nerves, but we could not determine this macroscopically.

Gliding Rodentia

Gliding evolved at least twice within Rodentia: in scaly tailed squirrels (Anomaluridae) and in flying squirrels (Petauristinae, Sciuridae). Several authors have described the propatagial muscles of the petauristines (Leche, 1886; Peterka, 1936; Bryant, 1945; Gupta, 1966; Johnson-Murray, 1977) and the airfoil muscles of *Anomalurus* are discussed by Johnson-Murray (1980).

The propatagial muscles of petauristines are generally described as originating from the lateral side of the face. They give rise to a fusiform muscle in the leading edge of the propatagium and insert by means of a tendon on the first digit. Our dissection results for *Glaucomys* agree with those of previous authors for the propatagial segment of the complex, but they disagree with respect to the facial origin of the muscle.

We observed that a muscle extended subcutaneously between root of the ear and the mandibular symphysis (Fig. 4). This muscle, called mandibuloauricularis by Bryant (1945), is innervated by cranial nerve VII. Extending between the zygomatic arch and mandibuloauricularis are a few weak muscle fibers that Bryant (1945) describes as the anterior fibers of sphincter colli profundus. These fibers insert on the dorsal side of the middle segment of mandibuloauricularis, and the propatagial muscle originates from the ventral side of the same segment. All three muscles are extremely weak in this area, and continuity between sphincter colli profundus and the propatagial muscles could easily be assumed if particular attention was not given to the contact with mandibuloauricularis.

It is likely that previous authors referred to the propatagial muscles of petauristines (Peterka, 1936; Bryant, 1945, Johnson-Murray, 1977) and anomalurids (Johnson-Murray, 1980) as platysma because of the continuity between the propatagial muscles and muscles innervated by cranial nerve VII. We have now established that cervical spinal nerves III and IV innervate the neck segment of the propatagial complex in petauristines and that cranial nerve VII is not involved. The antebrachial segment of the propatagial muscles is innervated by a branch from the radial nerve in petauristines.

Gliding Marsupialia

Gliding flight has evolved independently three times in marsupials: feathertail gliders (*Acrobates,* Burramyidae), sugar gliders (*Petaurus,* Petauridae), and greater glider phalangers (*Petauroides,* Pseudocheiridae). The propatagial musculature of these taxa was described in detail by Johnson-Murray (1987), who noted that the propatagium of these taxa is smaller than in other gliders. The propatagial muscles of the dissected *Petaurus* specimen (USNM 240270) were extremely weak. We noted some diffuse fibers rostral to the elbow, but these were not connected to any muscle fibers near the neck or shoulder. These fibers may originate from the superficial dorsal fascia of the arm or may be part of the cutaneous muscle sheet of the plagiopatagium. The cephalic vein extends with the propatagial muscle complex of *Petaurus* in the leading edge of the wing. Johnson-Murray (1987) reported that the propatagial muscles in all three gliding marsupials originated from the ventral midline of the neck. We cannot explain this discrepancy.

Birds

George and Berger (1966) and Raikow (1985) published a general account of the propatagial complex of birds. We studied a specimen of the American robin (*Turdus migratorius,* Muscicapidae) in which the muscle consists of four distinct bellies (Fig. 1). Three of these attach to the same terminal tendon, and originate respectively from the skin of the root of the neck and clavicle (M. cucullaris pars propatagialis), the rostral edge of the coracoid (M. tensor patagii longus), and the rostral part of the sternal keel (M. pectoralis propatagialis longus). The body of the last slip is embedded in the mass of the pectoral muscles. The three muscle bellies join at the shoulder and their terminal tendon extends in the leading edge of the propatagium and inserts on the rostral edge of the distal radius. Part of this tendon is highly elastic.

A second propatagial muscle complex consists of a single muscle belly. It originates from the rostral edge of the coracoid (M. tensor patagii brevis), and its terminal tendon extends parallel to the humerus and inserts proximally on the aponeurosis of the forearm extensors.

The arrangement of the propatagial muscle complex as seen in *Turdus* is more or less typical for birds (see Raikow, 1985). The segment of the distal tendon of many birds is divided into many parallel tendons that insert at the forearm. In other birds, slips from the pectoral muscles join one or both propatagial muscles. The complex may also lack cucullaris propatagialis. The propatagial complex of birds originates only from neck and chest and is never muscular distal to the shoulder. Large veins are never associated with the propatagial muscles of birds.

We have not determined the innervation of the propatagial muscles in *Turdus*. Fisher and Goodman (1955) found that spinal nerves 15 through 18 innervated tensor patagii longus and brevis, as well as pectoralis propatagialis longus in *Grus americana*. Pectoralis propatagialis brevis is absent in *Grus*. Buri (1900) studied the propatagial muscle complex of "*Micropus melba*" (apparently *Apus melba*, Apodidae). He identified three muscle bellies in the complex derived from deltoid, biceps, and cucullaris, respectively. The belly derived from cucullaris was innervated by cranial nerve X–XI, the other bellies by spinal nerves.

Discussion

Function

Most authors (e.g., Vaughan, 1970; Strickler, 1978; Altenbach, 1979) have suggested that the propatagial muscles are involved in flight. The presence of a similar muscle in mammalian gliders, birds, and pterosaurs (Wellnhofer, 1975) suggests that the muscle is important in aerial locomotion, but experimental evidence is lacking.

In addition to its probable role in flight, the propatagial muscle complex of bats may also perform other functions. Schumacher (1931) observed that induced contraction of the antebrachial belly of the propatagial muscle complex caused blood from the cephalic vein to drain toward the shoulder. This observation and the relations between vein and muscle led Schumacher (1931) and Kallen (1977) to suggest that the propatagial muscle complex was important in venous drainage of the forelimb in resting bats. The tendon of the complex in the microchiropteran *Saccopteryx* divides near the elbow, and the two resulting strands extend on either side of the orifice of the propatagial gland. This suggests that the muscle controls secretion of the gland in *Saccopteryx*. In the microchiropteran *Myzopoda*, the terminal tendon is closely associated with the adhesive pad at the wrist (Schliemann, 1970) and might operate it.

Pettigrew *et al.* (1989) suggested that the propatagial muscle complex evolved independently in Mega- and Microchiroptera as a result of selection for enhanced aerial locomotion. But the importance of this structure in phylogeny reconstruction is not preempted by its function in flight. Although the mere presence of the complex might be linked to aerial locomotion, its complexity and diversity among the aerial taxa merit further study.

Structural Variation

At a histological level, the propatagial complex of bats is composed of three tissue types: muscle, collagen, and elastic fibers. The arrangement of these tissues within the complex is variable not only between Mega- and Microchiroptera, but also within each of these taxa (Strickler, 1978). Parts of the propatagial complex are apparently extremely plastic evolutionarily, as noticed previously by Pettigrew *et al.* (1989).

For example, elastic tissue is usually present in the propatagial complex rostral to the elbow in bats and birds, possibly to accommodate stretching in flexion and extension of the flight membrane near the elbow. On the other hand, elastic tissue at the elbow is absent in *Pteropus* (where the complex is muscular) and *Rhinopoma* (where it is collagenous).

Another example of variability in the propatagial complex concerns the distal (brachial and antebrachial) muscle bellies. The size of the distal belly varies inter- or intraspecifically within *Pteropus*. Only a small muscle belly was present at the forearm in specimens dissected by Schumacher (1931) and Strickler (1978), whereas there was a large, elongate muscle belly that extended for nearly the entire length of arm and forearm in our specimen and that of Mori (1960). In other closely related bats, three muscle bellies may be present rostral to the arm and forearm (Strickler, 1978).

On the other hand, certain aspects of the propatagial complex are stable within higher taxa of bats and could be important in phylogenetic reconstruction. Previous authors have noted several consistent differences between the propatagial complex of Mega- and Microchiroptera. Strickler (1978) stated that the propatagial complex of the pteropodids is unique among bats in having a muscular pectoral attachment, whereas in Microchiroptera this attachment consists of collagenous fibers only. On a histological level, however, it is clear that a muscular attachment also occurs in Microchiroptera (Thewissen and Babcock, 1991, Fig. 1). The pectoral contributions of Mega- and Microchiroptera are similar in their structural components, and the only difference is that the pectoral slip of Megachiroptera is not embedded in the pectoral mass at the gross level.

The propatagial complexes of Mega- and Microchiroptera may differ in the facial origin. Although always present in Megachiroptera, it is almost always absent in Microchiroptera, except as an anomaly in certain individuals. We noted it in a specimen of *Eptesicus,* and Schumacher (1931) found it in a specimen of *Vespertilio.* It is possible that the facial slip in these specimens represents an atavistic trait, reminiscent of an ancestral condition. It could also represent the facial muscle of a quadruped.

Systematic Implications

Mega- and Microchiroptera are very similar in two aspects of the propatagial complex, but different from mammalian gliders and birds: the presence of an occipital belly, and the close relation between the propatagial muscle complex and the cephalic vein (Table I). The propatagial muscle complex of both birds and mammalian gliders lacks an occipital origin, suggesting that this origin is not necessary for aerial locomotion. Within bats, this belly is present in all Mega- and Microchiroptera studied by us and by Strickler (1978). One possible exception is the microchiropteran *Myzopoda,* for which

Table I. Character Distributions for Propatagial Characters among Flyers and Gliders, Based on Our Observations

	Occipital origin present	Cephalic vein in leading edge	Innervation		
			VII	Cerv.	VII/cerv.
Megachiroptera					
Pteropus	+	+			+
Haplonycteris	+	+			+
Macroglossus	+	+			
Microchiroptera					
Rhinopoma	+	+			+
Myotis	+	+			+
Eptesicus	+	+			
Tadarida	+	+	+		
Palaeochiropteryx		+			
Dermoptera					
Cynocephalus	−	−	+[a]	+[a]	
Rodentia					
Glaucomys	−	−	+[b]	+	
Marsupialia					
Petaurus	−	+			
Aves					
Turdus	−	−			

[a]Cranial nerve VII and cervical spinal nerves innervate different propatagial muscles. In bats, they innervate sections of the same muscle complex.
[b]Muscles innervated by cranial nerve VII are restricted to the face and do not extend into the propatagium. They are loosely attached to the propatagial muscles by connective tissue.

Schliemann (1970) suggests that the occipital belly is absent. This taxon could be unique with respect to this character.

The cephalic vein of bats, including fossil *Palaeochiropteryx*, is located in the propatagium just caudal to the propatagial muscle complex. The vein is located in the propatagium in only one mammalian glider (*Petaurus*) and not in birds. It is thus probably unrelated to aerial locomotion. These two characters, occipital origin and location of the vein, are shared-derived for Mega- and Microchiroptera.

Additional strong evidence for the homology between the propatagial complexes of Micro- and Megachiroptera comes from their innervation. Cranial nerve VII usually does not innervate muscles caudal to the clavicle, while in bats and flying lemurs muscles innervated by cranial nerve VII reach as far distally as the forearm. Although this latter pattern is uncommon in mammals, it is not unique. Huber (1930) noted that muscles innervated by cranial nerve VII extend onto the forelimb in monotremes. It has never been shown that the propatagial muscles of nonarchontan mammalian gliders are innervated by cranial nerve VII, and the gliders that we investigated display different patterns.

The dual pattern of innervation by cranial nerve VII and by cervical spinal nerves in *Rhinopoma, Myotis, Haplonycteris,* and *Pteropus* is unique among mammals. It is relatively common for different parts of a muscle to receive innervation from adjacent sections of the central nervous system, but we are not aware of any other example of motor connections between cranial nerve VII and cervical spinal nerves. The absence of dual innervation of the propatagial muscles in any mammalian glider suggests that this kind of innervation is not a prerequisite for aerial locomotion. It also suggests that the phylogenetic heritage of mammals does not predetermine innervation of propatagial muscles by these sources. Consequently, we consider the dual innervation of the propatagial muscles in bats a strong indication that these muscles were present in the last common ancestor of all bats, and that this ancestor had an airfoil.

If the propatagial muscles of bats are primitively innervated by cranial nerve VII and cervical spinal nerves, then the innervation of *Tadarida* must be derived. This is supported by a pure parsimony argument. Phylogenetic trees of microchiropterans (Smith, 1976; Van Valen, 1979) usually support vespertilionoid monophyly. This superfamily includes vespertilionids (*Myotis*) and molossids (*Tadarida*), but not rhinopomatids (*Rhinopoma*). Given that dual innervation occurs in both *Myotis* and *Rhinopoma,* it is more parsimonious to infer that it also occurred in the ancestral microchiropteran and that reduction occurred in the lineage leading to *Tadarida*. Branches of cranial nerve VII and cervical spinal nerves form a plexus in some bats (*Myotis* and Megachiroptera), but not in others (*Rhinopoma*). It is unclear what the primitive arrangement is. The similarity of the innervation of the propatagial complex in Mega- and Microchiroptera suggests that the complex as a whole is homologous between these clades.

Position of Dermoptera

The innervation of the propatagial muscles also bears on the relations between bats and flying lemurs. Dermoptera is placed within Archonta by most authors (Novacek and Wyss, 1986; Pettigrew *et al.*, 1989), and it is often thought to be closely related to bats (Leche, 1886; Novacek and Wyss, 1986; Novacek, 1992; Szalay and Lucas, this volume), although a recent hypothesis claims close ties with primates (Beard, 1991). Evidence from the propatagial muscles is consistent with the first view (Thewissen and Babcock, 1991): the propatagial complex of *Cynocephalus* is an adequate structural intermediate between the muscles of the primitive quadrupedal mammal and a flying bat. There are two simple muscle sheets in *Cynocephalus,* one innervated by cranial nerve VII, the other by cervical spinal nerves. Tissues forming these muscle sheets could merge in ontogeny and differentiate to form the propatagial complex of bats. The resulting muscle complex would retain the original innervation of the two muscles, resulting in the dual innervation of the propatagial muscles in the ancestral bat.

Conclusions

Propatagial muscles occur in all bats, birds, and the gliding mammals that we investigated, suggesting that these muscles evolved many times as a result of selection for aerial locomotion. Mere presence of propatagial muscles is therefore a suspect character for phylogenetic inference; more detailed similarities must be sought to determine homologies of these muscles. The presence of an occipital belly and the propatagial location of the cephalic vein are aspects of the propatagial complex that are probably unrelated to aerial locomotion and are shared between Mega- and Microchiroptera. These traits are homologous in the sense that Jardine (1969) and Van Valen (1982) emphasized: structural similarities that may or may not follow similar ontogenetic pathways.

The similarity of innervation of the propatagial muscles in Mega- and Microchiroptera supports homology at a different level. Innervation is usually a faithful guide to ontogenetic origin, and thus supports homology as defined by Cracraft (1981) and Roth (1984): tissues with a similar developmental history. The innervation pattern in Chiroptera suggests a unique ontogenetic trajectory for the propatagial complex as a whole and corroborates the hypothesis that it was present in the last common ancestor of Mega- and Microchiroptera.

We agree with Pettigrew *et al.* (1989) that a propatagial muscle complex is necessary for aerial locomotion and that it evolved several times in aerial amniotes. However, Mega- and Microchiroptera share several detailed similarities of the propatagial complex that are absent in other flying and gliding vertebrates. These characters support bat monophyly. Our results as presented here and by Thewissen and Babcock (1991, 1992) are consistent with results from other studies (Wible and Novacek, 1988; Baker *et al.*, 1991; Simmons *et al.*, 1991; Luckett, this volume; Adkins and Honeycutt, this volume; Stanhope *et al.*, this volume).

ACKNOWLEDGMENTS

We thank Drs. J. H. Casseday and V. L. Roth (both of Duke University) for supplying us with specimens, and Drs. J. Habersetzer (SMF), G. Storch (SMF), R. W. Thorington (USNM), and P. K. Tucker (UMMZ) for access to museum specimens. We thank Drs. K. K. Smith and N. B. Cant for use of their laboratories and Drs. M. Cartmill and K. K. Smith for comments on the manuscript. We benefited from discussions with Drs. J. D. Pettigrew, and J. R. Wible. We are deeply indebted to Drs. J. L. Johnson-Murray and A. R. Wyss for sending us their unpublished manuscripts.

References

Altenbach, J. S. 1979. Locomotor morphology of the vampire bat, *Desmodus rotundus. Am. Soc. Mammal. Spec. Publ.* **6.**

Baker, R. J., Novacek, M. J., and Simmons, N. B. 1991. On the monophyly of bats. *Syst. Zool.* **40:**216–231.

Beard, K. C. 1991. Vertical postures and climbing in the morphotype of Primatomorpha: Implications for locomotor evolution in primate history, in: *Origines de la bipedie chez les hominidés,* pp. 79–87. Cahiers de Paléoanthropologie, Paris.

Bryant, M. B. 1945. Phylogeny of Nearctic Sciuridae. *Am. Midl. Nat.* **33:**257–390.

Buri, R. O. 1900. Zur Anatomie des Flügels von *Micopus melba* und einigen anderen Coracornithes, zugleich Beitrag zur Kenntnis des systematischen Stellung der Cypselidae. *Jena. Z. Naturwiss.* **33**(N. S. 26)**:**361–610.

Cracraft, J. 1981. The use of functional and adaptive criteria in phylogenetic systematics. *Am. Zool.* **21:**21–36.

Fisher, H. I., and Goodman, D. C. 1955. The myology of the whooping crane, *Grus americana. Ill. Biol. Monogr.* **24:**1–127.

George, J. C., and Berger, A. J. 1966. *Avian Myology.* Academic Press, New York.

Gregory, W. K. 1910. The orders of mammals. *Bull. Am. Mus. Nat. Hist.* **27:**1–524.

Gupta, B. B. 1966. Notes on the gliding mechanism in the flying squirrel. *Occas. Pap. Mus. Zool. Univ. Mich.* **645.**

Huber, E. 1930. Evolution of facial musculature and cutaneous field of trigeminus. Part I. *Q. Rev. Biol.* **5:**133–188.

Jardine, N. 1969. The observational and theoretical components of homology: A study based on the morphology of the dermal skull roofs of rhipidistian fishes. *Biol. J. Linn. Soc.* **1:**327–361.

Jazuta, K. 1937. Zur vergleichenden Anatomie der Hautmuskulatur bei Säugetiere. 4. Mitteilung: Halsmuskulatur bei den Chiropteren. *Anat. Anz.* **84:**26–31.

Johnson-Murray, J. L. 1977. Myology of the gliding membranes of some petauristine rodents (genera: *Glaucomys, Pteromys, Petinomys,* and *Petaurista*). *J. Mammal.* **59:**374–384.

Johnson-Murray, J. L. 1980. Comparative and functional morphology of selected genera of non-gliding and gliding mammals (Phalangeridae, Sciuridae, Anomaluridae, and Cynocephalidae). Unpublished Ph.D. thesis, University of Massachusetts.

Johnson-Murray, J. L. 1987. The comparative myology of the gliding membranes of *Acrobates*, *Petauroides* and *Petaurus* contrasted with the cutaneous myology of *Hemibelideus* and *Pseudoicheirus* (Marsupialia: Phalangeridae) and with selected gliding Rodentia (Sciuridae and Anomaluridae). *Aust. J. Zool.* **35**:101–113.

Kallen, F. C. 1977. The cardiovascular system of bats: Structure and function, in: W. A. Wimsatt (ed.), *Biology of Bats*, Vol. III, pp. 290–483. Academic Press, New York.

Kyou-Jouffroy, F. 1971. Musculature peaucière. II.—Les peauciers du tronc et des membres et leurs derivés, in: P.-P. Grassé (ed.), *Traité de Zoologie, Anatomie, Systematique, Biologie*, Vol. XVI, Part III, pp. 626–678. Masson, Paris.

Leche, W. 1886. Über die Säugethiergattung *Galeopithecus:* Eine morphologische Untersuchung. *K. Svenska Vet. Akad. Handl.* **21**:1–92.

MacAlister, A. 1872. The myology of Cheiroptera. *Philos. Trans. Soc. London* **162**:125–1710. (1873).

Mori, M. 1960. Muskulatur des *Pteropus edulis*. *Okijamas Fol. Anat. Jpn.* **36**:253–307.

Novacek, M. J. 1992. Mammalian phylogeny: Shaking the tree. *Nature* **356**:121–125.

Novacek, M. J., and Wyss, A. R. 1986. Higher-level relationships of the Recent eutherian orders: Morphological evidence. *Cladistics* **2**:257–287.

Peterka, H. E. 1936. A study of the myology and osteology of tree sciuruds with regard to adaptation to arboreal, glissant and fossorial habits. *Trans. Kans. Acad. Sci.* **39**:313–332.

Pettigrew, J. D. 1991a. Wings or brain? Convergent evolution in the origins of bats. *Syst. Zool.* **40**:199–216.

Pettigrew, J. D. 1991b. A fruitful, wrong hypothesis? Response to Baker, Novacek, and Simmons. *Syst. Zool.* **40**:239–243.

Pettigrew, J. D., Jamieson, B.G.M., Robson, S. K., Hall, L. S., McNally, K. I., and Cooper, H. M. 1989. Phylogenetic relations between microbats, megabats and primates (Mammalia: Chiroptera and Primates). *Philos. Trans. Soc. London Ser B* **325**:489–559.

Raikow, R. J. 1985. Locomotor system, in: A. S. King and J. McLelland (eds.), *Form and Function in Birds*, Vol. 3, pp. 57–148. Academic Press, New York.

Roth, V. L. 1984. On homology. *Biol. J. Linn. Soc.* **22**:13–29.

Schliemann, H. 1970. Bau und Funktion der Haftorgane von *Thyroptera* und *Myzopoda* (Vespertilionoidea, Microchiroptera, Mammalia). *Z. Wiss. Zool.* **181**:353–400.

Schumacher, S. 1931. Der "M. propatagialis proprius" und die "Tendo propatagialis" in ihren Beziehung zur V. cephalica bei den Fledermäusen. *Z. Anat. Entwicklungsgesch.* **94**:642–679.

Simmons, N. B., Novacek, M. J., and Baker, R. J. 1991. Approaches, methods, and the future of the chiropteran monophyly controversy: A reply to J. D. Pettigrew. *Syst. Zool.* **40**:239–243.

Smith, J. D. 1976. Chiropteran evolution. *Spec. Publ. Mus. Tex. Tech Univ.* **10**:49–69.

Smith, J. D., and Madkour, G. 1980. Penial morphology and the question of chiropteran phylogeny, in: D. E. Wilson and A. L. Gardner (eds.), *Proceedings of the 5th International Bat Research Conference*, pp. 347–365. Texas Tech Press, Lubbock.

Strickler, T. L. 1978. Functional osteology and myology of the shoulder in the Chiroptera. *Contrib. Vertebr. Evol.* **4**:1–198.

Thewissen, J.G.M., and Babcock, S. K. 1991. Distinctive cranial and cervical innervation of wing muscles: New evidence for bat monophyly. *Science* **251**:934–936.

Thewissen, J.G.M., and Babcock, S. K. 1992. The origin of flight in bats; to go where no mammal has gone before. *BioScience* **42**:340–345.

Ura, R. 1937. On the general differentiation of the superficial pectoral muscles, with special attention for the skin muscles of the trunk of mammals. *J. Tokyo Med. Assoc.* **51**:116–288, 338–390.

Van Valen, L. 1979. The evolution of bats. *Evol. Theor.* **4**:103–121.

Van Valen, L. 1982. Homology and causes. *J. Morphol.* **173**:305–312.

Vaughan, T. A. 1970. The muscular system, in: W. A. Wimsatt (ed.), *Biology of Bats*, Vol. 1, pp. 139–194. Academic Press, New York.

Wellnhofer, P. 1975. Die Rhamphorhynchoidea (Pterosauria) der Oberjura Plattenkalke Suddeutschlands. I. Allgemeine Skelettmorphologie. *Palaeontographica* **A148:**1–33.

Wible, J. R., and Novacek, M. J. 1988. Cranial evidence for the monophyletic origin of bats. *Am. Mus. Novit.* **2911:**1–19.

Ontogeny of the Tympanic Floor and Roof in Archontans

4

JOHN R. WIBLE and JAMES R. MARTIN

Introduction

To say that the middle ear and surrounding basicranium have played a critical role in current views of mammalian phylogeny is not an overstatement. Nearly every recent morphological treatment of phylogenetic relationships, be it higher or lower level, has included some characters from this complex region. The basicranium is certainly not a taxonomic touchstone, but it has been a helpful guide for elucidating affinities in many instances.

The relevance of basicranial characters is no more evident than in the analysis of the higher-level relationships of primates. Few morphological features distinguish the grouping of Recent primates and their last common ancestor, the Euprimates, from other eutherian (placental) mammals. Because several euprimate characters are from the basicranium, including an osseous auditory bulla (tympanic floor) completed by an outgrowth from the petrosal, the search for the "ancestors" of euprimates has typically included the search for potential synapomorphies of the basicranium.

Until recently, the taxon widely considered to be the sister group of Euprimates has been Plesiadapiformes, the so-called archaic primates, a di-

JOHN R. WIBLE and JAMES R. MARTIN • Department of Anatomical Sciences and Neurobiology, School of Medicine, University of Louisville, Louisville, Kentucky 40292.

Primates and Their Relatives in Phylogenetic Perspective, edited by Ross D.E. MacPhee. Plenum Press, New York, 1993.

verse array of extinct forms known from the Late Cretaceous to Late Eocene. Chief among the characters supporting this relationship is the presence of what has been interpreted to be a petrosal auditory bulla in some plesiadapiforms, specifically certain paromomyids and plesiadapids (Saban, 1963; Russell, 1964; Szalay, 1972). Not all authors accept this interpretation. Mac-Phee *et al.* (1983) note that some Recent mammals develop a nonpetrosal bulla that fuses seamlessly with the petrosal in the adult. A more telling challenge is the recent description of a suture separating the auditory bulla from the petrosal proper in the paromomyid *Ignacius* (Kay *et al.*, 1990). Because the bulla in *Ignacius* is also separated from the remaining basicranial ossifications, including the ectotympanic, it is identified as consisting of an entotympanic (Kay *et al.*, 1990), an independent element known to occur in various Recent eutherians (Novacek, 1977; MacPhee, 1979). Given that the bulla in *Ignacius* is not petrosal in origin, it seems likely that this is also the case in other plesiadapiforms (Kay *et al.*, 1990).

If some or all plesiadapiforms do not constitute the sister group of euprimates, as has been suggested recently by several authors (e.g., Wible and Covert, 1987; Beard, 1990b), then what taxa are the next most likely candidates? A variety of groups have been implicated as close relatives of euprimates, but during the last few years attention has focused mainly on the various members of the superordinal grouping Archonta. Originally defined by Gregory (1910), Archonta was resurrected by McKenna (1975), who included within it chiropterans, dermopterans, scandentians, and paromomyid and plesiadapid plesiadapiforms along with euprimates. Not without its critics (e.g., Novacek, 1980a, 1982; Cartmill and MacPhee, 1980), Archonta has received support in several recent analyses (e.g., Novacek and Wyss, 1986; Wible and Covert, 1987; Novacek *et al.*, 1988), though few characters explicitly support its monophyly. Within Archonta, euprimates have been linked variously with scandentians (Wible and Covert, 1987), with megachiropterans and dermopterans (Pettigrew *et al.*, 1989), and with a clade composed of dermopterans and various plesiadapiforms (Beard, 1990a, 1991).

Current knowledge of the basicranium and its developmental history in Recent archontans is well advanced for most euprimates and scandentians. Starck's (1975) benchmark summary of chondrocranial studies for these groups included developmental stages from 20 species of euprimates and 2 species of the scandentian *Tupaia*. He noted (p. 150) that the chondrocrania of these forms "do not show basic differences from the general eutherian pattern" and that "the presently known facts of chondrocranial morphology do not allow for many conclusions about phylogenetic relationships." Subsequent developmental studies of euprimate and scandentian basicrania including more stages and taxa (e.g., MacPhee, 1981; MacPhee and Cartmill, 1986; Zeller, 1986, 1987) have essentially reiterated Starck's claims, uncovering few characters that speak to the higher-level affinities of these groups. Where these studies and Starck's summary have made their impact is in the analysis of ingroup relationships (i.e., within Euprimates and within Scandentia) and in highlighting the importance of detailed developmental information for clearly defining adult character states used in reconstructing phylogeny.

Assessment of the basicranial evidence for the higher-level affinities of euprimates is limited not by what is known about euprimates and scandentians, but by what is not known about other potential outgroups, such as dermopterans and chiropterans. Some developmental studies on the basicranium of dermopterans and particular chiropterans have appeared (e.g., Klaauw, 1922), but detailed ontogenetic information on the characters traditionally used in phylogenetic analysis is much less complete for these groups. The goal of our contribution to this volume is to narrow this gap. Herein we report on the development of basicranial structures—the components of the tympanic floor and roof—in the dermopteran *Cynocephalus* and the megachiropteran *Pteropus*. The ontogenies reported for these structures are then compared with those in other archontans and eutherians.

Methods and Materials

Basicranial structures were studied in serially sectioned prenatal specimens of *Cynocephalus* and *Pteropus* (Table I). Specimens were housed either in the Duke University Comparative Embryological Collection, Durham, North Carolina (DUCEC) or the Department of Anatomical Sciences and Neurobiology, University of Louisville, Louisville, Kentucky (ULASN). The illustrations in Figs. 2, 4, and 6 were redrawn from computer-generated reconstructions

Table I. Sectioned Specimens[a]

Specimens	CRL (mm)	HL (mm)	ST (μm)	Stains
Dermoptera				
Cynocephalus variegatus				
DUCEC 804	88	28	15	Mallory
DUCEC 805	128	42	20	Azan
DUCEC 806	136	44	15/20	Azan
Cynocephalus sp.				
DUCEC 8310	107	28	15	Mallory
DUCEC 839	150	49	20/40	Azan
Megachiroptera				
Pteropus giganteus				
ULASN Pg 11	14.15	8.7	10	Masson
ULASN Pg 15	26.95	18.1	15	Masson
Pteropus sp.				
DUCEC 831	93.5		16	Mallory

[a]All specimens were sectioned in a frontal plane with one exception. The fetal colugo DUCEC 806 was bisected midsagittally, and one half was then sectioned in a frontal plane and the other was sagittally sectioned. Abbreviations: CRL, crown–rump length; HL, head length; ST, section thickness.

made with a three-dimensional reconstruction program from Jandel Scientific, Sausalito, California (PC3D, version 5.0). Every tenth section for *Cynocephalus variegatus* (DUCEC 804 and 806) and every fifth for *Pteropus* sp. (DUCEC 831) were projected and drawn. After aligning the sections using visual best-fit methods (Gaunt and Gaunt, 1978), we traced the outlines of relevant structures with a Jandel digitizing tablet. The PC3D program, running on an IBM PS2, stacked the outlines from each section to produce a three-dimensional representation that could be viewed in any specified orientation. The sections in Figs. 3, 5, and 7 were photographed with a Wild (M400) photomacroscope.

Descriptions

The developing ear region of a generalized fetal eutherian is shown in Fig. 1. The cavum tympani, the air-filled diverticulum of the pharynx that completely fills in the middle-ear (tympanic) cavity in the adult, is situated lateral to the promontorium of the petrosal, the cochlear housing (Fig. 1C). The floor and roof of the primordial middle-ear cavity consist largely of connective-tissue fibers organized in sheetlike membranes (dashed lines in Fig. 1C). The membrane that forms the primordial floor, the fibrous membrane of the tympanic cavity (MacPhee, 1981), is stretched between the overlying chondrocranial elements and the ectotympanic bone, which is the only other constant occupant of the developing floor. The ectotympanic, the intramembranous ossification that supports the tympanic membrane, is usually a narrow ring that is incomplete laterally in fetal mammals (Fig. 1B). The membrane contributing to the primordial tympanic roof, the spheno-obturator membrane (Gaupp, 1908), occupies the gap in the chondrocranium anterolateral to the promontorium. Different terms are applied to this gap depending on how far rostralward it extends. Following MacPhee (1981), we refer to the gap circumscribed by the petrosal, squamosal, alisphenoid, and basisphenoid as the piriform fenestra (Fig. 1A). A variety of soft-tissue structures (e.g., nerves, vessels) pass through the fibrous membrane of the tympanic cavity and spheno-obturator membrane to either enter or exit the middle ear.

During later developmental stages, bone and cartilage from a variety of sources fill in the fibrous membrane of the tympanic cavity and spheno-obturator membrane to varying extents in different mammals. Among the major elements that contribute to the closure (partial or complete) of the tympanic floor and roof are the surrounding basicranial ossifications (i.e., alisphenoid, basisphenoid, basioccipital, exoccipital, petrosal, pterygoid, and squamosal), the ectotympanic, the cartilage of the auditory tube, the tympanohyal (the proximal part of Reichert's cartilage), and independent elements called entotympanics (Kampen, 1905; Klaauw, 1931; Novacek, 1977). The adult form of the tympanic floor and roof is achieved by expansion of the

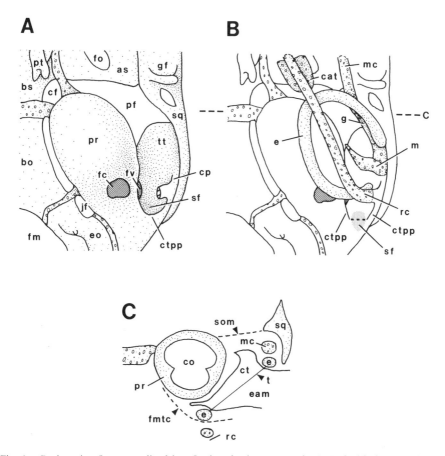

Fig. 1.   Basicrania of a generalized late-fetal eutherian mammal. (A) Left side in ventral view. Soft tissues, middle-ear ossicles, and tympanic floor elements removed. (B) As above, but with tympanic floor elements in place. (C) Frontal section at level of dashed line labeled C in panel B, illustrating the connective-tissue membranes (dashed lines) forming the primordial roof of the middle-ear cavity. Open circles represent cartilage. Two arrangements of the caudal tympanic process of the petrosal (ctpp) are depicted. In A, the process surrounds the fossa for the origin of the stapedius muscle; in B, the process is not continuous behind the fossa, but is divided into two parts. Because the primordial tympanic wall is stretched between the caudal tympanic process and the ectotympanic, the stapedius muscle in A is enclosed within the middle-ear cavity, but in B a part of the muscular origin (regular light stipple) extends beyond the posterior boundary of the middle ear (represented by the dashed line between the two parts of the caudal tympanic process). Abbreviations as in Table II.

contributing skeletal elements and pneumatization (inflation) of the cavum tympani in late prenatal and postnatal stages. Both the identification of the contributing skeletal elements and the pattern of pneumatization are important for defining the adult character state (MacPhee and Cartmill, 1986; MacPhee *et al.*, 1989).

   A cautionary note regarding the term *entotympanic* is indicated. Entotympanics are a category of independent cartilages that appear in the tympanic

Table II. Abbreviations

abc	Anterior basicapsular commissure
ac	Auricular cartilage
as	Alisphenoid
bf	Basicapsular fissure
bo	Basioccipital
bp	Basal plate
bs	Basisphenoid
cat	Cartilage of auditory tube
"cat"	Fused cartilage of the auditory tube, rostral and caudal entotympanics
cc	Cochlear capsule (pars cochlearis of auditory capsule)
ce	Caudal entotympanic
cf	Carotid foramen
co	Cochlea
cp	Crista parotica
ct	Cavum tympani
ctpp	Caudal tympanic process of petrosal
dre	Dorsolateral part of rostral entotympanic
e	Ectotympanic
eam	External acoustic meatus
eo	Exoccipital
ep	Process on ectotympanic
er	Epitympanic recess
es	Element of Spence
esq	Primordial epitympanic sinus of squamosal
ew	Epitympanic wing
fc	Fenestra cochleae (round window)
ff	Foramen faciale
fm	Foramen magnum
fmtc	Fibrous membrane of tympanic cavity
fo	Foramen ovale
fs	Facial nerve sulcus
fv	Fenestra vestibuli (oval window)
g	Gonial
gf	Glenoid fossa
gpn	Greater petrosal nerve
hf	Hypoglossal foramen
i	Incus
ic	Internal carotid artery
jf	Jugular foramen
m	Malleus
mc	Meckel's cartilage
me	Mastoid eminence
pa	Parietal
pbc	Posterior basicapsular commissure
pe	Petrosal
pf	Piriform fenestra
pfc	Prefacial commissure
pgp	Postglenoid process
pr	Promontorium of petrosal
pt	Pterygoid

(continued)

Table II. (*Continued*)

rc	Reichert's cartilage
re	Rostral entotympanic
sc	Secondary cartilage
scg	Superior cervical ganglion
sf	Stapedius fossa
som	Spheno-obturator membrane
sq	Squamosal
t	Tympanum (tympanic membrane)
th	Tympanohyal
tm	Tensor tympani muscle
tt	Tegmen tympani
u	Utricle
V	Trigeminal ganglion
VII	Facial nerve
vc	Vascular canal
vre	Ventromedial part of rostral entotympanic

floor and roof in various eutherians (Klaauw, 1922; Novacek, 1977; MacPhee, 1979). Following Klaauw (1922), entotympanics are divided into two types, based on their approximate site of formation, either rostral or caudal. Our use of the terms *rostral entotympanic* and *caudal entotympanic* is for descriptive purposes and does not necessarily imply homology. Determining the homologies of these cartilages requires phylogenetic analysis (which is beyond the scope of this contribution), because homologous features are those that characterize monophyletic groups (Patterson, 1982).

Cynocephalus

Reconstructions of the developing tympanic region were made for two of the five prenatal colugos studied. In the younger reconstructed specimen, DUCEC 804 (Fig. 2), the chondrocranium is largely unossified, with ossification centers representing the alisphenoid, basisphenoid, basioccipital, and exoccipital. A petrosal ossification has not yet appeared, but is represented by its cartilaginous precursor, the auditory capsule, which has a cochlear part (for the cochlear duct and saccule) and a canalicular part (for the semicircular canals and utricle). The cavum tympani is not much inflated and does not occupy the entire middle-ear cavity (see Fig. 3A). Forming from the dorsolateral aspect of the cavum tympani is an epitympanic sinus, an accessory air space that grows into one of the neighboring intramembranous ossifications, the squamosal (Figs. 2A, 3A). The epitympanic sinus is continuous with the epitympanic recess, the space that includes the dorsal parts of the malleus and incus (Fig. 3B). In adult *C. volans*, this squamosal epitympanic sinus is about four times the volume of the cavum tympani proper (Hunt and Korth, 1980).

In the older reconstructed specimen, DUCEC 806 (Fig. 4), the basi-

A

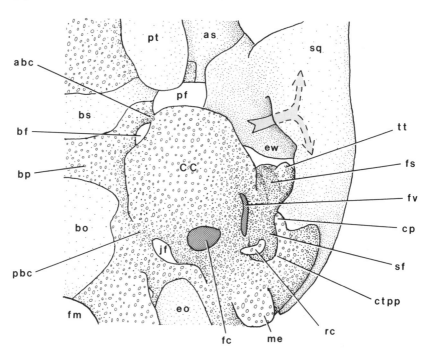

B

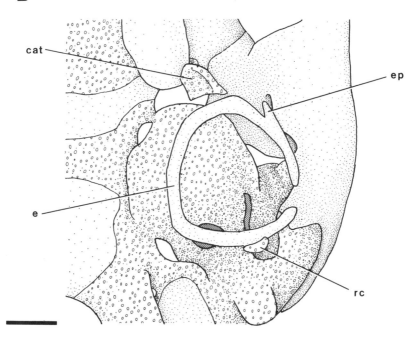

cranium is largely ossified; on the petrosal, for example, only the postero-ventral tip of the mastoid eminence is still cartilaginous. The epitympanic sinus has inflated within the squamosal, increasing the size of its cavity both anteriorly and posteriorly (Fig. 4B), and the cavum tympani proper has inflated to fill nearly the entire middle-ear cavity (Fig. 5A). The petrosal is isolated completely from the basioccipital via the basicapsular (basicochlear) fissure, which is continuous anteriorly with the piriform fenestra and posteriorly with the jugular foramen (Fig. 4A). By contrast, in the younger specimen anterior and posterior basicapsular commissures enclose the basicapsular fissure and connect the auditory capsule and basal plate (Fig. 2A). Whereas the pterygoid and alisphenoid are separate ossifications in the younger specimen, they are fully fused in DUCEC 806.

Tympanic Floor

DUCEC 804 and 8310. In the two youngest specimens, the major elements in the tympanic floor are the fibrous membrane of the tympanic cavity and the ectotympanic (Fig. 3A). The outer attachments of the fibrous membrane are as follows (refer to Fig. 2): medially, the basal plate; posteriorly, the auditory capsule, including the low process behind the stapedius fossa (the caudal tympanic process of the petrosal); and anteriorly, the anterior pole of the cochlear capsule and the cartilage of the auditory tube. From these attachments the fibrous membrane runs to the ectotympanic, becoming continuous with that bone's periosteum. Fibers extend laterally beyond the ectotympanic, but are diffuse.

The ring-shaped ectotympanic is incomplete laterally, with both its anterior and posterior crura in contact with the squamosal (Fig. 2B). From the anterior crus a process extends forward, underlying the exit point of Meckel's cartilage from the developing tympanic cavity (Fig. 2B). Along the anteromedial aspect of the ectotympanic, that bone's medial edge is expanded slightly, such that it makes a small contribution to the anteromedial wall of the tympanic cavity.

As shown in Fig. 1B, Reichert's cartilage usually contacts the auditory capsule at the crista parotica in mammals (de Beer, 1937). It does not in any of the colugos studied here. DUCEC 804 and 8310 have a small segment of Reichert's cartilage that lies near (but does not contact) the crista parotica (Fig. 2A). This rod-shaped element lies outside the fibrous membrane and therefore does not contribute to the tympanic floor. A comparable segment of

---

Fig. 2.   Left basicranium of fetal *Cynocephalus variegatus* (DUCEC 804) in ventral view. (A) Cartilages (open circles) and endochondral and intramembranous ossifications of the tympanic region with the ectotympanic and cartilage of the auditory tube removed. Stippled arrow passes into and shows the extent of the epitympanic sinus of the squamosal. (B) As above but with the ectotympanic and cartilage of the auditory tube in place. Scale bar = 1.0 mm. Abbreviations as in Table II.

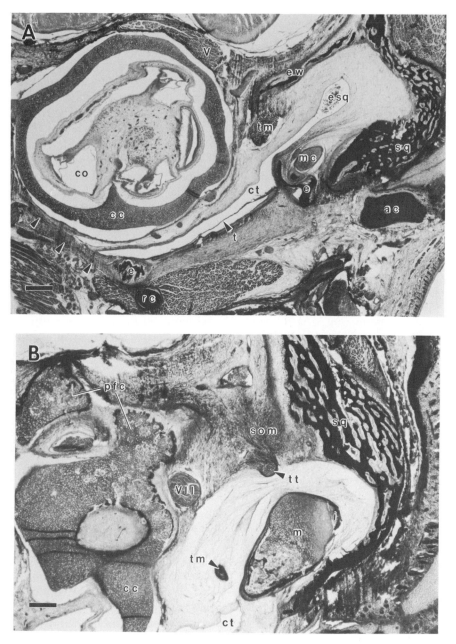

Fig. 3. Frontal sections through the basicranium of fetal *Cynocephalus* sp. (DUCEC 8310). (A) Section 1438, through the cochlear capsule (cc) in front of the tegmen tympani. Arrowheads between the cochlear capsule and ectotympanic (e) point to the fibrous membrane of the tympanic cavity. The cavum tympani (ct) and its dorsolateral diverticulum, the primordial epitympanic sinus of the squamosal (esq), do not completely fill in the middle-ear space. Scale bar = 0.25 mm. (B) Section 1488, through the cochlear capsule at the rostral tip of the tegmen tympani (tt). The tegmen tympani lies in the fibrous tympanic roof, the spheno-obturator membrane (som), and posterior to this section, is attached to the prefacial commissure (pfc). The space dorsal to the malleus (m) and incus (i) represents the epitympanic recess. Scale bar = 0.33 mm. Abbreviations as in Table II.

Reichert's cartilage does not occur in the other colugos studied and so is not depicted in Fig. 4.

DUCEC 805, 806, and 839. In the remaining specimens the tympanic floor is nearly completely closed by bone and cartilage. Most of the growth responsible for the floor closure is on the ectotympanic, which is now a conch-shaped element applied to the ventral surface of the middle ear (Fig. 4C). The outer contour of the previously ring-shaped ectotympanic has expanded dorsally along the fibrous membrane to fill in much of the anterior, medial, and posterior walls of the middle ear, and the inner contour has expanded laterally to form a nearly complete floor (Fig. 5A). The site of attachment of the tympanic membrane (crista tympanica) to the ectotympanic, however, is essentially in the same position as in the earlier stages. As a result, the tympanic membrane lies at roughly the same subhorizontal angle that it does in the earlier stages (compare Figs. 3A and 5A); a horizontal position is reported for the tympanum in adult *C. volans* by Hunt and Korth (1980). The lateral expansion of the ectotympanic floors the external ear and forms the ventral rim of an osseous external acoustic meatus (Figs. 4C,D, 5A). The rostrally directed process on the ectotympanic's anterior crus present in the earlier stages (Fig. 2B) has also expanded and now forms the anterior wall of the conduit into the squamosal epitympanic sinus.

The ectotympanic is not the only element in the anterior, medial, and posterior walls of the middle ear. Within the fibrous membrane of the tympanic cavity, in the gap between the ectotympanic and the ventral surface of the petrosal is a mass of cartilage (Figs. 4B, 5A). In DUCEC 805, this is a precartilaginous mass (rounded mesenchymal cells and connective-tissue fibers) that appears to be continuous anteriorly with the more advanced cartilage of the auditory tube. However, because sections were lost for DUCEC 805, it is uncertain whether only one precartilaginous mass or several centers of precartilage (or cartilage) are present. In any case, in the remaining specimens (DUCEC 806 and 839), there is one cartilaginous mass that at its anteromedial end includes the cartilage of the auditory tube. Along the medial and posteromedial tympanic walls, this mass is wider than the underlying ectotympanic and is therefore partly visible in ventral view with the ectotympanic in place (Fig. 4C). Between this cartilage and the anterior pole of the promontorium is a tiny, anteromedially directed canal transmitting the internal carotid nerves and a minute remnant of the internal carotid artery, which is supplied through the circle of Willis. The internal carotid nerves run in a transpromontorial position on the ventral surface of the promontorium (Fig. 5A), whereas the remnant of the artery involutes at the caudal end of the carotid canal and does not enter the middle ear.

Klaauw (1992) is the only author to have described separate cartilages, in addition to the cartilage of the auditory tube, in the colugo tympanic floor. In a study of three prenatal specimens of *C. volans* (32-, 33-, and 40-mm head lengths), Klaauw (1922) found that (1) the youngest specimen had only a tubal cartilage (as in DUCEC 804 and 8310; Fig. 2B); (2) the intermediate specimen added a separate cartilage behind that, which Klaauw called a rostral ento-

A

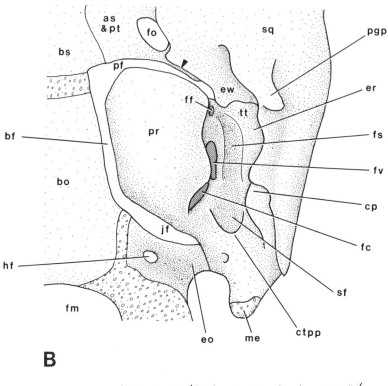

B

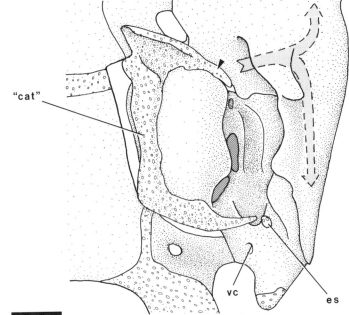

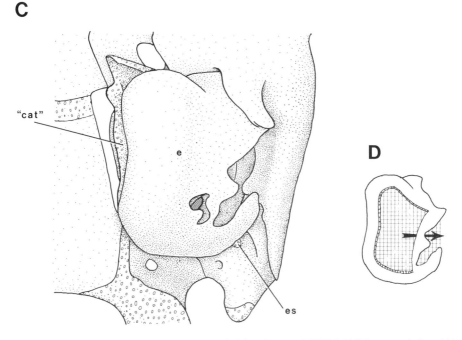

Fig. 4.   Left basicranium of late-fetal *Cynocephalus variegatus* (DUCEC 806) in ventral view. (A) Cartilages (open circles) and endochondral and intramembranous ossifications of the tympanic region with the ectotympanic, cartilage of the auditory tube, entotympanics, and element of Spence removed. Alisphenoid and pterygoid are fused. Arrowhead points to epitympanic wing of alisphenoid. (B) As above but with fused cartilage of the auditory tube + entotympanics ("cat") and the element of Spence (es) in place. Stippled arrow passes into and shows the extent of the epitympanic sinus of the squamosal. Arrowhead points to cartilage in tympanic roof that floors greater petrosal nerve. (C) As above but with ectotympanic (e) in place. Fenestration in ectotympanic results from incomplete ossification and transmits no structures. A similar opening is present in the macerated skull of a juvenile *C. variegatus* from the Smithsonian Institution (USNM 307553). (D) Isolated ectotympanic with window cut to show the attachments of the subhorizontal tympanum (grid pattern). Arrow passes through the external acoustic meatus. Proximal segment of Reichert's cartilage found in DUCEC 804 (see Fig. 2) is absent in DUCEC 806. Scale bar (for A, B, and C) = 1.6 mm. Abbreviations as in Table II.

tympanic; and (3) in the oldest specimen, the first two cartilages fused and a third was added behind that, which Klaauw called a caudal entotympanic. In their dissection of a neonatal *C. volans*, Hunt and Korth (1980) found only one cartilage in the gap between the ectotympanic and the petrosal, which they believed represented the fused state of the three separate cartilages identified by Klaauw (1922). Because DUCEC 806 and 839 are comparable to the neonatal *C. volans* described by Hunt and Korth (1980) in that one cartilaginous mass is present, we can neither confirm nor dismiss the ontogeny described by Klaauw (1922). However, until evidence to the contrary is forthcoming, we assume that the tympanic floor of *C. variegatus* develops like that of *C. volans* as

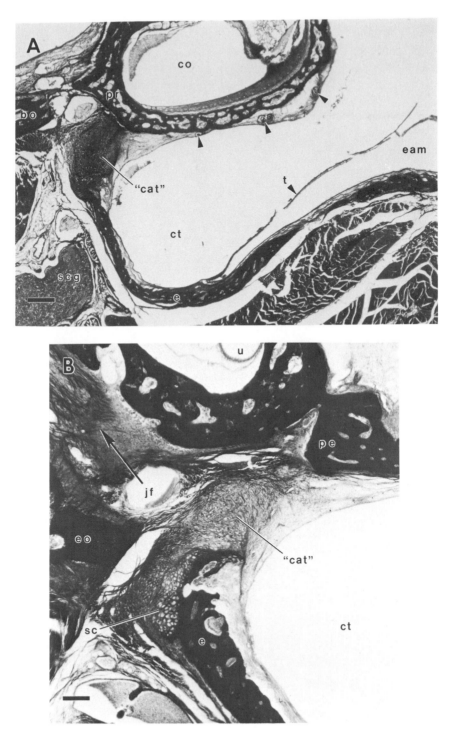

described by Klaauw (1922). Hunt and Korth (1980) further report that the entire cartilaginous mass in the neonatal *C. volans* (i.e., cartilage of the auditory tube plus rostral and caudal entotympanics) is ossified in the adult.

The ectotympanic and the elements corresponding to the rostral and caudal entotympanics of Klaauw (1922) that form in the medial and posterior tympanic wall behind the cartilage of the auditory tube exhibit an unusual relationship. In DUCEC 805, the surface of the ectotympanic that faces the precartilaginous mass of the rostral and caudal entotympanics is composed not of intramembranous bone, but of secondary cartilage. This is the case whether the precartilaginous mass abuts the ectotympanic or is separated from it by undifferentiated mesenchymal cells. In DUCEC 806, the secondary cartilage that is found on some areas of the ectotympanic and the cartilage of the rostral and caudal entotympanics are not distinctly separable, but appear to be encased within the same connective-tissue wrapping (Fig. 5B). Secondary cartilage is absent from the ectotympanic of the oldest specimen (DUCEC 839) and the boundary between the intramembranous ectotympanic and cartilaginous entotympanics is distinct. A similar relationship between the ectotympanic and entotympanic is reported for a fetal stage of the hyracoid *Dendrohyrax dorsalis* by Fischer (1989). The surface of the ectotympanic that abuts the (rostral) entotympanic in that form consists of secondary cartilage, and a boundary between the two bullar elements is marked only by a change in the structure of the cartilage. However, the appearance of secondary cartilage in the colugo ectotympanic may be part of a more widespread phenomenon. Gaupp (1907) noted an unusual preponderance of secondary cartilage in the intramembranous ossifications of the colugo, including the mandible, maxilla, squamosal, pterygoid, vomer, jugal, and frontal. From the sections of the colugo basicranium available for this study, we confirm the presence of secondary cartilage in the mandible, squamosal, pterygoid, and jugal.

In a cross section of the oldest *C. volans* studied by Klaauw (1922, Fig. 16), two additional tiny cartilages appear in the posterolateral tympanic wall, between the ectotympanic and the crista parotica. In a subsequent publication, Klaauw (1923) described a single cartilage in a comparable position in the two younger specimens of *C. volans* and suggested that these cartilages represent the element of Spence, an inconstant skeletal element associated with the chorda tympani nerve at the rear of the tympanic cavity. In DUCEC 806, a

---

Fig. 5.   Sections through the basicranium of late-fetal *Cynocephalus variegatus* (DUCEC 806). (A) Frontal section 658, through the promontorium (pr) in front of the tegmen tympani. The tympanic floor is formed by the greatly expanded ectotympanic (e) and the fused cartilage of the auditory tube + rostral and caudal entotympanics ("cat"). The cavum tympani (ct) is expanded and nearly completely fills the middle-ear space. Arrowheads point to branches of the internal carotid nerve running in a transpromontorial position. Scale bar = 0.33 mm. (B) Sagittal section 300, through the back of the ear region just medial to the round window. Arrow points in the direction of the jugular foramen (jf). The secondary cartilage (sc) on the ectotympanic grades into that of the "cat." Scale bar = 0.6 mm. Abbreviations as in Table II.

small, rod-shaped cartilaginous element of Spence in contact with the ecto-tympanic medially forms the anteromedial border of stylomastoid foramen, the point of exit of the facial nerve from the tympanic cavity (Fig. 4B,C). This element projects into the middle ear and the chorda tympani nerve arises from the facial nerve immediately in front of it. A smaller version of the element of Spence appears in DUCEC 804, but its incidence in the remaining specimens was not ascertainable (because of missing sections). The fate of the element of Spence in the adult is unknown.

Tympanic Roof

DUCEC 804 and 8310. In the two youngest specimens, two elements, one in front of the other, nearly close the tympanic roof lateral to the cochlear capsule (Fig. 2A). A narrow gap spanned by connective-tissue fibers (the spheno-obturator membrane) separates these two elements from each other and from the cochlear capsule. The anterior element is the broad epitympanic wing of the squamosal (Fig. 3A); the posterior is the tegmen tympani, the part of the auditory capsule that projects forward from the crista parotica (Fig. 2A). Connected dorsomedially with the cochlear capsule by the prefacial commissure, the tegmen tympani is short and narrow and tapers to a small, round process at its anterior end (Fig. 3B) that lies in a plane ventral to the epitympanic wing of the squamosal. A tegmen tympani has been described previously in a prenatal *C. volans* (42-mm head length) by Parker (1886). However, this structure forms later than the other parts of the auditory capsule, because it is wholly absent from much younger specimens of *C. variegatus* (Henckel, 1929) and *C. volans* (Halbsguth, 1973) (28- and 63-mm crown–rump lengths, respectively).

The cartilage of the auditory tube lies immediately in front of the cochlear capsule (Fig. 2B) and forms the roof over the auditory tube, the communication between the cavum tympani and pharynx. The lateral lip of the tubal cartilage is turned ventrally and nearly contacts the anteriormost edge of the ectotympanic. Two muscles arise from the tubal cartilage—the tensor tympani posteriorly from the dorsolateral surface, and the tensor veli palatini anteriorly from the ventrolateral surface. The levator veli palatini muscle arises directly from the auditory tube in front of the origin of the tensor palati. The tubal cartilage is not mentioned in the studies of younger prenatal colugos by Henckel (1929) and Halbsguth (1973).

DUCEC 806 and 839. The epitympanic wing of the squamosal and the tegmen tympani remain the major elements in the tympanic roof lateral to the promontorium of the petrosal, the ossified cochlear capsule (Fig. 4A). Moreover, the proportions of their contributions are essentially as in the earlier stages. The tegmen tympani is ossified as part of the petrosal and its anteromedial edge lies ventral to the epitympanic wing of the squamosal (Fig. 4A).

Two additional elements contribute to the tympanic roof, closing the piriform fenestra between the epitympanic wing of the squamosal and the anterior pole of the promontorium. The alisphenoid has grown posterolaterally, producing an epitympanic wing that is just visible dorsal to the posteromedial edge of the squamosal's epitympanic wing and enclosing a foramen ovale for the mandibular division of the trigeminal nerve (arrowhead in Fig. 4A). Ventral to these elements and filling the gap between them and the petrosal is a cartilaginous mass, continuous anteromedially with the fused primordia of the tubal cartilage and rostral and caudal entotympanics (arrowhead in Fig. 4B). This mass was not described by Klaauw (1992), and we cannot determine whether it forms as an extension of the tubal cartilage or from a separate center. Running in the gap between this mass on the one hand, and the promontorium and the alisphenoid and squamosal epitympanic wings on the other, is the greater petrosal nerve, a branch of the facial nerve. The tensor tympani muscle lies immediately ventrolateral to this mass and the epitympanic wing of the squamosal and appears to have attachments to both.

Sections that included the elements of the tympanic roof were not available for DUCEC 805.

Pteropus

Reconstructions of the developing tympanic region were made for the most advanced prenatal specimen, *Pteropus* sp. DUCEC 831 (Fig. 6). Its basicranium is largely ossified: narrow zones of cartilage separate the endochondral ossifications of the basal plate, and the tegmen tympani and the canalicular part of the auditory capsule remain cartilaginous. Selected features of the basicranium of this specimen have been discussed previously (Wible and Novacek, 1988; Wible, 1992), but all of the components of the tympanic floor and roof are described and illustrated here for the first time.

The other two specimens, both of which are *P. giganteus*, are not as advanced developmentally. The smaller one, ULASN Pg 11, is at an early stage in chondrocranial formation, with the squamosal and gonial the only intramembranous ossifications present in the basicranium. The larger one, ULASN Pg 15, has some endochondral ossification in the basal plate, but the auditory capsule is wholly cartilaginous.

Tympanic Floor

ULASN Pg 11. The tympanic floor in the youngest specimen is composed largely of mesenchymal cells, most of which are undifferentiated and diffuse, but with one area of condensation representing the primordial ectotympanic. The only skeletal element contributing to the floor is Reichert's cartilage, which is connected to the crista parotica by a precartilaginous mass and runs forward and medially beneath the cochlear capsule (as in Fig. 1B).

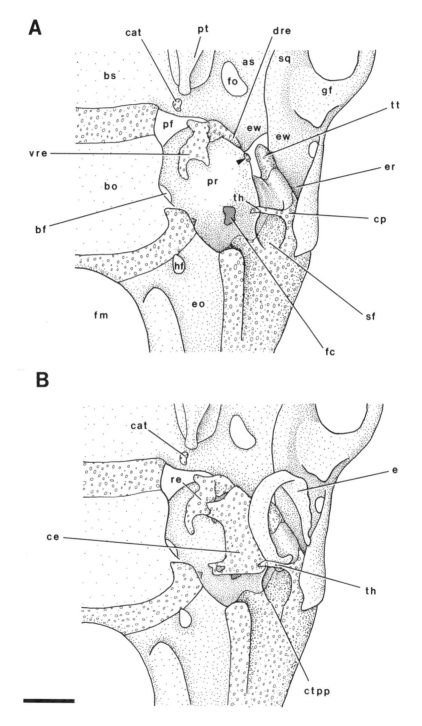

ULASN Pg 15. In the second specimen, the major elements in the tympanic floor are the ectotympanic and the fibrous membrane of the tympanic cavity. The ectotympanic, to which the fibrous membrane is attached, is essentially a simple ring, open laterally, with some lateral expansion on the anterior crus. The ectotympanic is highly inclined (at roughly a 60° angle to the horizontal) and contributes to the lateral as well as the ventral wall of the tympanic cavity. The fibrous membrane forms the anterior, medial, and posterior tympanic walls and extends lateral to the ectotympanic, beneath the external acoustic meatus. Posteriorly, the fibrous membrane is attached to the caudal tympanic process of the petrosal, as described above for *Cynocephalus*, but the membrane's relationship to fossa for the stapedius muscle differs. In the colugo, the caudal tympanic process of the petrosal surrounds the fossa for the stapedius muscle posteriorly (Figs. 2A, 4A), including the muscular origin within the middle ear. In contrast, in ULASN Pg 15 (and DUCEC 831), the caudal tympanic process is divided into medial and lateral parts that are not continuous behind the stapedius fossa (as in Figs. 1B, 6B), the lateral part being the posterior prolongation of the crista parotica. The fibrous membrane is stretched between the two parts of the caudal tympanic process, but a portion of the stapedius fossa between the two extends outside the middle ear, behind the fibrous membrane (see Fig. 1B). Anteriorly, two muscles—the tensor and levator veli palatini—take their origin from the lateral and ventral surface of the fibrous membrane near the junction of the auditory tube and cavum tympani. A cartilage of the auditory tube is lacking at this stage of development.

Most of the distal part of Reichert's cartilage has degenerated into a ligament. It is represented by a short proximal segment (tympanohyal) that makes a small contribution to the tympanic floor behind the posterior crus of the ectotympanic (as in Fig. 6). Extending medially from its attachment on the crista parotica, the tympanohyal ends beneath the lateral margin of the fenestra cochleae. Just beyond the crista parotica, the tympanohyal penetrates the fibrous membrane and its distal end lies outside the tympanic cavity.

DUCEC 831. Within the fibrous membrane of the tympanic cavity are three independent cartilages, representing the caudal and rostral entotympanics and cartilage of the auditory tube described in prenatal *Pteropus edulis* and *Rousettus amplexicaudatus* by Klaauw (1922) (Fig. 6B). Structurally, the rostral entotympanic is the least mature with intercellular matrix that contains collagenous fibers; the caudal entotympanic and cartilage of the auditory tube are composed of mature hyaline cartilage resembling that, for example, of the tympanohyal.

Fig. 6.   Left basicranium of late-fetal *Pteropus* sp. (DUCEC 831) in ventral view. (A) Cartilages (open circles) and endochondral and intramembranous ossifications of the tympanic region with the ectotympanic and caudal entotympanic removed. Arrowhead points to precartilage ventral to tensor tympanic muscle. (B) As above but with the ectotympanic (e) and caudal entotympanic (ce) in place. Scale bar = 1.6 mm. Abbreviations as in Table II.

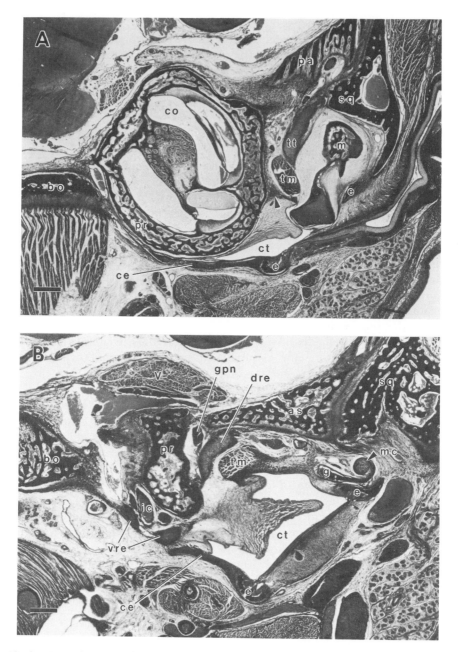

Fig. 7.   Frontal sections through the basicranium of late-fetal *Pteropus* sp. (DUCEC 831). (A) Section 1477, through the promontorium (pr) near the rostral end of the tegmen tympani (tt). The latter projects ventromedially into the middle ear and forms the medial wall of the epitympanic recess, the space dorsal to the malleus (m). The horizontally oriented caudal entotympanic (ce) is the major element in the tympanic floor. Arrowhead points to precartilage ventromedial to the tensor tympani muscle (tm). (B) Section 1367, through the anterior pole of the promon-

The caudal entotympanic is an elongate, thin element lying in a near horizontal plane that fills in the tympanic floor between the promontorium of the petrosal and the posterior crus of the ectotympanic (Figs. 6B, 7A). It is separated throughout its length from the promontorium by a gap that is narrow posteriorly but broader anteriorly (cf. Fig. 7A,B). Posteriorly, the caudal entotympanic nearly covers the fenestra cochleae (Fig. 6B). Most of its medial edge lies ventral to the midpoint of the promontorium in the sagittal plane (Fig. 7A). Anteriorly, the caudal entotympanic forms a floor beneath the auditory tube and the levator veli palatini muscle arises from its dorsal surface. The lateral edge of the caudal entotympanic abuts the ectotympanic (Fig. 7), except posterolaterally, where the two are separated by an opening through which the tympanohyal, facial nerve, and stapedius muscle pass (Fig. 6B). A small fenestration in the posteromedial corner of the caudal entotympanic transmits no structures.

The rostral entotympanic occupies the gap between the caudal entotympanic and the anterior pole of the promontorium (Figs. 6, 7B). It is composed of two distinct parts—ventromedial and dorsolateral (Figs. 6A, 7B). The former contributes to the medial wall of the cavum tympani and auditory tube; the latter is in the tympanic roof and is described with the tubal cartilage separately below. The two parts of the rostral entotympanic converge anteriorly, but are connected only at their anteriormost extent by a narrow bridge. It is possible that the rostral entotympanic formed from two separate centers that fused at their anterior ends: Klaauw (1922) described two independent cartilages in a comparable location in a prenatal *P. edulis*. The main mass of the ventromedial part of the rostral entotympanic is in front of the anterior pole of the promontorium, dorsomedial to the caudal entotympanic (Fig. 7B). Projecting posteroventrally from the main mass are medial and lateral processes, both of which come to lie in the same plane as the caudal entotympanic (Fig. 6A). Running forward, first in the gap between the medial and lateral process and then dorsal to the main mass are the internal carotid artery and nerve (Fig. 7B). These run in a perbullar position (Wible, 1986) with the ventromedial part of the rostral entotympanic forming the floor of the carotid canal and the anterior pole of the promontorium forming the roof.

Reports on the fate of the entotympanics in adult *Pteropus* exist, but a distinction between rostral and caudal elements is not always made. Neither element's contribution to the tympanic floor enlarges significantly from the condition described above. The rostral entotympanic apparently ossifies, though whether it is wholly or partially ossified is unknown; the caudal entotympanic may remain cartilaginous or may ossify to varying degrees with one

---

torium. The internal carotid artery (ic) and nerves run in a canal between the promontorium and the ventromedial part of the rostral entotympanic (vre); the greater petrosal nerve (gpn) runs in a canal between the promontorium, the dorsolateral part of the rostral entotympanic (dre), and the epitympanic wing of the alisphenoid (as). Scale bar (for A and B) = 0.2 mm. Abbreviations as in Table II.

or more centers present (Kampen, 1905; Novacek, 1980b; Wible, 1984). Fusion of the rostral and caudal entotympanics to each other and to surrounding cartilages has been reported for some prenatal megachiropterans. The caudal entotympanic was found to fuse: (1) anteriorly with the ventromedial part of the rostral entotympanic (*Rousettus amplexicaudatus*, Klaauw, 1922; *R. aegyptiacus*, Jurgens, 1963; *R. leschenaulti*, Martin, 1991); (2) posterolaterally with the tympanohyal (*Pteropus* sp., Kampen, 1915; *R. amplexicaudatus*, Klaauw, 1922; *R. aegyptiacus*, Jurgens, 1963; Wible, 1984; *R. leschenaulti*, Martin, 1991); and (3) posteriorly with the auditory capsule, behind the fenestra cochleae (*R. amplexicaudatus*, Klaauw, 1922; *R. leschenaulti*, Martin, 1991). Incidences of fusion of the rostral entotympanic to various elements of the tympanic roof are discussed in the following section.

The tympanohyal continues to make a small contribution to the posterolateral wall of the tympanic cavity, behind the caudal entotympanic (Fig. 6B). The tympanohyal and caudal entotympanic are in close proximity, but a distinct gap filled by connective-tissue fibers separates the two. The ectotympanic in DUCEC 831 is not as highly inclined (at approximately a 45° angle to the horizontal, Fig. 7B) as in ULASN Pg 15, but is expanded both medially and laterally to some extent.

Tympanic Roof

ULASN Pg 11. In the youngest specimen, the tympanic roof is largely composed of undifferentiated mesenchymal cells. The only skeletal elements dorsal to the cavum tympani are Meckel's cartilage and its associated intramembranous ossification, the gonial (see Fig. 1B); a tegmen tympani is wholly lacking at this stage. The tegmen tympani was also reported as absent in an early prenatal stage of *P. seminudus* (= *Rousettus leschenaulti*) by Starck (1943).

ULASN Pg 15. The spheno-obturator membrane closes the large piriform fenestra in front of the cochlear capsule. Surrounding the piriform fenestra are, anteriorly, the ala temporalis and alisphenoid (which forms a complete foramen ovale), medially, the alicochlear commissure (which forms the lateral margin of a separate carotid foramen), and laterally, the squamosal with its small epitympanic wing. The piriform fenestra extends posteriorly, lateral to the cochlear capsule, to the level of the geniculate ganglion of the facial nerve, where the tegmen tympani is encountered. The tegmen tympani (see Fig. 6A) extends forward from the crista parotica, abuts the squamosal laterally, and is connected with the cochlear capsule posteromedially via the prefacial commissure. A small gap in front of the prefacial commissure separates the tegmen tympani and cochlear capsule. The tegmen tympani is a slender process oriented at a 60° angle to the horizontal and forming the medial wall of the epitympanic recess (as in Fig. 7A), the space that includes the dorsal parts of the malleus and incus (and out of which the gonial and Meckel's cartilage extend anteriorly). The cartilage of the tegmen tympani is not as mature as that of the surrounding chondrocranium, where large lacunae are already in evidence. A cartilage of the auditory tube is not yet present.

DUCEC 831. The large piriform fenestra present in ULASN Pg 15 is almost completely filled in with bone and cartilage in DUCEC 831, though a narrow gap still separates the alisphenoid and petrosal (Fig. 6A). The near closure of the piriform fenestra results from the expansion of the epitympanic wing of the alisphenoid posteriorly, the epitympanic wing of the squamosal medially, and the tegmen tympani anteriorly. The various skeletal elements that form the boundaries of the piriform fenestra do not all lie in the same tissue plane (also true of ULASN Pg 15). In the plane of the spheno-obturator membrane are the epitympanic wing of the alisphenoid and the prefacial commissure of the petrosal. In a more ventral plane are the tegmen tympani and the dorsolateral part of the rostral entotympanic (Fig. 7). The epitympanic wing of the squamosal abuts the epitympanic wing of the alisphenoid anteriorly and the tegmen tympani posteriorly and, therefore, lies in both the dorsal and ventral planes. An alicochlear commissure is lacking; consequently, the carotid foramen is not an entity separate from the piriform fenestra.

The tegmen tympani remains a slender process at roughly a 60° angle to the horizontal and forms the medial wall of the epitympanic recess (Fig. 7A). Whereas the cochlear capsule, prefacial commissure, and crista parotica are ossified, the tegmen tympani is cartilaginous (Figs. 6A, 7A). Two zones of cartilage are identifiable within the tegmen tympani, but the boundary between them is not distinct. Posteriorly, the cartilage is more mature with large separate lacunae, while anteriorly there is a mass of collagenous fibers and mesenchymal cells resembling that of the rostral entotympanic. Attached to the immature cartilage of the tegmen tympani ventromedially is the tensor tympani muscle (Fig. 7A) and abutting it ventrolaterally is the gonial. The tegmen tympani contacts neither the tensor tympani muscle nor the gonial in ULASN Pg 15; therefore, this anterior part of the tegmen tympani must form later.

The tegmen tympani of *Pteropus* and other megachiropterans is the subject of recent controversy. Wible and Novacek (1988, p. 6) noted that the tegmen tympani of megachiropterans and microchiropterans differs from that of other eutherians in that it "tapers to a slender process that projects ventrally into the middle ear cavity medial to the epitympanic recess." Though such a tegmen tympani was illustrated only for an adult vespertilionid microchiropteran, *Myotis myotis*, it was stated that the megachiropterans studied exhibited the same pattern. King (1991) was unable to confirm this in pre- and postnatal specimens of *Pteropus* and figured and described the tegmen tympani to be a robust, horizontal bony element closing the roof of the middle ear. Wible (1992) reviews the criteria by which the tegmen tympani is identified: continuity with the crista parotica and position lateral to the geniculate ganglion of the facial nerve (de Beer, 1937). Using evidence from DUCEC 831, Wible (1992) affirms that the tegmen tympani of *Pteropus* sp. is as described by Wible and Novacek (1988). Furthermore, he traces the controversy to King's (1991) misidentification of the epitympanic wing of the alisphenoid as the tegmen tympani.

As mentioned above, the dorsolateral part of the rostral entotympanic contributes to the tympanic roof (Figs. 6A, 7B). From the anteriormost extent of the rostral entotympanic, the dorsolateral part runs dorsally in front of the anterior pole of the promontorium. It then sends a small rod-shaped process posteriorly beneath the narrow gap between the epitympanic wing of the alisphenoid and the promontorium (Fig. 7B). Running in this gap is the greater petrosal nerve. The dorsolateral part of the rostral entotympanic provides origin for some fibers of the tensor veli palatini muscle, as does the sphenoid bone behind the pterygoid. The tensor palati is the rostral continuation of the tensor tympani muscle; the two are separated by a narrow connective-tissue band. In the prior specimen, ULASN Pg 15, the band separating the tensor palati and tensor tympani muscles is very broad.

Two additional small, rod-shaped cartilages are present in DUCEC 831. Ventral to the sphenoid and dorsal to the auditory tube at its junction with the pharynx is the cartilage of the auditory tube (Fig. 6A). Along the anterolateral aspect of the promontorium of the petrosal is an even smaller precartilaginous mass (Fig. 6A), which is composed of collagenous fibers and mesenchymal cells. This precartilage (arrowhead in Fig. 7A) lies ventral to and provides origin for the tensor tympani muscle. It lies within a thick band of connective-tissue fibers that is attached anteriorly to the rostral entotympanic and dorsally to the anterior end of the tegmen tympani. This fibrous band forms a sling through which the tensor tympani muscle passes. As suggested by Wible and Novacek (1988), this sling may act as a pulley directing the action of the muscle. A separate tubal cartilage and a larger cartilage beneath the tensor tympani muscle are described for a prenatal *P. edulis* by Klaauw (1922).

What are the fates of the various independent cartilages in the tympanic roof of *Pteropus*? Klaauw (1922) reported that the cartilage of the auditory tube fuses with the rostral entotympanic in prenatal *P. edulis* and *Rousettus amplexicaudatus* and that the cartilage ventral to the tensor tympani muscle fuses with the rostral entotympanic and tegmen tympani in *P. edulis*. However, the fate of these components in the adult state is poorly understood. An ossification in the position of the rostral entotympanic is reported in adult *Pteropus* (Kampen, 1905; Novacek, 1980b), but it is not known whether this includes the entire cartilaginous mass at the anterior pole of the promontorium. To resolve this issue, juveniles of *P. giganteus* and *R. leschenaulti* are currently being processed for serial sectioning.

Other Recent Archontans

Tympanic Floor

Euprimates

The major element in the adult tympanic floor is an outgrowth from the petrosal bone (MacPhee, 1981; MacPhee and Cartmill, 1986).

MacPhee and Cartmill (1986) recognize two phases in the ontogeny of the tympanic floor, an early phase during which all euprimates exhibit essentially the same pattern and a late phase during which different patterns of pneumatization are expressed. In the early phase, periosteal outgrowths (the petrosal plate) arise from a thin strip on the ventral surface of the auditory capsule (the midcapsular arc) and expand along the fibrous membrane of the tympanic cavity to the ectotympanic. The anterior part of the midcapsular arch (the rostral tympanic process of the petrosal) is on the cochlear capsule, and the posterior part (the caudal tympanic process of the petrosal) is on the canalicular part of the auditory capsule. The latter process encloses the fossa for the stapedius muscle within the tympanic cavity in strepsirhines, but part of the fossa is excluded in *Tarsius* and possibly other haplorhines. In the late phase, sutural tissue forms between the petrosal plate and ectotympanic in most euprimates, resulting in a phaneric (visible) condition for the ectotympanic in the adult. In lemuriforms, however, these floor elements decouple and the petrosal plate expands laterally beneath the ectotympanic, enclosing it entirely within the bulla, an aphaneric (not visible) condition.

Scandentia

The major element in the adult tympanic floor is the greatly expanded entotympanic; the caudal tympanic process of the petrosal, which does not surround the stapedius fossa posteriorly, makes a small contribution posterolaterally (Cartmill and MacPhee, 1980; Zeller, 1986).

In *Tupaia,* which is known from relatively complete ontogenetic series, the entotympanic forms within the fibrous membrane shortly before birth and "is in direct continuity with the anlage of the tubal cartilage from the beginning" (Zeller, 1987, p. 40). This combined tubal cartilage + entotympanic has a broad medial and dorsal plate from which processes extend in postnatal stages (Spatz, 1966; MacPhee, 1981). From the medial plate is a posterior spur that contributes to the medial and posterior tympanic walls and fuses with the cartilaginous caudal tympanic process of the petrosal. The dorsal plate and its process contribute to the tympanic roof and are discussed separately. The posterior spur of the medial plate expands laterally beneath the ectotympanic, enclosing it within the bulla and producing an aphaneric condition. Ossification of the entotympanic begins near the anterior end, but a small tubal cartilage at the pharyngeal aperture of the auditory tube is retained in the adult (MacPhee, 1981). The entotympanic and caudal tympanic process of the petrosal are fused in the adult (Spatz, 1966; MacPhee, 1981).

In the only prenatal specimen of *Ptilocercus* to be described, the entotympanic is also said to develop in primary connection with the tubal cartilage (Zeller, 1986). However, whereas only one center of chondrification occurs in *Tupaia,* two separate centers are present in *Ptilocercus* (Zeller, 1986).

Microchiroptera

The major elements in the adult tympanic floor are the ectotympanic, which is usually not much expanded, and the entotympanic, which may remain cartilaginous or ossify (Kampen, 1905; Klaauw, 1931; Novacek, 1980b). A low caudal tympanic process of the petrosal with medial and lateral parts that do not surround the stapedius fossa is present.

The ontogeny of the tympanic floor has been adequately studied only for the vespertilionid *Myotis myotis* (Frick, 1954; Wible, 1984). Two entotympanics, rostral and caudal, develop within the fibrous membrane. The rostral entotympanic forms around the transpromontorial internal carotid artery and nerve beneath the carotid foramen, at the anterior pole of the cochlear capsule. Its shape and position are reminiscent of the rostral entotympanic of *Pteropus* sp. described above; however, the laterally and posteriorly directed processes extending from the main mass are shorter. The caudal entotympanic appears slightly before the rostral in the posterior and posteromedial tympanic walls. Expansion of the floor elements, in particular the ectotympanic, results in the nearly complete closure of the tympanic floor in the adult, which was described from serial sections by Wible (1984). The caudal entotympanic expands forward ventral to the posterior edge of the rostral entotympanic, which remains restricted to the area beneath the carotid foramen. Though both entotympanics are cartilaginous in the serially sectioned adult studied by Wible (1984), they are ossified in some museum specimens (Wible, personal observation).

Rostral and caudal entotympanics are reported for prenatal specimens of the vespertilionid *Pipistrellus* and the molossid *Tadarida* by Wible and Novacek (1988). The only other report of a prenatal microchiropteran with an entotympanic concerns the vespertilionid *Miniopterus schreibersi* (Fawcett, 1919). However, only one element, equivalent in position to the caudal entotympanic described for *Myotis myotis,* is present. Given the paucity of descriptions, it is not known whether the number of entotympanics varies among microchiropterans or whether the specimen studied by Fawcett was too young to show a rostral entotympanic.

Tympanic Roof

Euprimates

The tympanic roof is always complete in euprimates and is formed largely by the tegmen tympani and continuations of the petrosal bulla that extend into the plane of the roof; epitympanic wings of the alisphenoid and squamosal usually also have some contribution (MacPhee and Cartmill, 1986).

A characteristic of the euprimate chondrocranium is a tegmen tympani that is elongate and broad (Starck, 1975; MacPhee, 1981). Along its anteromedial surface, the tegmen tympani is separated from the cochlear capsule by a gap, which is filled in by periosteal outgrowths from the petrosal (an epitympanic wing) in later stages (reported in strepsirhines by MacPhee, 1981, but presumably present in haplorhines also). On the posteroventral surface of the tegmen tympani in strepsirhines, a sulcus and foramen for the stapedial artery (Diamond, 1991) are often found; these are transformed into an incomplete osseous septum (epitympanic crest) within which the stapedial artery runs, bordering the anterior margin of the epitympanic recess (MacPhee, 1981). The euprimate tubal cartilage is elongate and in some prenatal strepsirhines underlies (but is separated from) the rostral margin of the cartilaginous tegmen tympani (MacPhee, 1981).

Scandentia

The major elements in the adult tympanic roof are the entotympanic, tegmen tympani, and epitympanic wing of the alisphenoid (Le Gros Clark, 1926; MacPhee, 1981). In addition, in adult *Tupaia,* the entotympanic covers a small epitympanic wing of the petrosal that extends forwards from the anterior pole of the promontorium (MacPhee, 1981). The transpromontorial internal carotid artery and nerve leave the middle ear for the endocranium anteriorly via a canal between the epitympanic wing of the petrosal and the entotympanic (MacPhee, 1981).

The ontogeny of the tympanic roof is described for *Tupaia* by Spatz (1966), MacPhee (1981), and Zeller (1987). The tegmen tympani is elongate and broad, ending on a parallel with the anterior pole of the cochlear capsule, and with a prominent foramen for the stapedial artery (Zeller, 1986). It is separated from the cochlear capsule by a wide gap, which is closed by the posteriorly expanded epitympanic wing of the alisphenoid. In postnatal stages, the dorsal plate of the combined tubal cartilage + entotympanic described above and the leading edge of the tegmen tympani fuse, forming a broad entotympanic + tegminal commissure. The entotympanic and tegmen tympani ossify from separate centers (the latter as part of the petrosal), but are lightly fused to each other in the adult in the area adjacent to the foramen for the stapedial artery (Spatz, 1966; MacPhee, 1981). The stapedial artery becomes suspended in an osseous canal beneath the tegmen tympani that resembles the epitympanic crest of some strepsirhines (MacPhee, 1981). A small epitympanic wing of the squamosal forms the side wall of the epitympanic recess (MacPhee, 1981).

The tegmen tympani in the prenatal specimen of *Ptilocercus* described by Zeller (1986) differs from that in *Tupaia* in several regards. The tegmen tympani is not separated from the cochlear capsule, but is broadly attached to form a wide prefacial commissure. Moreover, an arterial foramen pierces the tegmen tympani, but it transmits the ramus superior of the stapedial, and not the stapedial proper as in *Tupaia*. It is not known whether an entotympanic + tegminal commissure forms in *Ptilocercus*. As in *Tupaia*, the tegmen tympani in *Ptilocercus* is large and has a vertical lamella resembling the epitympanic crest of some strepsirhines. The tensor tympani muscle is lacking in *Ptilocercus* (Zeller, 1986), but there are conflicting reports regarding its occurrence in tupaiines (cf. Saban, 1963; MacPhee, 1981).

Microchiroptera

The major elements in the adult tympanic roof are the tegmen tympani and the epitympanic wings of the alisphenoid and squamosal. However, these are usually separated to varying degrees by a patent piriform fenestra (Novacek, 1980b).

In the chondrocrania of microchiropterans for which reports exist, including representatives of Rhinolophidae, Vespertilionidae, and Molossidae (Fawcett, 1919; Sitt, 1953; Frick, 1954; Wible, 1984; Wible and Novacek, 1988), the tegmen tympani (tuberculum tympani of some authors) is a thin anteroventrally projecting process on the anterolateral aspect of the auditory capsule. Its development is closely tied to that of the cartilage of the auditory tube in the vespertilionid *Myotis myotis* (Wible, 1984). In that form, the tegmen tympani is a late-forming portion of the chondrocranium. Extending forward from its anterior tip between the gonial and the tensor tympani muscle is a zone of connective-tissue fibers and mesenchymal cells. At the rostral end of this mass is a precartilaginous zone dorsal to the auditory tube, the tubal cartilage anlage. In later stages, a third center of chondrification appears in the mass connecting the tubal cartilage and the tegmen tympani. These various cartilages subsequently fuse, but a distinct histological boundary separates the tegmen tympani with its more mature cartilage cells from the immature cartilage rostral to it. In the adult, the tegmen tympani is ossified as part of the petrosal and extending forward from it is a single tubal cartilage. The tubal cartilage and tegmen tympani are also fused in prenatal specimens of the vespertilionid *Pipistrellus* and the molossid *Tadarida* (Wible and Novacek, 1988), and these elements, in turn, are fused with the rostral entotympanic in prenatal specimens of the phyllostomid *Carollia* (Wible, personal observation).

Comparisons

Fifteen basicranial characters derived from the above descriptions are listed in Table III, and the distribution of these characters in various archontans is shown in Table IV. Because only characters present in more than one taxon are included, several autapomorphies of the colugo (e.g., the subhorizontal ectotympanic, the enlarged epitympanic wing of the squamosal) are not listed. In the following comparisons, the abbreviation CS refers to character and character state numbers in Tables III and IV.

Megachiroptera and Microchiroptera

As noted previously by Wible and Novacek (1988), the ontogeny of the tympanic floor and roof is remarkably similar in megachiropterans and microchiropterans. Both groups of bats have two independent cartilages in the

Table III. Basicranial Character List[a]

1. Tympanic floor
   - 0—Largely membranous (MacPhee, 1981)
   - 1—Fully ossified, ectotympanic major element
   - 2—Fully ossified (chondrified), entotympanic major element
   - 3—Fully ossified, petrosal plate major element (forms anterior, medial, and posterior walls)*
2. Entotympanic
   - 0—Absent (Novacek, 1986)
   - 1—Main element forms anteromedial to cochlear capsule continuous with cartilage of auditory tube
   - 2—Main element forms posteromedial to cochlear capsule
3. Ectotympanic
   - 0—Phaneric (MacPhee *et al.*, 1988)
   - 1—Aphaneric
4. Ectotympanic shape
   - 0—Simple (annular or horseshoe shaped), not expanded greatly medially or laterally (Novacek, 1986)
   - 1—Expanded significantly relative to ontogenetically early condition
5. Caudal tympanic process of petrosal
   - 0—Surrounds stapedius fossa (Wible, 1990)
   - 1—Does not surround stapedius fossa
6. Basioccipital
   - 0—No contact with medial bullar wall (see text)
   - 1—Contacts entotympanic element in medial bullar wall
7. Tegmen tympani
   - 0—Forms broad roof over mallear–incudal articulation (Wible, 1991)
   - 1—Enlarged to roof entire middle-ear ossicular chain
   - 2—Reduced, tapered to a short, round process*
   - 3—Reduced, tapered to an elongate, round process*
8. Epitympanic crest (canal for stapedial artery) on tegmen tympani
   - 0—Absent (Wible, 1987)
   - 1—Present
9. Arterial foramen in tegmen tympani
   - 0—Absent (Wible, 1987)
   - 1—Present, for stapedial artery
   - 2—Present, for ramus superior of stapedial artery
10. Cartilage of auditory tube + entotympanic + tegminal commissure
    - 0—Absent (see text)
    - 1—Present
11. Epitympanic wing of petrosal
    - 0—Present (MacPhee, 1981)
    - 1—Absent
12. Epitympanic wing of alisphenoid
    - 0—Moderately large, expanded posteriorly at least to the level of the promontorium's anterior pole (MacPhee *et al.*, 1988)
    - 1—Small, does not reach to the level of the anterior pole
13. Internal carotid artery
    - 0—Well developed and in transpromontorial course (Wible, 1986)
    - 1—Perbullar course (within petrosal)*
    - 2—Perbullar course (between entotympanic and petrosal)
    - 3—Insignificant or obliterated during ontogeny (accompanying nerve in transpromontorial course)

(continued)

Table III. (*Continued*)

---

14. Osseous carotid canal leading to carotid foramen
     0—Absent (Wible, 1986)
     1—In petrosal
     2—Between entotympanic and petrosal
15. Greater petrosal nerve
     0—Partial or complete canal floored by petrosal (Wible, 1990)
     1—Partial or complete canal floored by entotympanic and/or cartilage of the auditory tube

---

*a*Sources for the primitive eutherian state (0) follow the state's description.
*Derived states that are unique to the taxa listed in Table IV among eutherians.

tympanic floor: the larger element, the caudal entotympanic, forms first in the posterior and posteromedial walls (CS 2.2), and subsequent to that another, the rostral entotympanic, forms a partial or complete canal for the internal carotid artery and nerve beneath the carotid foramen (CS 14.2) and a floor beneath the greater petrosal nerve (CS 15.1). The overall positions and shapes of these elements are essentially the same in the two bat groups. Additionally, both megachiropterans and microchiropterans have a caudal tympanic process of the petrosal that does not surround the stapedius fossa (CS 5.1). In the tympanic roof, both groups of bats have a tegmen tympani that tapers to an elongate, anteroventrally projecting process providing attachment area for the tensor tympani muscle (CS 7.3). This process fuses with the rostral entotympanic and/or cartilage of the auditory tube, at least in *Pteropus edulis* (Klaauw, 1922), *Myotis myotis, Pipistrellus, Tadarida,* and *Carollia* (CS 10.1). Finally, an epitympanic wing of the petrosal is lacking (CS 11.1).

Whereas the form of the chiropteran tegmen tympani is unique among investigated Recent mammals (Wible and Novacek, 1988), two independent cartilages forming posteroventrally and anterodorsally in the tympanic floor are also found in dermopterans (Klaauw, 1922), macroscelideans (MacPhee, 1981), carnivorans (Hunt, 1974; Wible, 1984), and the xenarthran *Dasypus* (Reinbach, 1952). Hyracoids were described as having both rostral and caudal entotympanics by Klaauw (1922), but the caudal element is a process of the petrosal according to Fischer (1989). Among the taxa having both elements, the rostral entotympanic is in proximity to the internal carotid artery and nerve (beneath the carotid foramen) in chiropterans, dermopterans, and carnivorans. (This may also occur in some xenarthrans, but the development of the entotympanics is poorly known for this group.) However, while the internal carotid artery and nerve run between the rostral entotympanic and the promontorium in megachiropterans and dermopterans (and through the rostral entotympanic in microchiropterans), these structures reach the carotid foramen by passing between the rostral and caudal entotympanics, between the rostral entotympanic and the ectotympanic, or through the rostral entotympanic in carnivorans (Hunt, 1987). A caudal tympanic process of the petrosal that is not continuous behind the stapedius fossa occurs in some other

Table IV. Taxon–Character List[a]

Taxon	1	2	3	4	5	6	7	8	9	10	11	12	13	14	15
											Character				
Lemuriformes	3	0	1	0	0	0	1	1	1	0	0	1	0,3	1	0
Lorisiformes	3	0	0	1	0	0	1	0	0	0	0	1	3	1	0
Adapidae	(3)	(0)	1	0	?	0	(1)	1	(1)	(0)	?	(1)	0	(1)	?
Omomyidae	(3)	(0)	0	1	?	0	(1)	0	0	(0)	?	(1)	0	(1)	?
Tarsius	3	0	0	1	1	0	1	0	0	0	?	1	1	1	0
Anthropoidea	3	0	0	1	?	0	1	0	0	0	?	1	1	1	0
Tupaiinae	2	1	1	0	1	1	1	1	1	1	0	0	0	2	1
Ptilocercinae	2	1	1	0	1	0	1	1	2	?	1	0	0	2	1
Cynocephalus	1	1	0	1	0	1	2	0	0	0	1	0	3	2	1
Megachiroptera	1,2	2	0	0,1	1	0	3	0	0	1	1	0	2	2	1
Microchiroptera	1,2	2	0	0,1	1	0	3	0	0	1	1	0	0,3	2	1
Paromomyidae	2	?	0	1	?	1	?	0	0	?	?	?	3	?	?
Plesiadapidae	(2)	?	(0)	(1)	?	?	?	0	0	?	?	?	3	?	?

[a]Characters and character states described in Table III. States in parentheses are those for which a parsimonious inference is permitted; states known to exist in one member of a highly corroborated sister dyad (i.e., lemuriformes + adapids, haplorhines + omomyids, paromomyids + plesiadapids) are inferred to be present in its sister taxon, if there is no contrary evidence. ? = unknown condition.

eutherians (e.g., scandentians, *Tarsius*, and lipotyphlans; MacPhee, 1981), and a petrosal epitympanic wing is lacking in dermopterans and *Ptilocercus* (the incidence of this structure in nonarchontan eutherians has not been studied beyond what was reported by MacPhee, 1981).

Among the differences distinguishing megachiropterans from microchiropterans is the course of the internal carotid artery and nerve through the rostral entotympanic. These structures follow a perbullar pathway wholly excluded from the middle-ear space in megachiropterans (CS 13.2), but are in a transpromontorial position within the middle ear in microchiropterans (Wible, 1986) (CS 13.0). However, Wible (1984) has shown that the perbullar pathway of megachiropterans develops from an ontogenetically earlier transpromontorial position.

Euprimates and Scandentia

Euprimates, in particular lemuriforms (and adapids), share more resemblances in the developing tympanic floor and roof with tree shrews than either does with other Recent archontans (or other eutherians) (Cartmill and MacPhee, 1980; MacPhee, 1981; Wible and Covert, 1987). Resemblances of tree shrews and lemuriforms include: an inclined, aphaneric ectotympanic (CS 3.1) that is not expanded (CS 4.0); and an enlarged tegmen tympani that roofs the entire middle-ear ossicular chain (CS 7.1), has an epitympanic crest housing the stapedial artery (CS 8.1), and encloses an arterial foramen. This foramen transmits the stapedial artery in lemuriforms and *Tupaia* (Zeller, 1987; Diamond, 1991) (CS 9.1) and the ramus superior in *Ptilocercus* (Zeller, 1986) (CS 9.2). An enlarged tegmen tympani is shared by tree shrews and all euprimates, not just lemuriforms.

A wholly aphaneric condition for the ectotympanic is rare among eutherians, occurring in several extinct groups (leptictids, brachyericine insectivorans; MacPhee *et al.*, 1988) along with lemuriforms and tree shrews. An enlarged tegmen tympani is found in, in addition to lemuriforms and tree shrews, in hyracoids (Fischer, 1989), macroscelideans, lagomorphs, and rodents (Wible and Covert, 1987). Of these, an arterial foramen pierces the tegmen tympani in only lemuriforms, tree shrews, and the lagomorph *Oryctolagus* (Frick and Heckmann, 1955; Wible, 1984). However, an epitympanic crest enclosing the stapedial artery is lacking in *Oryctolagus* (Wible, personal observation). Arterial foramina pierce the tegmen tympani and canals enclose the stapedial artery in other taxa (Wible, 1987), but the combination of an enlarged tegmen tympani having both an arterial foramen and an epitympanic crest is unique to lemuriforms and tree shrews among eutherians.

The major difference between euprimates and tree shrews is the composition of the tympanic floor: an independent cartilage (CS 1.2) versus periosteal outgrowths from the petrosal (CS 1.3). Consequently, the aphaneric ectotympanics in lemuriforms and tree shrews develop in different ways. In lemuriforms, after sutural tissues form between the petrosal and the ectotympanic, the elements decouple and the petrosal grows around the ectotympanic; in tree shrews, sutural tissue never appears between the entotympanic and ectotympanic (MacPhee, 1981). Another remarkable difference is the reduced size of the epitympanic wing of the alisphenoid in euprimates (CS 12.1); tree shrews exhibit the state considered to be primitive for eutherians (CS 12.0).

Chiroptera, Dermoptera, and Scandentia

As noted elsewhere (e.g., Cartmill and MacPhee, 1980; MacPhee and Cartmill, 1986), no basicranial characters considered to be derived relative to the eutherian morphotype are shared among Recent archontans. However, MacPhee and Cartmill (1986, p. 263) observed that all Recent archontans except Euprimates have one bullar feature in common, "the possession of at least one entotympanic or entotympanic-like structure." Some additional resemblances regarding the independent cartilage(s) in the noneuprimate archontan tympanic floor and roof can be added. These various cartilages form in close proximity to four structures: (1) the tubal cartilage—the two are fused in some chiropterans, dermopterans, and scandentians (see CS 2.1); (2) the tegmen tympani—the two are fused in some chiropterans and *Tupaia* (CS 10.1); (3) the internal carotid artery and nerve, forming at least the floor of the carotid canal beneath the carotid foramen (CS 14.2); and (4) the greater petrosal nerve, forming the floor of this nerve's canal (CS 15.1).

Among Recent eutherians, entotympanics have also been identified in developmental stages for macroscelideans, carnivorans, xenarthrans, hyracoids, and chrysochlorids (Klaauw, 1922; Reinbach, 1952; Wible, 1984; Fischer, 1989; MacPhee and Novacek, 1993). In addition, Kampen (1905) described an independent ossification in the medial tympanic wall in the pholidotans *Manis tricuspis* and *M. gigantea* and the perissodactyl *Rhinoceros sumatrensis*, but the development of these structures is unknown. None of the entotympanics that occur in these nonarchontan eutherians exhibits all four of the relationships described above for the elements in archontans. The (rostral) entotympanic fuses with the tubal cartilage in macroscelideans (MacPhee, 1981) and hyracoids (Fischer, 1989), but does not form in proximity to the tegmen tympani or floor the carotid canal or greater petrosal nerve. In *Rhinoceros*, the entotympanic is fused with the tegmen tympani, but does not contribute to a carotid canal (Kampen, 1905). Finally, the entotympanics contribute to a carotid canal in carnivorans (Hunt, 1974) and apparently in some xenarthrans (Kampen, 1905), but do not form in close proximity to the tubal cartilage or tegmen tympani (Wible, 1984, personal observation).

Additional resemblances in the ontogeny of the tympanic floor and roof are shared among some, but not all, of the noneuprimate archontans. Bats, colugos, and *Ptilocercus* lack an epitympanic wing of the petrosal (CS 11.1). Bats and colugos have two independent cartilages corresponding in position to rostral and caudal entotympanics. Though not elongate, the colugo tegmen tympani resembles that of bats in that in its cartilaginous state it tapers to a slender process (CS 7.2). Bats and tree shrews have a stapedius fossa that is partially excluded from the middle ear space (CS 5.1); and some bats and *Tupaia* have cartilage connecting the tegmen tympani and tubal cartilage (CS 10.1). Finally, in colugos and tree shrews, the formation of the cartilage in the tympanic floor proceeds in a caudal direction, starting from the tubal cartilage + (rostral) entotympanic primordium at the anterior pole of the cochlear capsule (CS 2.1): in bats the direction is reversed (CS 2.2).

Plesiadapiformes and Dermoptera

Traditionally held to be the sister group of Euprimates, Plesiadapiformes has recently been linked with Dermoptera based on evidence from the postcranium (Beard, 1990a,b, 1991) and basicranium (Kay *et al.*, 1990). The basicranial characters supporting this relationship are a "collar-shaped ectotympanic, a partially involuted internal carotid system, a bulla composed partly of the entotympanic, and an entotympanic that contacts the basioccipital medially" (Kay *et al.*, 1990, p. 343). These authors note that the last two characters have been demonstrated for only the paromomyid *Ignacius*. However, the actual dimensions of the ectotympanic are also known only for *Ignacius*, because sutures delimiting that bone from the bulla in other plesiadapiforms are not apparent (MacPhee and Cartmill, 1986). As in colugos, the ectotympanic forms a floor for the external acoustic meatus in *Ignacius*, but we see no special resemblance beyond that. Judging from the illustrations in Kay *et al.*, the ectotympanic was inclined in *Ignacius* and perhaps a quarter the size of the entotympanic (CS 1.2). As shown above, the colugo's ectotympanic is sub-horizontal and roughly four times the size of the entotympanic (CS 1.1). The entotympanics of both *Ignacius* and colugos do contact the basioccipital medially [as in tupaiines (MacPhee, 1981), palaeanodonts, some xenarthrans (Patterson *et al.*, 1992), some carnivorans (Hunt, 1974) and perhaps plagiomenids (MacPhee *et al.*, 1989)] (CS 6.1), but again we see no special resemblance beyond that. The proportions of these elements are dramatically different. Regarding the remaining plesiadapiform + colugo basicranial character, partial involution of the internal carotid system is not an unusual feature among other archontans (CS 13.3) or eutherians (Wible, 1984, 1986, 1987). For now, we find the basicranial characters linking plesiadapiforms and colugos less convincing than those cited above shared by colugos, bats, and tree shrews.

Conclusions

The ontogeny of the tympanic floor and roof provides no characters distinguishing all Recent archontans from other eutherians. There are, however, characters distinguishing all Recent archontans except euprimates— entotympanic(s) that form in relationship to the tubal cartilage, tegmen tympani, internal carotid artery (and nerve), and greater petrosal nerve. While entotympanics occur in some non-archontan eutherians, they do not exhibit this suite of relationships.

Special relationships between euprimates, dermopterans, and megachiropterans proposed elsewhere (e.g., Pettigrew *et al.*, 1989; Beard, 1990a, 1991) are not supported by any characters of tympanic floor and roof ontogeny. Megachiropterans share a unique pattern of tympanic floor and roof formation with microchiropterans, further supporting chiropteran monophyly as have recent analyses of DNA sequence changes (e.g., Adkins and Honeycutt, 1991; Mindell *et al.*, 1991). The few basicranial resemblances linking dermopterans and plesiadapiforms noted by Kay *et al.* (1990) are not convincing or occur in other archontans. If euprimates share a special relationship with any archontans, it is with scandentians based on the basicranial evidence. Both have an enlarged tegmen tympani that roofs the entire middle-ear ossicular chain and there are further unique resemblances in the tegmen tympani of lemuriforms and scandentians.

While the evidence of tympanic floor and roof ontogeny supports various groupings within Archonta, to resolve whether there actually is a clade of noneuprimate archontans, a clade of euprimates and scandentians, or even a clade of archontans requires another giant leap—phylogenetic analyses that include characters from diverse anatomical and biochemical systems among all of the relevant extant and extinct taxa.

Acknowledgments

The authors gratefully acknowledge the following individuals and institutions for the specimens described herein: Drs. M. Cartmill and J. G. M. Thewissen, Department of Biological Anthropology and Anatomy, Duke University, Durham, North Carolina; Dr. Kunwar P. Bhatnagar, Department of Anatomical Sciences and Neurobiology, University of Louisville, Louisville, Kentucky; and Dr. K. B. Karim, Department of Zoology, Institute of Science, Nagpur, India. Comments by Ross MacPhee and one anonymous reviewer greatly improved the manuscript. We thank Louise Bond and Lucinda Schultz for technical assistance and Cathie Caple for photographic assistance. The research reported here was supported in part by National Science Foundation Grant BSR 89-96278 and a project completion grant from the University of Louisville to J. R. Wible.

References

Adkins, R. M., and Honeycutt, R. L. 1991. Molecular phylogeny of the superorder Archonta. *Proc. Natl. Acad. Sci. USA* **88:**10317–10321.

Beard, K. C. 1990a. Flying lemurs, primates, and fossils. *Am. J. Phys. Anthropol.* **81:**192.

Beard, K. C. 1990b. Gliding behaviour and palaeoecology of the alleged primate family Paromomyidae (Mammalia, Dermoptera). *Nature* **345:**340–341.

Beard, K. C. 1991. Postcranial fossils of the archaic primate family Microsyopidae. *Am. J. Phys. Anthropol. Suppl.* **12:**48–49.

Cartmill, M., and MacPhee, R. D. E. 1980. Tupaiid affinities: The evidence of the carotid arteries and cranial skeleton, in: W. P. Luckett (ed.), *Comparative Biology and Evolutionary Relationships of Tree Shrews*, pp. 95–132. Plenum Press, New York.

De Beer, G. R. 1937. *The Development of the Vertebrate Skull.* Oxford University Press (Clarendon), London.

Diamond, M. K. 1991. Homologies of the stapedial artery in humans, with a reconstruction of the primitive stapedial artery configuration of Euprimates. *Am. J. Phys. Anthropol.* **84:**433–462.

Fawcett, E. 1919. The primordial cranium of *Miniopterus schreibersi* at the 17 millimetre total length stage. *J. Anat.* **53:**315–350.

Fischer, M. S. 1989. Zur Ontogenese der Tympanalregion der Procaviidae (Mammalia: Hyracoidea). *Gegenbaurs Morph. Jb.* **135:**795–840.

Frick, H. 1954. *Die Entwicklung und Morphologie des Chondrokraniums von Myotis Kaup.* Georg Thieme Verlag, Stuttgart.

Frick, H., and Heckmann, U. 1955. Ein Beitrag zur Morphogenese des Kaninchenschädels. *Acta Anat.* **24:**268–314.

Gaunt, W. A., and Gaunt, P. N. 1978. *Three Dimensional Reconstruction in Biology.* University Park Press, Baltimore.

Gaupp, E. 1907. Demonstration von Präparaten, betreffend Knorpelbildung in Deckknochen. *Verh. Anat. Ges.* **21**:251–252.

Gaupp, E. 1908. Zur Entwicklungsgeschichte und vergleichenden Morphologie des Schädels von *Echidna aculeata* var. *typica. Semon's Zool. Forschungsreisen Australien* **6**(2):539–788.

Gregory, W. K. 1910. The orders of mammals. *Bull. Am. Mus. Nat. Hist.* **27**:1–524.

Halbsguth, A. 1973. Das Cranium eines Foeten des Flattermaki *Cynocephalus volans (Galeopithecus volans)* (Mammalia. Dermoptera) von 63 mm SchStlg. Inaug. Diss. Med., Frankfurt am Main.

Henckel, K. O. 1929. Die Entwicklung des Schädels von *Galeopithecus temmincki* Waterh. und ihre Bedeutung für die stammesgeschichtliche und systematische Stellung der Galeopithecidae. *Gegenbaurs Morph. Jb.* **62**:179–205.

Hunt, R. M., Jr. 1974. The auditory bulla in Carnivora: An anatomical basis for reappraisal of carnivore evolution. *J. Morphol.* **143**:21–76.

Hunt, R. M., Jr. 1987. Evolution of the aeluroid Carnivora: Significance of auditory structure in the nimravid cat *Dinictis. Am. Mus. Novit.* **2886**:1–74.

Hunt, R. M., Jr., and Korth, W. K. 1980. The auditory region of Dermoptera: Morphology and function relative to other living mammals. *J. Morphol.* **164**:167–211.

Jurgens, J. D. 1963. Contributions to the descriptive and comparative anatomy of the cranium of the Cape fruit-bat. *Ann. Univ. Stellenbosch* **38**:3–37.

Kampen, P. N. van. 1905. Die Tympanalgegend des Säugetierschädels. *Gegenbaurs Morph. Jb.* **34**:321–722.

Kampen, P. N. van. 1915. De phylogenie van het entotympanicum. *Tijdschr. Ned. Dierkd. Ver.* **14**:xxiv.

Kay, R. F., Thorington, R. W., Jr., and Houde, P. 1990. Eocene plesiadapiform shows affinities with flying lemurs not primates. *Nature* **345**:342–344.

King, A. J. 1991. Re-examination of the basicranial anatomy of the Megachiroptera. *Acta Anat.* **140**:313–318.

Klaauw, C. J. van der. 1922. Uber die Entwickelung des Entotympanicums. *Tijdschr. Ned. Dierkd. Ver.* **18**:135–174.

Klaauw, C. J. van der. 1923. Die Skelettstückchen in der Sehne des Musculus stapedius und nahe dem Ursprung der Chorda tympani. *Z. Anat. Entwicklungsgesch.* **69**:32–83.

Klaauw, C. J. van der. 1931. On the auditory bulla in some fossil mammals. *Bull. Am. Mus. Nat. Hist.* **62**:1–352.

Le Gros Clark, W. E. 1926. On the anatomy of the skull of the pen-tailed tree-shrew (*Ptilocercus lowii*). *Proc. Zool. Soc. London* **1926**:1179–1309.

McKenna, M. C. 1975. Towards a phylogenetic classification of the Mammalia, in: W. P. Luckett and F. S. Szalay (eds.), *Phylogeny of the Primates: A Multidisciplinary Approach*, pp. 21–46. Plenum Press, New York.

MacPhee, R. D. E. 1979. Entotympanics, ontogeny and primates. *Folia Primatol.* **27**:245–283.

MacPhee, R. D. E. 1981. Auditory regions of primates and eutherian insectivores: Morphology, ontogeny and character analysis. *Contrib. Primatol.* **18**:1–282.

MacPhee, R. D. E., and Cartmill, M. 1986. Basicranial structures and primate systematics, in: D. R. Swindler and J. Erwin (eds.), *Comparative Primate Biology*, Vol. 1, pp. 219–275. Liss, New York.

MacPhee, R. D. E., and Novacek, M. J. 1993. Definition and relationships of Lipotyphla, in: F. S. Szalay, M. J. Novacek, and M. C. McKenna (eds.), *Mammal Phylogeny: Placentals.* Springer-Verlag, Berlin (in press).

MacPhee, R. D. E., Cartmill, M., and Gingerich, P. D. 1983. New Paleogene primate basicrania and definition of the order Primates. *Nature* **301**:509–511.

MacPhee, R. D. E., Novacek, M. J., and Storch, G. 1988. Basicranial morphology of early Tertiary erinaceomorphs and the origin of primates. *Am. Mus. Novit.* **2921**:1–42.

MacPhee, R. D. E., Cartmill, M., and Rose, K. D. 1989. Craniodental morphology and relationships of the supposed Eocene dermopteran *Plagiomene* (Mammalia). *J. Vertebr. Paleontol.* **9**:329–349.

Martin, J. R. 1991. The development of the auditory region of the rousette fruit bat *Rousettus leschenaulti* (Mammalia, Megachiroptera). M.S. thesis, University of Louisville.

Mindell, D. P., Dick, C. W., and Baker, R. J. 1991. Phylogenetic relationships among megabats, microbats, and primates. *Proc. Natl. Acad. Sci. USA* **88:**10322–10326.

Novacek, M. J. 1977. Aspects of the problem of variation, origin and evolution of the eutherian auditory bulla. *Mammal Rev.* **7:**131–149.

Novacek, M. J. 1980a. Cranioskeletal features in tupaiids and selected Eutheria as phylogenetic evidence, in: W. P. Luckett (ed.), *Comparative Biology and Evolutionary Relationships of Tree Shrews*, pp. 35–93. Plenum Press, New York.

Novacek, M. J. 1980b. Phylogenetic analysis of the chiropteran auditory region, in: D. E. Wilson and A. L. Gardner (eds.), *Proc. 5th Int. Bat Res. Conf.* pp. 347–365. Texas Tech Press, Lubbock.

Novacek, M. J. 1982. Information for molecular studies from anatomical and fossil evidence on higher eutherian phylogeny, in: M. Goodman (ed.), *Macromolecular Sequences in Systematics and Evolutionary Biology*, pp. 3–41. Plenum Press, New York.

Novacek, M. J., and Wyss, A. 1986. Higher-level relationships of the recent eutherian orders: Morphological evidence. *Cladistics* **2:**257–287.

Novacek, M. J. 1986. The skull of leptictid insectivorans and the higher-level classification of eutherian mammals. *Bull. Am. Mus. Nat. Hist.* **183:**1–112.

Novacek, M. J., Wyss, A. R., and McKenna, M. C. 1988. The major groups of eutherian mammals, in: M. J. Benton (ed.), *The Phylogeny and Classification of Tetrapods*. Oxford University Press (Clarendon), London.

Parker, W. K. 1886. On the structure and development of the skull of the Mammalia. Part III. Insectivora. *Philos. Trans. R. Soc. London* **176:**121–275.

Patterson, B., Segall, W., Turnbull, W. D., and Gaudin, T. J. 1992. The ear region in xenarthrans (=Edentata: Mammalia). Part II. Pilosa (sloths, anteaters), palaeanodonts, and a miscellany. *Fieldiana, Geol., n.s.* **24:**1–79.

Patterson, C. 1982. Morphological characters and homology, in: K. A. Joysey and A. E. Friday (eds.), *Problems of Phylogenetic Reconstruction*, pp. 21–74. Academic Press, New York.

Pettigrew, J. D., Jamieson, B. G. M., Robson, S. K., Hall, L. S., McAnally, K. I., and Cooper, H. M. 1989. Phylogenetic relations between microbats, megabats, and primates (Mammalia: Chiroptera and Primates). *Philos. Trans. R. Soc. London Ser. B* **325:**489–559.

Reinbach, W. 1952. Zur Entwicklung des Primordialcraniums von *Dasypus novemcinctus* Linné (*Tatusia novemcincta* Lesson) II. *Z. Morphol. Anthropol.* **45:**1–72.

Russell, D. E. 1964. Les mammifères paléocènes d'Europe. *Mem. Mus. Natl. Hist. Nat. Paris Ser. C* **13:**1–321.

Saban, R. 1963. Contribution à l'étude de l'os temporal des Primates. *Mem. Mus. Natl. Hist. Nat. Paris Ser. A* **29:**1–378.

Sitt, W. 1943. Zur Morphologie des Primordialcraniums und des Osteocraniums eines Embryos von *Rhinolophus rouxii* von 15 mm Scheitl-Steiß-Länge. *Gegenbaurs Morph. Jb.* **88:**268–342.

Spatz, W. B. 1966. Zur Ontogenese der Bulla tympanica von *Tupaia glis* Diard 1820 (Prosimiae, Tupaiiformes). *Folia Primatol.* **4:**26–50.

Starck, D. 1943. Beitrag zur Kenntnis der Morphologie und Entwicklungsgeschichte des Chiropterancraniums. Das Chondrocranium von *Pteropus seminudus*. *Z. Anat. Entwicklungsgesch.* **112:**588–633.

Starck, D. 1975. The development of the chondrocranium in primates, in: W. P. Luckett and F. S. Szalay (eds.), *Phylogeny of the Primates: A Multidisciplinary Approach*, pp. 127–155. Plenum Press, New York.

Szalay, F. S. 1972. Cranial morphology of the early Tertiary *Phenacolemur* and its bearing on primate phylogeny. *Am. J. Phys. Anthropol.* **36:**59–76.

Wible, J. R. 1984. The ontogeny and phylogeny of the mammalian cranial arterial pattern. Ph.D. dissertation, Duke University.

Wible, J. R. 1986. Transformations in the extracranial course of the internal carotid artery in mammalian phylogeny. *J. Vertebr. Paleontol.* **6:**313–325.

Wible, J. R. 1987. The eutherian stapedial artery: Character analysis and implications for super-ordinal relationships. *Zool. J. Linn. Soc.* **91**:107–135.

Wible, J. R. 1990. Late Cretaceous marsupial petrosal bones from North America and a cladistic analysis of the petrosal in therian mammals. *J. Vertebr. Paleontol.* **10**:183–205.

Wible, J. R. 1991. Origin of Mammalia: The craniodental evidence reexamined. *J. Vertebr. Paleontol.* **11**:1–28.

Wible, J. R. 1992. Further examination of the basicranium of the Megachiroptera: A reply to A. J. King. *Acta Anat.* **143**:309–316.

Wible, J. R., and Covert, H. H. 1987. Primates: Cladistic diagnosis and relationships. *J. Hum. Evol.* **16**:1–22.

Wible, J. R., and Novacek, M. J. 1988. Cranial evidence for the monophyletic origin of bats. *Am. Mus. Novit.* **2911**:1–19.

Zeller, U. 1986. Ontogeny and cranial morphology of the tympanic region of the Tupaiidae, with special reference to *Ptilocercus. Folia Primatol.* **47**:61–80.

Zeller, U. 1987. Morphogenesis of the mammalian skull with special reference to *Tupaia,* in: H.-J. Kuhn and U. Zeller (eds.), *Morphogenesis of the Mammalian Skull.* Verlag Paul Parey, Hamburg.

Developmental Evidence from the Fetal Membranes for Assessing Archontan Relationships

5

W. PATRICK LUCKETT

Introduction

During the past decade, increased attention has been devoted to a search for the nearest relatives of Primates, from molecular, paleontological, and comparative anatomical perspectives (Goodman *et al.,* 1982; Novacek, 1982, 1990; Novacek and Wyss, 1986; Miyamoto and Goodman, 1986; Shoshani, 1986; Pettigrew, 1986; Wible and Covert, 1987; Pettigrew *et al.,* 1989; Czelusniak *et al.,* 1990). Many paleontologists and evolutionary morphologists continue to find support for a cladistic relationship among Primates, Scandentia, Dermoptera, and Chiroptera as the superorder Archonta, although the morphological evidence for this association remains sparse (Novacek, 1990). In contrast, recent analyses of amino acid sequence data suggest that primates are more closely related to rodents and lagomorphs (Miyamoto and Goodman, 1986), or to a larger unresolved cluster of Rodentia, Lagomorpha, Scandentia, Chiroptera, Lipotyphla, Pholidota, and Carnivora (Czelusniak *et al.,*

W. PATRICK LUCKETT • Department of Anatomy, University of Puerto Rico, Medical Sciences Campus, San Juan, Puerto Rico 00936. *Current address:* Institute des Sciences de l'Evolution, Laboratoire de Paléontologie, Université Montpellier II, Montpellier, France.

Primates and Their Relatives in Phylogenetic Perspective, edited by Ross D.E. MacPhee. Plenum Press, New York, 1993.

1990). Mitochondrial DNA and immunodiffusion distance data (Adkins and Honeycutt, this volume; Cronin and Sarich, 1980; Sarich, 1993) consistently group Dermoptera as the sister taxon of Primates, and strongly support the monophyly of Chiroptera, contrary to assertions by Pettigrew *et al.* (1989) based on selected neuroanatomical findings.

Paleontological studies have resulted in new debates about whether the archaic Paleocene plesiadapiforms are cladistically and adaptively members of the order Primates, or whether some "plesiadapiforms" exhibit closer phyletic affinities with the order Dermoptera (Szalay *et al.*, 1987; Beard, 1990; this volume; Kay *et al.*, 1990). In addition, controversy continues about whether Eocene omomyids or adapids are nearer to the ancestry of Anthropoidea (Gingerich, 1984; Rosenberger *et al.*, 1985; Szalay *et al.*, 1987; Simons and Rasmussen, 1989), with disagreements engendered in part by the relative weight given to assessments of the limited dental, cranial, or postcranial remains.

Among living Primates, there is increased support from molecular and morphological studies that *Tarsius* and Anthropoidea share a sister-group relationship as members of the monophyletic suborder Haplorhini, in contrast to the Strepsirhini. Resistance to this hypothesis continues from Schwartz and Tattersall (1985, 1987; see also Schwartz, 1986), who believed that selected dental and postcranial attributes support the affinities of *Tarsius* with lorisoids, rather than with Anthropoidea. More disagreement exists concerning whether *Tarsius* shares special affinities with Eocene omomyids, and whether extant lemuriforms are affiliated with Eocene adapids (for differing views, see Gingerich, 1984; Szalay *et al.*, 1987; Andrews, 1988). A consideration of these controversies is critical for evaluating the significance of shared similarities between *Tarsius* and Anthropoidea.

Most controversies about evolutionary relationships within the order Primates, as well as those concerning the superordinal affinities of Primates, can be attributed to disagreements concerning the homology or homoplasy of shared derived attributes among taxa being compared. Incongruences between different hypotheses are commonly the result of differing weights given to particular character complexes, and the degree to which reasonable hypotheses of character transformations are proposed and tested. Many evolutionary analyses are restricted to the selective enumeration of shared derived characters assessed on the basis of outgroup comparisons, with little or no attempt made to evaluate the adaptive significance of these features within the context of a functionally integrated system.

Role of Fetal Membranes in Assessing Mammalian Phylogeny

The fetal membranes and placenta comprise a complex and functionally interrelated organ system that is essential for the normal prenatal development of mammals. Although a true mammalian placental relationship is restricted to therians, the same extraembryonic or fetal membranes (amnion, chorion, yolk sac, and allantois) develop in monotremes, as well as in all other amniotes (Mossman, 1937, 1987; Amoroso, 1952; Luckett, 1977). The mammalian fetal membranes develop in continuity with the embryonic germ layers and exhibit earlier functional differentiation during ontogeny than do tissues of the embryo proper; this is correlated with the fact that they perform the essential functional roles of nutrition, excretion, respiration, and protection throughout the prenatal phase of ontogeny.

The functional primacy of the fetal membranes during ontogeny is evident from both morphological and experimental studies, which demonstrate that the trophoblast (the epithelial layer of the chorion) is the earliest tissue to differentiate during eutherian development (see Gardner, 1985, 1989, for recent reviews). Trophoblast forms the outer epithelium through which most fetal–maternal exchange is mediated during all stages of pregnancy; as such, it is the functionally specialized component of the choriovitelline and chorioallantoic placentas in intimate contact with maternal tissues.

As emphasized by Mossman (1937, 1987), the developmental, genetic, and functional complexity of the fetal membranes makes them an excellent system for assessing evolutionary relationships within Mammalia, and especially within Eutheria. In his comprehensive analysis of fetal membrane data from all mammalian orders, Mossman was able to demonstrate that these developmental features have remained relatively conservative at the generic, familial, and superfamilial levels during mammalian evolution. He suggested that this conservatism was caused, in part, by the minimal selective effect of "external" influences, such as climate, habitat, diet, body size, or locomotor patterns, on the developing fetal membranes and fetus within the relatively stable and uniform "internal" environment of the uterus. This does not imply that the fetal membranes have evolved under "neutral" selective conditions; instead, it means that obvious adaptive or selective factors remain poorly known. The concept of "developmental constraints" doubtlessly applies here (Smith *et al.*, 1985), even though we remain largely ignorant of the nature or extent of these constraints. One example is that the definitive placenta in *all* eutherians consists of the chorion vascularized by mesoderm from the allantois (i.e., a chorioallantoic placenta), regardless of the number of histological layers that separate maternal and fetal blood within the placental barrier. Therefore, whatever evolutionary changes occur during the development of the amnion, yolk sac, or allantois in different taxa, they must be compatible with (or constrained by) this essential functional interrelationship. Similar constraints can be inferred for some of the other mammalian fetal membranes (for further discussion, see Luckett, 1993).

Fetal membrane developmental patterns may vary considerably at the subordinal and ordinal levels among eutherians, or they may exhibit only moderate differences within some orders. Because the entire ontogenetic pattern is often available for each fetal membrane trait, as well as for the complete interrelated complex, these features are particularly amenable to analysis for homologous versus convergent similarities, and for the detection of character transformations. Assessment of these crucial aspects during character analysis should be an essential prerequisite for the testing of phylogenetic hypotheses (Bock, 1977).

Fetal Membrane Developmental Patterns of Archontans

Detailed studies on the morphogenesis of the fetal membranes and placenta in all major groups of Primates have been described and illustrated previously (Hubrecht, 1896, 1899, 1902; Hill *et al.*, 1928; Wislocki, 1929; Hill, 1932; Starck, 1956; Hill and Florian, 1963; Luckett, 1974), and King (1986) has reviewed the available ultrastructural evidence on the chorioallantoic placental barrier in Strepsirhini and Anthropoidea. Development of placentation in several genera of tupaiids (order Scandentia) has also been extensively documented (Hill, 1965; Luckett, 1968; Kuhn and Schwaier, 1973), and the ultrastructure of the endotheliochorial placenta in *Tupaia* was illustrated by Kaufmann *et al.* (1985). Recent histological examination of a single gestation sac of *Ptilocercus lowii* from the last third of pregnancy reveals that its definitive fetal membranes and placenta are identical to those of tupaiines (Luckett, unpublished). Moreover, an evolutionary analysis of the primitive, derived, homologous, or convergent nature of shared similarities in these developmental features has been presented for Primates and Scandentia (Luckett, 1975, 1976, 1980a,b).

Because of the controversy surrounding the mono- or diphyletic origin of Chiroptera (Smith, 1980), an evolutionary analysis of fetal membrane traits was also carried out for both Mega- and Microchiroptera, and the reconstructed morphotype for each suborder was compared with that known for Primates, Dermoptera, and Scandentia (Luckett, 1980d). Information on the morphogenesis of the fetal membranes and placenta in Dermoptera has not been extensively reported (Hubrecht, 1919; Luckett, 1983; Mossman, 1987), although a summary of the major developmental features has been included in character analyses of archontans (Luckett, 1980d). These evolutionary assessments provided further corroboration for the hypothesis of chiropteran monophyly, but little or no direct support for special affinities among any of the four orders commonly included in Archonta. Several of the derived fetal membrane attributes of bats have been incorporated into larger syntheses of eutherian or archontan phylogeny (Novacek, 1982; Wible and Novacek, 1988).

My approach to evolutionary reconstruction is "character-centered," and not "taxon-centered," *sensu* Wheeler (1990), in that emphasis is placed on examining ontogenetic evidence for character homology and character state transformation within the class Mammalia. Therefore, when evaluating relatively primitive or derived aspects of fetal membrane developmental features within primates and other archontans, no assumptions are made about specific eutherian "outgroups" for determining homology or polarity. Instead, *all* other mammalian orders are considered during the evolutionary analysis of fetal membrane attributes, and, when applicable, all other amniotes (see Luckett, 1977, in press). Phylogenetic analyses of relationships among supraspecific taxa always involve hypotheses about character state transformations between taxa; such transformations cannot be observed *directly* in either ontogenetic or stratophenetic sequences, but remain as hypotheses to be tested by all available means (including ontogeny, morphology, stratophenetic position, and functional analysis).

Placental Development and Evolution in Primates

Evolutionary analyses of ontogenetic features of the fetal membranes and placenta for all major groups of Primates have been presented previously (Luckett, 1975, 1980a,b), and will only be summarized here. The morphotypic condition for most fetal membrane attributes of Strepsirhini is virtually identical to that reconstructed for the ancestral eutherian stock (see Luckett, 1977, in press). These developmental features include superficial, noninvasive implantation; amniogenesis by folding; development of a transitory choriovitelline placenta; reduced, vestigial nature of the free yolk sac in later ontogenetic stages; a large vesicular allantois; and a diffuse, definitive epitheliochorial placenta (Fig. 1). Both Platyrrhini and Catarrhini share a number of relatively derived eutherian traits in their morphotypes, including invasive implantation of the blastocyst at its embryonic pole; amniogenesis by cavitation; precocious differentiation of a mesodermal body stalk, associated with a tubular, rudimentary allantoic diverticulum; absence of a choriovitelline placenta; formation of both primary and secondary yolk sacs; and development of a double-discoidal hemochorial placenta (Fig. 2B). These shared, derived ontogenetic traits provide further corroboration for monophyly of Anthropoidea (Luckett, 1980a).

As with most other biological attributes, assessment of the fetal membranes and placenta in *Tarsius* is critical for elucidating the ontogenetic and phylogenetic transformations that have led to the uniquely derived anthropoidean pattern of development. All investigators who have examined fetal membrane morphogenesis in *Tarsius* (Hubrecht, 1896, 1899, 1902; Hill, 1932; Luckett, 1974) have emphasized the intermediate nature of many of its developmental characters, between those of Strepsirhini and Anthropoidea (Fig. 2A). Evolutionary analysis of these ontogenetic traits in all primate higher taxa reveals that the shared similarities between *Tarsius* and Strepsirhini, such as amniogenesis by folding, are primitive eutherian retentions, whereas attributes shared by *Tarsius* and Anthropoidea, including a rudimentary allantoic diverticulum and absence of a choriovitelline placenta, are shared, derived eutherian features (Luckett, 1975, 1980a). These analyses strongly corrobo-

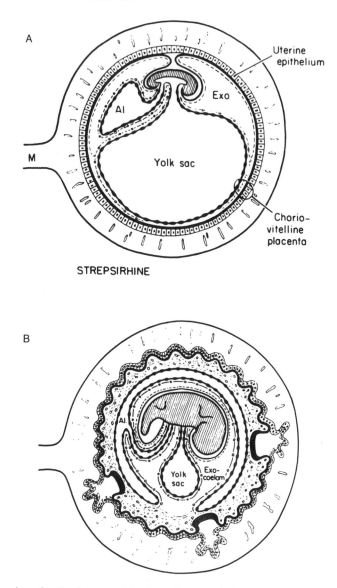

Fig. 1. Overview of major features of fetal membrane and placental development in the reconstructed strepsirhine morphotype (from Luckett, 1975). (A) Choriovitelline placental stage, showing amniogenesis and early allantoic development. (B) Chorioallantoic placental stage, with expanded allantois and reduced yolk sac. Al, allantois; Exo, exocoelom; M, mesometrium of uterus.

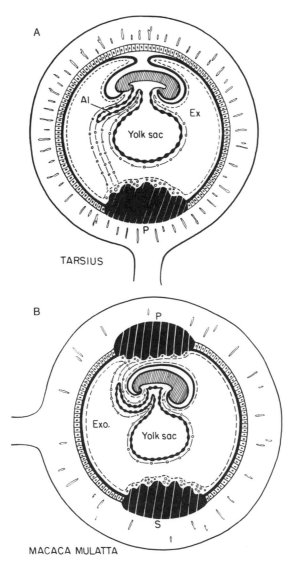

Fig. 2.   Diagram of fetal membranes and placenta in haplorhine primates (from Luckett, 1975).
(A) *Tarsius*. (B) *Macaca*, representing the reconstructed anthropoidean morphotype. Note the
similar nature of the allantois and yolk sac in both, but the differences in amniogenesis and
number of placental discs. See text for details. Al, allantois; Ex, exocoelom; P, primary placental
disc; S, secondary placental disc.

rate the hypothesis for a sister-group relationship between extant *Tarsius* and Anthropoidea as members of a monophyletic suborder Haplorhini.

The ontogeny of double-discoidal placentation in the anthropoidean morphotype (Fig. 2B) occurs in a uniquely derived (autapomorphic) manner, in that one disc develops at the initial implantation site (the primary disc), while the secondary disc forms later during ontogeny at the abembryonic pole (Luckett, 1974, 1982). This ontogenetic pattern differs from the convergently derived, double-discoidal condition of tupaiids (Fig. 3), in which both discs are primary in their initiation (for further details, see Luckett, 1980c). Consideration of the developmental, functional, and adaptive significance of the double-discoidal pattern in Anthropoidea suggests that it could have been derived phyletically from an ancestral condition similar to that which occurs in extant *Tarsius*. For this to have taken place in the common anthropoidean ancestor, the main developmental changes required from the *Tarsius* condition would have been: (1) formation of a simplex uterus and (2) precocious implantation of the early blastocyst by its embryonic pole to the anterior or posterior pole of the simplex uterus. The uterine morphology of *Tarsius*, with a long corpus and short uterine horns, is intermediately derived from the ancestral primate (and strepsirhine) pattern of long uterine horns and a short corpus. Viewed as a transformational hypothesis, no other primate (or eutherian) developmental pattern except that of *Tarsius* could serve as a model for the evolutionary origin of the unique anthropoidean fetal membrane complex. For a further discussion of the probable evolutionary transformations in haplorhine placentation, see Luckett (1982).

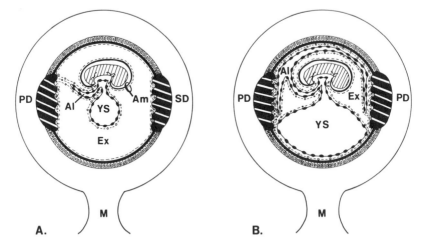

Fig. 3.   Convergent development of double-discoidal placentation in Anthropoidea (A) and Tupaiidae (B) (from Luckett, 1980c). Note differences in nature of the allantois and yolk sac, and the relation of the allantoic stalk to only the primary disc in Anthropoidea. For further details, see text. Al, allantois; Am, amnion; Ex, exocoelom; M, mesometrium; PD, primary placental disc; SD, secondary placental disc; YS, yolk sac cavity.

Morphotypic Condition of Strepsirhine Placentation

The hypothesis that a diffuse, noninvasive, epitheliochorial placenta may not represent the morphotype for Strepsirhini or Primates has been proposed by Martin (1975, 1990). Recently, Martin (1990, pp. 451, 457) suggested that "the endotheliochorial type of placenta is ancestral for placental mammals," and that "development of epitheliochorial placentation in strepsirhine primates can be explained as a divergent specialization away from a hypothetical endotheliochorial type in the ancestral primates, rather than as a purely primitive condition." These proposals were based in part on unsubstantiated theoretical considerations, and in part as the consequence of new interpretations of implantation and placentation in strepsirhines. Martin (1990, p. 455) suggested that invasive activity of the trophoblast in the galagids *Galagoides demidovii* (Gérard, 1932; Luckett, 1974) and *Galago senegalensis* (Butler, 1967) might represent a "primitive condition," rather than a secondary specialization, contrary to the opinion of others (Luckett, 1974; Butler, 1982). Further corroboration for this hypothesis was provided by his reinterpretation of recent studies on implantation and placentation in *Microcebus* (Reng, 1977; Strauss, 1978a,b), in which Martin (1990, p. 455) stated that this genus "has an initial invasive attachment comparable to that of *G. demidovii*" and retains an endotheliochorial zone within its definitive placenta. He then concluded that these new findings support his "prediction that residual endotheliochorial placentation would be found among the more primitive strepsirhines" (p. 455).

No evolutionary analysis was presented by Martin (1990) to assess the homologous or convergent nature of invasive implantation and placentation in *Microcebus, Galagoides,* or *Galago,* or to evaluate the primitive or derived nature of invasive trophoblastic activity within Strepsirhini. As reported previously (Luckett, 1974), it is probable that the transitory, abembryonic, invasive trophoblast of *G. senegalensis* is not homologous with the persisting, paraembryonic, invasive trophoblast of *G. demidovii* (Table I), when both their position and developmental fate are considered. Early implantation stages of *Microcebus* resemble the condition in *Loris* and *Nycticebus* (Table I), in that an expanded, bilaminar blastocyst undergoes a central, noninvasive attachment (Luckett, 1974). This was confirmed by Strauss (1978a) in his study of the same specimens of *Microcebus* from the Bluntschi Collection. Although Strauss (1978a) presented no precise description of the developmental stages of "implantation" studied, the "apposition" and "adhesion" phases discussed by him range from the early attaching, bilaminar blastocyst to embryos with 11–13 pairs of somites that possess a choriovitelline placenta (see Table I). Thus, contrary to Martin's (1990) assertion, *Microcebus* exhibits no evidence of invasive trophoblastic activity during implantation or early embryonic stages of development.

The suggestion (Reng, 1977; Strauss, 1978b) that later ontogenetic stages of placentation in *Microcebus* include a "syndesmochorial" or endothelio-

Table I. Comparison of Early Developmental Stages within Strepsirhini

Stage	*Microcebus*	*Galagoides*	*Galago*	*Nycticebus, Loris*
Early implantation	2.8-mm blastocyst; noninvasive[a]	0.35-mm blastocyst; invasive and paraembryonic	0.3-mm blastocyst; invasive and abembryonic	1.0- to 2.0-mm blastocyst; noninvasive
Primitive streak or early somite stage	2–4 somites; noninvasive	Early primitive streak; invasive	Early primitive streak; remnant of invasion	Early primitive streak; noninvasive
Choriovitelline placenta	11–13 somites; noninvasive	Localized invasive activity[b]	18 somites; noninvasive	11–19 somites; noninvasive

[a]Implantation is probably initiated in an earlier, unknown, stage.
[b]Invasive giant cell trophoblast persists throughout pregnancy.

chorial central zone, surrounded by a more extensive, diffuse epitheliochorial placenta, is an interesting hypothesis, and it may indeed be true. However, both the quality of tissue fixation and the resolution of the photographic documentation for this central "zone of penetration" are insufficient to corroborate this hypothesis, as noted by Mossman (1987, p. 198). My own examination of chorioallantoic placental development during the first half of pregnancy in *Microcebus* suggests that an epitheliochorial placenta, with slender branched villi, is formed during this period. The epithelial layer that separates cytotrophoblast from uterine stroma in the placenta was considered to be syncytiotrophoblast by Reng (1977); however, it is equally possible that this is a layer of modified uterine epithelial symplasma. Until a more careful study is undertaken for placental development in *Microcebus,* including an ultrastructural analysis, it will be difficult to resolve the nature of its placental barrier.

It is evident that developmental relationships of the endotheliochorial zone within the diffuse epitheliochorial placenta of *Galagoides* are not homologous with those of any portion of the chorioallantoic placenta of *Microcebus.* Trophoblastic giant cells denude the uterine epithelium in a localized paraembryonic region during blastocyst implantation in *Galagoides* (Gérard, 1932; Luckett, 1974), and this giant cell zone persists during subsequent developmental stages (Table I). The definitive placenta of both *Galago crassicaudatus* and *G. senegalensis* is diffuse and epitheliochorial (Strahl, 1899; Gérard, 1932; Butler and Adam, 1964; King, 1984), as it is in other described lorisiforms and lemuriforms, and it seems probable that the localized endotheliochorial zone of placentation in *Galagoides* is derived secondarily within the Galagidae. Although Reng (1977) and Strauss (1978a,b) suggested that *Microcebus* shows a close "placentological relationship" with *Galagoides,* there is no indication that they meant this in an evolutionary, rather than phenetic, context. In summary, Martin's (1990) hypothesis that the epitheliochorial placentation of extant strepsirhines was derived secondarily from an endotheliochorial condition in ancestral primates is clearly unwarranted.

A final comment concerns Mossman's (1937) suggestion that endotheliochorial placentation might represent the ancestral eutherian condition. This was based in part on his belief that a transition from endotheliochorial to epitheliochorial placentation occurs during the ontogeny of the American mole, *Scalopus aquaticus* (Mossman, personal communication). However, a more detailed analysis of fetal membrane morphogenesis in *Scalopus* (Prasad *et al.*, 1979), including an ultrastructural examination of the placental barrier, revealed that there is no loss of uterine epithelium during ontogeny. Consequently, Mossman (1987, p. 168) now believes that the epitheliochorial placenta and other fetal membrane attributes of *Scalopus* and some other talpids represent the "most primitive" developmental pattern among recent eutherians. Most of these features are identical to the reconstructed eutherian morphotype proposed by Luckett (1977).

Placental Development in Dermoptera

An extensive study by Hubrecht of implantation and amniogenesis in *Cynocephalus* was published posthumously (1919), based on his collection of 182 female reproductive tracts from Indonesia. This embryological material is maintained at the Hubrecht Laboratory in Utrecht, and a brief report was presented on the major features of fetal membrane development in this collection (Luckett, 1983). Because of the question of possible affinities of Dermoptera with Primates or Chiroptera, a synopsis of fetal membrane morphogenesis in *Cynocephalus variegatus* is presented here.

Implantation and Amniogenesis

In the preimplantation blastocyst, the polar trophoblast overlying the embryonic mass remains intact, and the inner cell mass is differentiated into epiblast and hypoblast. The latter has spread peripherally to line at least half of the blastocyst cavity as extraembryonic endoderm. As demonstrated by Hubrecht (1919), the earliest implanted blastocysts are attached to the antimesometrial pole of the uterine horn by the polar trophoblast (Fig. 4). The invasive, multilayered trophoblast soon invades the uterine epithelium and destroys it, so that the trophoblast approaches the underlying dilated maternal capillaries.

In slightly later implantation stages, a primordial amniotic cavity develops by cavitation within the elongated epiblast (Fig. 5). This primordial cavity persists in later ontogenetic stages as the definitive amniotic cavity. At about the same time, the caudal end of the embryonic epiblast has differentiated precociously into an early primitive streak, and extraembryonic mesoderm derived from it forms a short stalklike attachment to the overlying tro-

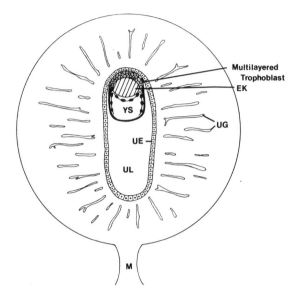

Fig. 4. Early implantation stage in the dermopteran *Cynocephalus variegatus*. Initial attachment occurs antimesometrially by the polar trophoblast overlying the embryonic knot. EK, embryonic knot; UE, uterine epithelium; UG, uterine glands; UL, uterine lumen; other abbreviations as in previous figures.

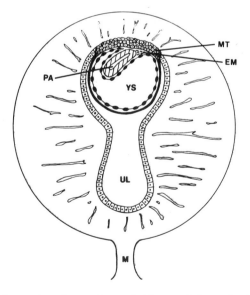

Fig. 5. Stage of amniogenesis and precocious primitive streak formation in *Cynocephalus*. Extraembryonic mesoderm is arising from the caudal end of the tubular amnioembryonic disc. EM, extraembryonic mesoderm; MT, multilayered trophoblast; PA, primordial amniotic cavity; other abbreviations as in previous figures.

phoblast. This was considered by Hubrecht (1919) as a body stalk, homologous with that of *Tarsius*. Subsequently, an exocoelom differentiates within the extraembryonic mesoderm in presomite stages, so that the embryo proper becomes separated from the preplacental trophoblast by this mesoderm-lined cavity (Fig. 6). Although initially a limited area of contact remains between the short body stalk and preplacental trophoblast (Hubrecht, 1919), the continued expansion of the exocoelom in presomite stages results in the disappearance of the transitory body stalk. This developmental relationship differs considerably from that seen in haplorhine primates, in which the precociously differentiated body stalk persists as the site of development of allantoic blood vessels to vascularize the developing placenta (cf. Figs. 2 and 6).

Yolk Sac and Allantoic Vesicle

Blood islands and primitive blood vessels differentiate within the yolk sac splanchnopleure in late presomite stages, and the embryonic one-third of the yolk sac has a functioning vitelline circulation by the time the embryo has attained ten pairs of somites. A choriovitelline placenta is lacking in this and later developmental stages, however, because the expanded exocoelom separates the yolk sac splanchnopleure from the chorion (Fig. 7). An avascular bilaminar omphalopleure, consisting of trophoblast and extraembryonic endoderm, occupies the abembryonic pole of the yolk sac at this stage.

An allantoic diverticulum is first detected as a dorsocaudal outgrowth

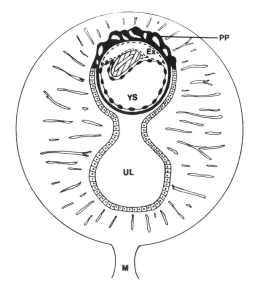

Fig. 6.   Presomite embryonic stage of *Cynocephalus*, with development of the exocoelom within the extraembryonic mesoderm, and with formation of the preplacenta at the initial implantation site. PP, preplacenta; other abbreviations as in previous figures.

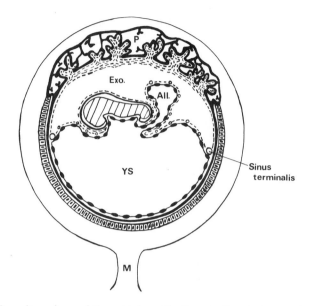

Fig. 7.   An 18-somite embryo of *Cynocephalus,* with allantoic diverticulum projecting into exocoelom. P, preplacenta with distended maternal vascular sinuses; other abbreviations as in previous figures.

from the hindgut in a 10-somite embryo, and it has formed a globular allantoic vesicle, surrounded by vascularized mesoderm, that projects freely into the exocoelom in an 18-somite embryo (Fig. 7). With continued expansion, the medium-sized to moderately large allantoic vesicle extends across the exocoelom dorsal to the developing embryo and fuses with the chorion in embryos that have about 24–30 somites (Luckett, 1983). This results in the fusion of vascular allantoic mesoderm with avascular chorionic mesoderm and marks the onset of chorioallantoic placentation (Fig. 8). The allantoic vesicle remains medium-sized and covers the entire surface of the placental disc until at least midpregnancy, when the embryo has attained a greatest length (GL) of 18–20 mm (Fig. 8B). Unfortunately, the nature of the allantoic vesicle could not be determined in late stages of pregnancy.

The developmental fate of the yolk sac is not completely clear, although a bilaminar omphalopleure still persists in 6- to 7-mm GL embryos. In the most mature of these, an embryo with paddle-shaped forelimb buds, the bilaminar omphalopleure is reduced in extent compared with earlier stages, and it now occupies only the abembryonic one-fourth of the gestation sac (cf. Figs. 7 and 8A). An extensive area of avascular smooth chorion and a smaller region of vascular chorioallantoic membrane separate the bilaminar omphalopleure from the placental disc. In a later conceptus from midpregnancy, bearing an 18- to 20-mm GL fetus, the smooth chorioallantoic membrane occupies the entire wall of the gestation sac beyond the margins of the placental disc, and

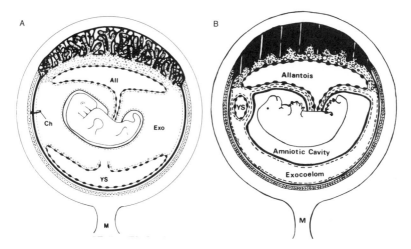

Fig. 8.   Later developmental stages of the yolk sac, allantois, and placental disc in *Cynocephalus*. (A) 7-mm embryo, with reduced yolk sac and expanded exocoelom. (B) 18- to 20-mm fetus, with greatly expanded exocoelom, and possible yolk sac remnant. Ch, smooth chorion.

there is no remnant of a bilaminar omphalopleure. Indeed, there is no distinct trace of the yolk sac at this stage, although a small bilaminar, epithelium-lined vesicle, lying in the exocoelom near one margin of the placental disc, may represent the last remnant of a reduced, free yolk sac (Fig. 8B). The vitelline vessels are reduced within the umbilical cord, and there is no distinct trace of a vitelline duct in the cord.

My interpretation (Luckett, 1983) of the fate of the yolk sac in *Cynocephalus* has been questioned recently by Mossman (1987). He suggested that it is unlikely that the bilaminar omphalopleure present during early limb-bud stages would disappear in later stages. However, such an ontogenetic transformation occurs in every eutherian taxon that possesses a reduced, free yolk sac in late stages, so that it should not be unexpected in dermopterans. The lack of evidence for the transformation of a temporary bilaminar omphalopleure to a reduced, free yolk sac by further expansion of the exocoelom in *Cynocephalus* is the consequence of a gap in the developmental series between 7- and 18-mm GL embryos. Histological examination of these intermediate stages would further clarify the morphogenesis of the yolk sac in this order.

Chorioallantoic Placenta

Invasive activity of the polar trophoblast into the antimesometrial uterine stroma leads to the development of a trophoblastic preplacenta during implantation stages. Maternal vascular sinusoids become surrounded by the invading trophoblast, and their endothelial lining appears to be partly disrupted during primitive streak stages (see Hubrecht, 1919). By early somite stages, the maternal endothelium appears to be lost completely within the shallow preplacenta, so that dilated maternal sinusoids are surrounded by a double layer of trophoblast. With the addition of a layer of exocoelomic mesoderm to the surface of the preplacenta, it can now be designated as a chorionic preplacenta (Fig. 6).

Shallow to moderately developed, chorionic mesodermal "villi" extend into the embryonic surface of the preplacenta in 14- to 18-somite embryos (Fig. 7), but these lack blood vessels. This observation is corroborated by examination of an older 24- to 26-somite embryo, in which fusion of the allantoic vesicle with the chorion has been initiated. The highly vascular mesoderm of the allantoic vesicle remains distinct from the avascular chorionic mesoderm in the area of recent fusion, and allantoic vessels do not extend into the mesodermal villi of the placenta. In older embryos with 30–50 somites, the allantoic capillaries extend into the mesodermal villi, so that a functional chorioallantoic placenta is established (Fig. 8A). At this stage the placental barrier exhibits a hemodichorial relationship, with syncytiotrophoblast, cytotrophoblast, embryonic connective tissue, and embryonic capillary endothelium separating maternal and embryonic blood in the placental disc. In more mature 30- to 35-mm GL fetuses, the cytotrophoblastic layer is reduced or lost in places, so that the placental barrier shows a hemomonochorial condition in some regions. In late fetal stages, the definitive placental barrier is primarily labyrinthine and hemomonochorial.

Placental Development in Chiroptera

Our knowledge of fetal membrane morphogenesis in chiropterans has been enriched during the past 30 years by the studies of Gopalakrishna and his students, and they have provided an excellent summary of their work, while reviewing comparative data from most families of bats (Gopalakrishna, 1958; Gopalakrishna and Karim, 1979, 1980). These authors have discussed shared similarities and differences among families, but there has been little attempt to assess these ontogenetic features in phylogenetic perspective. An evolutionary analysis was presented for the major developmental aspects of placentation in chiropterans, based on original observations from several families, as well as from a review of the literature (Luckett, 1980d). This analysis corroborated the hypothesis of chiropteran monophyly, and studies published on bat placentation since 1980 have provided no evidence to contradict this hypothesis. The excellent investigations by Rasweiler (1990, 1991) on fetal membrane development in molossids can serve as a model for future studies on all phases of placentation in bats.

A brief overview is presented here for the major aspects of fetal membrane development in Mega- and Microchiroptera, so that these ontogenetic features can be compared with those of other eutherians, especially Primates and Dermoptera.

Implantation and Amniogenesis

In all chiropterans studied to date, the polar trophoblast remains intact at the time of implantation. The initial attachment of the bilaminar blastocyst to the endometrium is circumferential or diffuse in most families, including pteropodids and rhinopomatids (Figs. 9A, 10A). This doubtlessly represents the primitive chiropteran and the primitive eutherian condition. In all bats, however, this initial attachment is followed by invasion of the trophoblast into

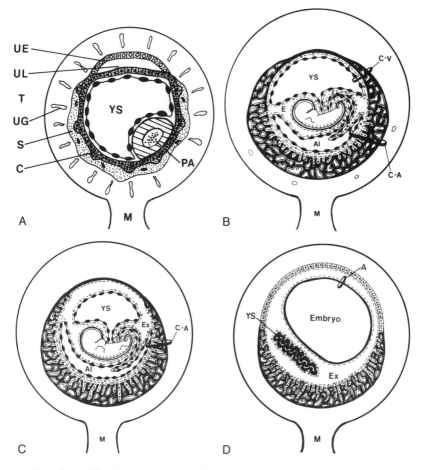

Fig. 9.   Overview of fetal membrane and placental development in the pteropodid *Rousettus*. (A) Implantation and early amniogenesis. (B) Choriovitelline and early chorioallantoic placentation. (C) Later limb bud stage, with chorioallantoic placenta and free yolk sac. (D) Definitive fetal membranes in late pregnancy, with mesometrial placental disc and free, glandlike yolk sac. A, amnion; C, cytotrophoblast; C-A, chorioallantoic placenta; C-V, choriovitelline placenta; S, syncytiotrophoblast; other abbreviations as in previous figures. (Adapted and modified in part from Karim, 1976, and Luckett, 1980d.)

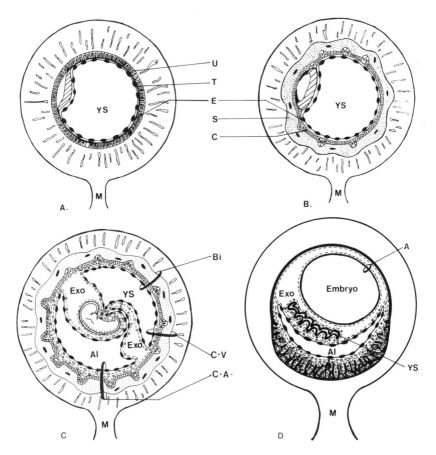

Fig. 10. Overview of fetal membrane and placental development in the rhinopomatid *Rhino-poma*, approximating the morphotypic condition for Microchiroptera. (A) Implantation. (B) Early amniogenesis. (C) Choriovitelline and early chorioallantoic placentation. (D) Definitive fetal membranes, with mesometrial placental disc and free, glandlike yolk sac. Bi, bilaminar omphalopleure; E, endoderm of yolk sac; other abbreviations as in previous figures. (Adapted in part from Srivastava, 1952.)

the endometrium, with loss of the uterine epithelium (Figs. 9A, 10B), an advance over the primitive eutherian pattern. In most bat families with circumferential implantation, the inner cell mass or embryonic disc does not bear a constant orientation to one pole of the uterus, a condition that differs from all other eutherian orders. This has resulted in orientation described as mesometrial, orthomesometrial, or even antimesometrial during different phases of early postimplantation development, even within the same species, as noted by Gopalakrishna and Karim (1980). The reasons for this variation are unclear, and several authors have noted that for some of these taxa, disc orientation seems to be directed more consistently toward the tubo-uterine junction than it is toward a particular endometrial pole. For a further discussion of this unusual condition, see Rasweiler (1979). In a few microbat families, the orientation of the embryonic disc is consistently antimesometrial, and this condition appears to be derived within Microchiroptera. Vespertilionids, thyropterids, and desmodontids are characterized by this pattern.

In all chiropterans, the inner cell mass and its overlying polar trophoblast come into intimate contact with the uterine endometrium during implantation. Consequently, a primordial amniotic cavity develops within the inner cell mass of all bats (Fig. 9A). In phyllostomoids and pteropodids, this primordial cavity persists to become the definitive amniotic cavity (Fig. 9B,C), whereas it has a different fate in most other families. In these, the epiblastic roof of the cavity becomes disrupted and opens beneath the overlying trophoblast, so that a transitory trophoepiblastic cavity separates the embryonic epiblast from the trophoblast (Fig. 10A,B). The definitive amniotic roof is formed later by the upfolding of the epiblastic margins of the embryonic disc, and the expanding exocoelom will subsequently separate the amnion from the overlying chorion. Compared with the eutherian morphotypic condition, this process of amniogenesis in most bat families is intermediately derived, between the primitive pattern of somatopleuric folding, and the most derived process of amniogenesis by cavitation.

Yolk Sac and Choriovitelline Placenta

Following primitive streak formation, extraembryonic mesoderm spreads peripherally between trophoblast and endoderm, and, with the subsequent differentiation of vitelline vessels, a choriovitelline placenta is established during early somite stages in all bat families (Figs. 9B, 10C). Differences exist in the degree to which mesoderm spreads to the abembryonic pole of the conceptus in later stages, and three different patterns can occur. In pteropodids and many microbat families, including rhinopomatids, rhinolophids, and molossids, vascular mesoderm spreads over the entire surface of the yolk sac (Fig. 9B,C), and with the expansion of the exocoelom, the yolk sac is removed from its contact with the inner chorionic surface. This developmental pattern resembles the primitive eutherian condition, and is probably primitive for Chiroptera (for a dissenting view, see Mossman, 1987). The subsequent developmental fate of the free yolk sac in these bat families differs, however, from all other known eutherians.

As first described in pteropodids, the free, reduced yolk sac does not disappear, as it does in other eutherians such as strepsirhine primates and artiodactyls. Instead, it persists, with hypertrophy of both the endodermal lining cells and the surrounding mesothelium, resulting in a "glandlike" appearance of the yolk sac (Fig. 9D). In molossids, it has been demonstrated both histochemically and ultrastructurally that the hypertrophied endodermal cells accumulate glycogen and lipid droplets during the second half of pregnancy (Stephens and Easterbrook, 1968; Rasweiler, 1990). These stored nutrients are rapidly depleted shortly before parturition, and it has been suggested that they serve as an energy source for the large fetus just prior to birth. This glandlike modification of the yolk sac was initially believed to be a pteropodid specialization (Mossman, 1937), but is has subsequently been identified in Rhinopomatidae (Fig. 10D), Emballonuridae, Rhinolophidae, and Molossidae (Srivastava, 1952; Gopalakrishna, 1958; Stephens and Easterbrook, 1968; Rasweiler, 1990). This developmental pattern of the yolk sac is unique within Eutheria, and evolutionary analysis suggests that it is an autapomorphous feature of the Chiroptera (Luckett, 1980d).

In a few microchiropteran families (Phyllostomidae, Desmodontidae, Noctilionidae), extraembryonic mesoderm fails to spread to the peripheral region of the yolk sac, so that the abembryonic pole remains bilaminar throughout gestation. Wimsatt (1954) suggested that the persisting bilaminar omphalopleure in these bats may have an absorptive function, as it does in other therians. Although several authors have proposed that retention of the bilaminar omphalopleure may be a primitive condition for both Chiroptera and Eutheria (Wimsatt, 1954; Mossman, 1987), evolutionary analysis of the yolk sac and all other fetal membrane features in eutherians suggests instead that this is a derived (and pedomorphic) feature, functionally correlated with its specialization for nutrient absorption (Luckett, 1977, 1980d). As such, it appears to be an additional shared, derived feature to support the monophyly of Phyllostomoidea.

An intermediately derived pattern of yolk sac morphogenesis character-izes vespertilionids, megadermatids, and thyropterids. In these families, ex-traembryonic mesoderm spreads peripherally to form a completely trilaminar omphalopleure, but expansion of the exocoelom is limited. This results in the incomplete separation of the trilaminar omphalopleure from its contact with the chorion abembryonically. The vascular wall of the yolk sac undergoes hypertrophy and collapse against the persisting trilaminar omphalopleure, and it exhibits similar ultrastructural specialization for absorption and secre-tion in vespertilionids (Enders *et al.*, 1976), as it does in the free, glandlike yolk sac of molossids.

Allantoic Vesicle

The endodermal allantoic vesicle grows into the exocoelom and expands to fuse with the chorionic preplacenta during somite stages of development in most chiropteran families. The allantoic vesicle becomes moderately large during limb bud stages, and occupies about half the circumference of the chorionic sac in pteropodids, rhinopomatids, and rhinolophids (Figs. 9B,C, 10C). There is a somewhat smaller, medium-sized allantoic vesicle in vesper-tilionids, megadermatids, and thyropterids, whereas a small, tubular, vestigial allantois characterizes known phyllostomoids. As for Eutheria in general, taxa with the largest allantoic vesicle probably approximate the primitive chiropteran condition. In most bat families, the allantoic vesicle is greatly reduced or disappears in late stages (Gopalakrishna and Karim, 1980).

Chorioallantoic Placenta

The circumferential, invasive implantation in most bats leads to development of a diffuse or horseshoe-shaped preplacenta (Figs. 9A, 10B). Following vascularization of the preplacenta by the allantoic vesicle, the early chorioallantoic placenta is also horseshoe-shaped (Figs. 9B,C, 10C) in many families. Further differentiation of the chorioallantoic placenta in pteropodids, rhinopomatids, rhinolophids, and megadermatids leads to an unusual reduction in extent of the placenta, so that the definitive chorioallantoic placenta assumes the form of a disc at the mesometrial pole of the uterus (cf. Figs. 9C,D, 10C,D). The preplacenta in vespertilionids is also horseshoe-shaped, but the incomplete portion occurs at the mesometrial rather than the antimesometrial pole. Subsequently, the placental disc becomes localized at the antimesometrial pole of the uterus in this family (see Gopalakrishna and Karim, 1980; Luckett, 1980d). This unusual pattern for reduction in extent of the definitive placental disc during ontogeny from a more diffuse condition is another autapomorphic feature for Chiroptera.

The histological layers that separate maternal from fetal blood in the placental labyrinth or zona intima have been studied for most families by light microscopy, histochemistry, or electron microscopy, and sometimes by all three techniques. In both Mega- and Microchiroptera, there is ontogenetic and ultrastructural evidence for the occurrence of labyrinthine endotheliochorial and labyrinthine hemochorial placentae (for a review, see Enders and Wimsatt, 1968; Gopalakrishna and Karim, 1980). The morphotypic condition for both suborders is a labyrinthine endotheliochorial placenta, with a prominent "interstitial membrane" separating the maternal endothelium from the syncytiotrophoblastic layer (Luckett, 1980d). Ultrastructural study of changing tissue relationships during ontogeny in the vespertilionid *Myotis* (Enders and Wimsatt, 1968) provides valuable evidence for the probable phylogenetic transformation of endotheliochorial to hemochorial placentation during chiropteran evolution.

Evolutionary Analysis of Archontan Fetal Membranes

The ontogenetic attributes of the fetal membranes and placenta in Primates, Dermoptera, Scandentia, and Chiroptera can be evaluated for their primitive, derived, homologous, or homoplastic similarities to each other, and to other eutherians, in order to determine whether such evidence can provide any corroboration for the hypothesis of Archonta monophyly. In addition, such a character analysis can serve to test hypotheses of special affinities between any two orders of suspected archontans. Because of the continuing controversy over the monophyly or diphyly of bats (Pettigrew, 1991; Baker *et al.*, 1991), the two chiropteran suborders will be treated separately in this analysis, as will the major subdivisions of extant Primates. Several molecular studies have suggested that rodents and lagomorphs are closely related to Primates (Miyamoto and Goodman, 1986; Czelusniak *et al.*, 1990; Stanhope *et al.*, this volume), and data on fetal membrane development in these two orders (see Luckett, 1985) are included in the present analysis in order to test this hypothesis.

Table II and Fig. 11 provide a summary of the relatively primitive and derived fetal membrane character states analyzed for "archontans," rodents, lagomorphs, and the reconstructed eutherian morphotype. In addition, developmental data on the composition of the auditory bulla are included for each order, because of the great value of this and other features of the middle-ear region for evolutionary analysis (see MacPhee and Cartmill, 1986; Wible and Novacek, 1988). For some ontogenetic traits analyzed here, such as the

Table II. Character Analysis of Developmental Features of the Fetal Membranes and Basicranium in Archontans and Other Eutherians

Primitive	Derived
1. Embryonic disc antimesometrial[a]	1. Embryonic disc orthomesometrial
2. Ossified bulla lacking, or not co-ossified with skull[b]	2. Petrosal auditory bulla
3. Choriovitelline placenta present	3. Choriovitelline placenta absent
4. Large allantoic diverticulum	4. Vestigial allantoic diverticulum
5. Epitheliochorial placenta	5. Hemochorial placenta
6. Amniogenesis by folding; no primordial amniotic cavity	6. Amniogenesis by cavitation
7. Polar trophoblast lost prior to implantation	7. Polar trophoblast persists
8. Embryonic disc oriented relative to mesometrium	8. Embryonic disc oriented toward tubo-uterine junction
9. Preplacenta similar to definitive placenta	9. Preplacenta and early chorioallantoic placenta horseshoe-shaped
10. Free yolk sac reduced or absent in late stages	10. Free yolk sac with enlarged, glandlike cells
11. Blastocyst attachment at para-embryonic pole[c]	11. Blastocyst attachment at embryonic pole
12. Ossified bulla lacking[b]	12. Entotympanic auditory bulla
13. Blastocyst attachment at para-embryonic pole[c]	13. Blastocyst attachment at abembryonic pole
14. Embryonic disc antimesometrial[a]	14. Embryonic disc mesometrial
15. Chorioallantoic placenta at site of initial attachment	15. Chorioallantoic placenta at opposite pole from attachment
16. Bilaminar omphalopleure transitory	16. Bilaminar omphalopleure persists
17. Ossified bulla lacking[b]	17. Ectotympanic auditory bulla

[a,b,c]For these features, several different derived character states are listed in the table; see text for details, and for intermediately derived states for some traits.

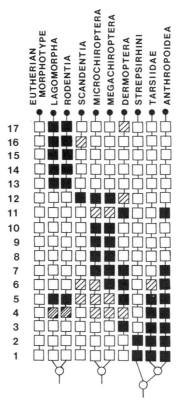

Fig. 11.   Character phylogeny of fetal membrane and basicranial traits in "archontans" and "glires." See Table II for characters evaluated. White squares = primitive character states; black squares = derived character states; diagonally lined squares = intermediately derived character states. Note: For characters 1, 2, 11, 12, 13, 14, and 17, several different derived character states, not forming parts of a transformation series, occur in eutherians, so that each derived state is illustrated separately (and listed separately in Table II). See text for detailed description and analysis.

size of the allantoic vesicle, more than one intermediately derived character state can be identified. The text should be consulted for further consideration of these intermediate states. It must be emphasized that the "character phylogeny" in Fig. 11 is a summary of the more extensive analysis presented here and in earlier reports (Luckett, 1974, 1975, 1980a,d), and it should not be considered a substitute for careful analysis of homology or homoplasy.

Analysis of fetal membrane data corroborates the traditional hypothesis of primate monophyly, in which Tupaiidae, Dermoptera, and Megachiroptera are excluded from the order. Assessments of both fetal membrane and molecular data support identical hypotheses of cladistic relationships among extant primate suborders and superfamilies (cf. Luckett, 1975, 1980a; Goodman, 1975; Goodman *et al.,* 1982; Czelusniak *et al.,* 1990), and an increasing number of multidisciplinary analyses of cranial, postcranial, and soft anatomical attributes provide additional corroboration for the strepsirhine–haplorhine dichotomy (MacPhee and Cartmill, 1986; Szalay *et al.*, 1987; Andrews, 1988; Beard *et al.*, 1988, 1991). In the following discussion, emphasis will be devoted to the manner in which fetal membrane data can contribute to resolution of the following controversies: (1) the phyletic position of Tarsiidae within Primates; (2) the monophyly or diphyly of Chiroptera, and the possible affinities of Megachiroptera with Primates; and (3) the possible affinities of Dermoptera with Chiroptera, Megachiroptera, or Primates.

Phyletic Position of Tarsiidae

Despite the overwhelming evidence for a close relationship between *Tarsius* and Anthropoidea, Schwartz (1978, 1986; also Schwartz *et al.*, 1978; Schwartz and Tattersall, 1987) continues to dismiss the shared, derived similarities between the two taxa either as convergences, or as traits that are actually autapomorphous for *Tarsius*. Each of the criticisms by Schwartz and colleagues regarding the fetal membrane evidence was rebutted previously (Luckett, 1982), but no mention or response to this rebuttal was offered by Schwartz and Tattersall (1987) in their most recent discussion of the data for affinities of *Tarsius*. Paradoxically, the "more robust theory of relatedness" offered by Schwartz and Tattersall (1987) places *Tarsius* as the sister-group of Lorisiformes within a reconstituted suborder "Prosimii," in contrast to the completely divergent hypothesis of Schwartz *et al.* (1978) that Tarsiiformes is the sister taxon of Paleocene Plesiadapiformes. Given the continuing erroneous interpretations of character transformations for fetal membrane data offered by Schwartz and colleagues, and their failure to consider the adaptive and evolutionary significance of intermediately derived character states for phylogenetic analyses, it is useful to reiterate my earlier refutation (Luckett, 1982) of their three main criticisms.

1. It was claimed (Schwartz and Tattersall, 1987) that several shared, derived attributes of *Tarsius* and Anthropoidea, such as precocious differentiation of a mesodermal body stalk, reduced allantoic diverticulum, development of a hemochorial placenta, and absence of a transitory choriovitelline placenta, are "all perfectly correlated." Thus, these authors argued that such features should be considered as a "single functional complex." However, such correlations can only be established by evaluating the developmental pattern of the fetal membranes in all major groups of eutherians. When this is done (Luckett, 1977, in press), it can be demonstrated that hemochorial placentation can also be associated with a medium-sized allantoic vesicle and the transitory occurrence of a choriovitelline placenta, as in some microchiropterans, tenrecids, and erinaceids. Moreover, in the dermopteran *Cynocephalus*, precocious differentiation of a mesodermal body stalk is associated with the subsequent development of a medium-sized allantoic vesicle (see Figs. 5–8).

Therefore, none of these developmental features is necessarily correlated within Eutheria, although the common occurrence of each of these derived attributes during the ontogeny of *Tarsius* and Anthropoidea suggests that they may form a functionally integrated character complex, strongly supporting the monophyly of Haplorhini.

2. While acknowledging that part of the process of amniogenesis in *Tarsius* is similar to that of Anthropoidea (formation of a primordial amniotic cavity within the epiblast), Schwartz (1978) claimed that the subsequent development of the definitive amnion by folding in *Tarsius* is independently derived, and not homologous to the pattern in Anthropoidea. As emphasized elsewhere (Luckett, 1974, 1980a), the condition in *Tarsius* is intermediately derived between the primitive eutherian pattern of Strepsirhini, lagomorphs, and soricids, in which no primordial cavity forms, and the most derived condition, in which the primordial cavity persists to become the definitive amniotic cavity (as in Anthropoidea, Dermoptera, and Megachiroptera). Only by considering patterns of amniogenesis as states of a transformational series is it possible to understand the functional interactions among amniogenesis, implantation, and primitive streak formation in eutherians (for further discussion, see Luckett, in press).

3. Because hemochorial placentation in *Tarsius* and Anthropoidea develops following different patterns of implantation, they have been considered by Schwartz and Tattersall (1987) to form convergently "purely out of functional necessity." Correlated with this, it has been claimed (Schwartz, 1978; Schwartz and Tattersall, 1987) that differences between *Tarsius* and Anthropoidea in amniogenesis and implantation are unlikely to be the result of occurrence of a simplex uterus in Anthropoidea, *contra* Luckett (1974, 1975), because some other eutherians with hemochorial placentation also develop "anthropoid" characteristics of implantation, amniogenesis, and placentation. Analysis of *all* ontogenetic aspects of the fetal membranes and placenta reveals, however, that an "anthropoid" pattern of placentation is found only in the morphotypic members of the Anthropoidea (secondarily modified in Hominoidea), whereas *individual* features of this developmental complex can occur by convergence in some other eutherians. Thus, cavitate amniogenesis and a hemochorial placenta appear in dasypodid edentatans, but, when their entire ontogenetic pattern is considered, it is evident that other features, such as the fundic site of implantation and the developmental fate and structure of the yolk sac, are derived in a different manner than in Anthropoidea (see Luckett, in press).

The above criticisms suggest that Schwartz and colleagues fail to appreciate the role that ontogenetic data can play in evaluating homology and polarity in transformational series for phylogenetic reconstruction. To characterize other eutherians as exhibiting an "anthropoid" pattern of placentation is an example of the approach that many (but not all) "cladists" take in their superficial analyses of selected morsels of data from the literature, especially for attributes that do not corroborate their own phylogenetic hypotheses.

Choosing a few isolated characters out of context, removed from the structural–functional complex to which they belong, is no substitute for a careful evolutionary analysis of *all* attributes of a single organ system (for an example of the former approach, see Tables 1–3 in Schwartz and Tattersall, 1987).

Monophyly or Diphyly of Chiroptera

An early review of the available evidence (Mossman, 1937) suggested that Mega- and Microchiroptera were widely separated on the basis of their fetal membrane development. Mossman's observation was cited to support a hypothesis of possible diphyly of Chiroptera by Jones and Genoways (1970). The latter authors were unaware, however, of the important study by Gopalakrishna (1958) on placental development in five families of Microchiroptera. On the basis of new observations, Gopalakrishna (1958, p. 176) noted that "Megachiroptera and the Microchiroptera are not as divergent as formerly believed" in their placental morphogenesis, and he emphasized the "close similarity" of these features between pteropodids and relatively "primitive" microbat families. Given the meager fetal membrane data for Microchiroptera available at the time to Mossman (1937), limited mainly to Vespertilionidae and Phyllostomidae, it is not surprising that he proposed a wide divergence between bat suborders, as acknowledged by Gopalakrishna (1958). As a consequence of the accumulation of new developmental data on bats, Mossman (1987, p. 186) recently concluded that "no clear-cut distinction in fetal membranes can as yet be made with certainty between the two suborders."

Evolutionary analysis of fetal membrane morphogenesis in most families of bats provides strong evidence for corroborating the hypothesis of chiropteran monophyly, as noted previously (Luckett, 1980d). In the previous assessment, no assumptions were made *a priori* regarding the monophyly or diphyly of bats; instead the morphotype was reconstructed for each suborder, and these ontogenetic morphotypes were compared with those for all other eutherian orders. Ontogenetic character precedence, commonality, and form–function considerations were all used in evaluating the primitive, derived, homologous, or homoplastic nature of shared similarities among fetal membrane features. Attributes were evaluated both among chiropteran families, and among archontans (see Tables 1, 2 and Figs. 4, 5 in Luckett, 1980d). The results of these analyses demonstrated the monophyly of Chiroptera, when compared with other archontan orders and to the eutherian morphotype, as well as the observation that Micro- and Megachiroptera are not readily distinguishable from each other in fetal membrane attributes (see Fig. 11).

Both these conclusions were also reached independently by Gopalakrishna and Karim (1980), on the basis of their *phenetic* analysis, although they interpreted their findings somewhat differently. These authors clustered Pteropodidae, Molossidae, Rhinopomatidae, Rhinolophidae, and Emballonuridae together in their dendrogram of suggested evolutionary relationships among chiropteran families, on the basis of shared fetal membrane similarities. This strongly influenced their conclusion that the traditional subordinal division of Chiroptera "is not justified" on developmental evidence. Gopalakrishna and Karim (1980) provided no assessment of the primitive or derived nature of shared fetal membrane similarities, nor did they reconstruct the ancestral developmental pattern for the order. Their opinion marked a complete turnabout from Mossman's diphyletic suggestion in 1937.

Evolutionary analysis suggests, however, that the shared similarities of fetal membrane features among pteropodids and the above-listed microchiropterans are caused primarily by the retention of ancestral chiropteran features (see Fig. 5 in Luckett, 1980d); as such, these have no bearing on assessing the subordinal allocation of these families. The only fetal membrane trait of pteropodids that differs from the inferred chiropteran morphotype is the occurrence of definitive amniogenesis by cavitation (Fig. 11); even this derived trait has evolved independently within Phyllostomoidea. Moreover, the ancestral chiropteran pattern for this character is intermediately derived (development of a transitory primordial amniotic cavity by cavitation), when compared with the primitive eutherian condition, and the chiropteran morphotype is retained by most microchiropteran families.

The differentiation of a free, glandlike yolk sac in later ontogenetic stages is a uniquely derived eutherian feature found only in Chiroptera. It probably represents a retention of the primitive chiropteran condition, which has been lost secondarily in phyllostomoids as a consequence of their pedomorphic formation of a functionally significant bilaminar omphalopleure. Another unique aspect of bat development is the formation of a relatively diffuse or horseshoe-shaped preplacenta and early chorioallantoic placenta, with the secondary reduction of this to a discoidal placenta in later stages. The mesometrial placental disc shared by pteropodids, rhinopomatids, and rhinolophids probably represents the primitive condition for Chiroptera.

The *combination* of primitive, derived, and autapomorphic eutherian features of fetal membrane morphogenesis that characterize the morphotypes of Mega- and Microchiroptera comprises an interrelated character complex that is not found in any other eutherian taxon. Some *individual* derived attributes, such as an endotheliochorial placenta in the morphotype of both suborders, can be found in some other higher taxa, including Scandentia and Carnivora. However, careful consideration of all developmental features associated with endotheliochorial placentation in the three orders, including fate of the polar trophoblast, amniogenesis, pattern of implantation, and fate of the definitive yolk sac, suggests that an endotheliochorial placenta has developed homologously in the chiropteran suborders, but convergently in carnivorans and scandentians. Such an ontogenetic approach for testing hypotheses of homology or convergence should carry greater weight than a simple appeal to parsimony based on character distribution and outgroup comparison.

Pettigrew *et al.* (1989, p. 543) claimed that fetal membrane homoplasy is "well known" within Eutheria, exemplified by the fact that an "interstitial membrane" has arisen convergently within the endotheliochorial placenta of carnivorans and chiropterans. From this they concluded that the latter feature "cannot be used with any great confidence as a synapomorphy to link the two groups of bats." Pettigrew *et al.* (1989) further argued that in "view of the limitations which foetal membrane characters exhibit in defining microchiropteran relations, it is perhaps expecting too much for these characters to be helpful in sorting out relationships between the Microchiroptera and Megachiroptera." Applying this kind of "logic" to systematics generally would eliminate the use of most biological traits, including occurrence of a laminated lateral geniculate nucleus and many other neuroanatomical features, from playing a useful role in evaluating phylogenetic relationships within Chiroptera or Eutheria. A further criticism to be leveled against the methodology of Pettigrew *et al.* (1989) is their brief discussion (then dismissal) of fetal membrane data taken out of context from a secondary literature source (Novacek, 1982), rather than considered in light of the entire analysis of fetal membrane development in archontans presented by Luckett (1980d). In doing so, they cite only a single feature (the interstitial membrane) that is clearly convergent between Chiroptera and Carnivora, while they conveniently ignore the other two features listed by Novacek (definitive yolk sac gland-like, and horseshoe-shaped preplacenta followed by a localized definitive placenta), even though these are uniquely derived eutherian attributes shared only by the two chiropteran suborders (see Fig. 11 and discussion above).

Many of the methodological deficiencies of Pettigrew and colleagues' (1989) investigations and analyses have been enumerated by Baker *et al.* (1991) and will not be reiterated here. Clearly, it is scientifically unsound to dismiss any type of biological data that can be brought to bear on assessing phylogenetic relationships, especially when it disagrees with one's own ideas about phylogeny. Pettigrew's (1991, p. 207) claim that "there appear not to be any unequivocal synapomorphies for bats apart from wing structure" reflects his willingness to summarily dismiss, ignore, or reinterpret other data that falsify his hypothesis of chiropteran diphyly, rather than conduct a careful and unbiased evaluation of the available evidence. For example, molecular biologists who have collected and analyzed immunological and amino acid sequence data from a wide range of eutherians are in agreement concerning the support for chiropteran monophyly provided by their assessments (Cronin and Sarich, 1980; Czelusniak *et al.*, 1990), while Pettigrew (1991) has dismissed the significance of the immunological evidence, and reinterpreted selected portions of the sequence data. Indeed, Pettigrew *et al.* (1989, p. 541) claimed that there is "a clear and strong association between megabats and primates" on the basis of their analysis of the published ß-hemoglobin sequences. However, examination of their figures for phenetic, cladistic, and DNA-triplet analyses of these data reveals that in *each case* the pteropodids clustered most closely with several (but not all) genera of microchiropterans, including *Rhinopoma*. This observation reflects the findings from evolutionary analysis of the chiropteran fetal membranes (Luckett, 1980d), in which the greatest number of developmental similarities were shared by pteropodids and rhinopomatids. These similarities appear to be the result of their joint possession of primitive chiropteran attributes for the fetal membranes and placenta, although some of these features are derived within Eutheria, as noted above.

While rejecting the numerous shared, derived attributes of wing and cranial morphology cited by Wible and Novacek (1988) and Baker *et al.* (1991) to corroborate chiropteran monophyly, Pettigrew (1991, p. 209) argued that "Smith and Madkour (1980) have made a reasonable case for the polarity" of their penial morphology characters, which also support a hypothesis of chiropteran diphyly. From both a logical and a biological perspective this conclusion seems premature. Careful analysis of the serial histological and vascular relationships of cavernous and accessory cavernous erectile tissues of the eutherian penis has been carried out only for two genera: the vespertilionid bat *Myotis* and the domestic dog *Canis* (Wimsatt and Kallen, 1952; Christensen, 1954). Broader but more superficial surveys have demonstrated the presence of accessory erectile tissue in many, but not all, families of lipotyphlous insectivorans and microchiropterans (Kaudern, 1911; Matthews, 1941; Smith and Madkour, 1980). Careful examination of the structural and vascular relationships of the accessory erectile tissues (bulbus glandis, pars longa glandis) in the dog suggested to Christensen (1954) that these were not homologous with the accessory tissues in *Myotis*.

As noted previously (Luckett, 1980c), the hypothesis that accessory erectile tissue is an ancestral eutherian feature, as proposed by Smith and Madkour (1980), remains to be tested by ontogenetic and histological analysis in a wide range of taxa. The absence of accessory cavernous tissue in some families of lipotyphlans and microchiropterans, including nycterids and miniopterines (Matthews, 1941), suggests that this tissue must have been lost secondarily in some taxa, if the hypothesis of Smith and Madkour (1980) is correct. If so, it is difficult to comprehend why a comparable secondary loss could not have occurred also in pteropodids.

In summary, the available fetal membrane developmental data corroborate the traditional hypothesis of chiropteran monophyly, in agreement with analyses of the molecular, cranial, and postcranial data (Cronin and Sarich, 1980; Wible and Novacek, 1988; Czelusniak *et al.*, 1990; Baker *et al.*, 1991; Adkins and Honeycutt, this volume; Stanhope *et al.*, this volume). Given the fact that the primary evidence for suggested homologies among the visual pathways of pteropodids, primates, and dermopterans has yet to be presented in detail by Pettigrew and colleagues, it is not surprising that the premature phylogenetic interpretations derived from these data have not been accepted uncritically by other neural scientists (Johnson and Kirsch, 1991; Kaas, 1993). Neurobiological data have traditionally been poorly represented in evolutionary analyses of the Mammalia; this is caused primarily by the lack of extensive comparative observations, as noted by Campbell (1980). Clearly, such data are of great potential value; however, as emphasized by Baker *et al.* (1991, p. 222), "neural characters must be appropriately sampled and analyzed before they can contribute to our understanding of mammalian phylogeny." Hopefully, a wider sampling of neurobiological features in microchiropterans, including phyllostomoids and rhinopomatids, will lead to a greater understanding of the homologous or convergent nature of shared similarities among archontans and other eutherians.

Affinities of Dermoptera

Although the initial stages of precocious primitive streak and extraembryonic mesoderm differentiation in *Cynocephalus* resemble the condition in *Tarsius* and Anthropoidea, as noted by Hubrecht (1919), the subsequent development of a large exocoelom removes the short mesodermal "body stalk" from any functional relationship with the preplacenta during presomite and early somite stages in dermopterans. In haplorhine primates, the precocious differentiation of the primitive streak is involved primarily with development of a mesodermal body stalk, and the latter remains in functional communication with the preplacental disc as the pathway for differentiation of the allantoic vascularization. By contrast, the dermopteran preplacenta receives its allantoic vascularization in the primitive eutherian manner, by the growth and expansion of the allantoic vesicle across the exocoelom in later somite stages. Both ontogenetic and functional considerations suggest that the pattern of "body stalk" formation and placental disc vascularization is nonhomologous in dermopterans and haplorhines.

The developmental pattern of the fetal membranes and placenta in *Cynocephalus* does not closely resemble that of any other eutherian; it combines a uniquely derived pattern of exocoelom formation with a primitive pattern of allantoic vesicle development. Mossman's (1987) characterization of dermopteran fetal membranes as being of the "soricoid type" is misleading, and it apparently referred only to his belief that dermopterans retained a soricoid pattern of the bilaminar omphalopleure. As noted above, this does not appear to be true for midgestation embryos. Other derived attributes, such as amniogenesis by cavitation and the development of a hemochorial placental disc, occur in a number of eutherian higher taxa, and, by themselves, contribute little to resolving the evolutionary affinities of extant dermopterans. Given the fact that living dermopterans are represented by a monogeneric family, it is difficult to evaluate to what degree this genus may represent the ontogenetic pattern of early members of the order.

The fetal membrane evidence does not provide any direct corroboration for a sister-group relationship between Chiroptera and Dermoptera (see Novacek, 1982, 1990), but neither does it falsify such a hypothesis. The suggestion for special affinities between Dermoptera and Primates receives even less support from the fetal membrane data. Given the doubts raised concerning suggested dermopteran affinities of the Eocene plagiomenids (MacPhee *et al.*, 1989), as well as the controversy over possible primate or dermopteran relations of some early Eocene "plesiadapiforms" (Beard, 1990; this volume; Kay *et al.*, 1990; Krause, 1991), it is unclear whether undoubted fossil representatives of the order Dermoptera can be recognized at present. More than ever, additional molecular data are needed from dermopterans and chiropterans, especially from DNA and protein sequences, as well as careful descriptions and analyses of the neurobiological data, in order to help evaluate the phylogenetic relationships of Dermoptera.

Summary and Conclusions

Because of the imprecision involved in analysis of any phylogenetic hypothesis, the present assessment of the fetal membrane evidence is offered as a test of several hypotheses of primate and "archontan" phylogeny. Developmental evidence from the fetal membranes and placenta provides corroboration for hypotheses of monophyly for many higher eutherian taxa, including Haplorhini, Chiroptera, and Scandentia. The highly conservative pattern of evolutionary change in some taxa, such as the Strepsirhini, sometimes limits the usefulness of fetal membrane data for phylogenetic reconstruction. Such limitations are to be expected if a mosaic pattern of primitive and derived attributes characterizes most taxa during mammalian phylogeny; this seems to be the case for most, if not all, organ systems analyzed. This mosaic nature of evolutionary change demonstrates the need for multidisciplinary analyses of different organ systems, including molecular data, to facilitate the reconstruction of mammalian phylogeny.

Because of the striking dichotomy in developmental patterns of the fetal membranes and placenta between Strepsirhini and Haplorhini, with retention of the reconstructed primitive eutherian morphotype in most strepsirhines, monophyly of the order Primates is only weakly corroborated by this ontogenetic evidence. It is for this reason that some students of comparative mammalian placentation, including Hubrecht (1908) and Mossman (1937, 1987), have questioned the inclusion of Strepsirhini within the order Primates. Mossman's (1987) argument that Strepsirhini should be clustered more closely with Perissodactyla, rather than with Haplorhini, on the basis of numerous shared fetal membrane similarities, appears to be based exclusively on shared primitive eutherian retentions. Many of these similarities are also shared with Pholidota, some Artiodactyla, and Talpidae, and Mossman (1987) acknowledged that the ontogenetic pattern of the fetal membranes and placenta in talpids probably approximated the primitive eutherian condition.

Such differing interpretations of the significance of fetal membrane data do not diminish their value for phylogenetic analysis; instead, they remind us of the necessity for evolutionary analysis of all types of biological data before using them for phylogenetic reconstruction. When other types of data are evaluated, such as basicranial and postcranial morphology, strong corroboration is provided for primate monophyly (at least for "euprimates") (see Szalay *et al.*, 1987; Martin, 1990; Wible and Martin, this volume).

The findings of the present study also emphasize the limitations for phylogenetic reconstruction when individual attributes from a morphofunctional organ system, such as the fetal membranes and placenta, are taken out of context and incorporated into broader, multidisciplinary syntheses of mammalian phylogeny. These systematic overviews, such as those of Novacek (1982, 1990), serve a valuable purpose in providing a broad perspective for the major patterns of mammalian evolution. When considered out of context by other authors, such as by Pettigrew *et al.* (1989) when discussing the fetal membrane data for chiropteran monophyly, the evidence for homology or convergence of shared, derived similarities can be diminished or trivialized. Evaluating all systematically significant attributes of a single organ system can increase the probability of correctly assessing the likelihood of homology versus homoplasy for individual features of the complex shared by two or more taxa. This was demonstrated in the present study by the analysis of homology or convergence for the endotheliochorial placenta of chiropterans, scandentians, and carnivorans. In other cases, derived features, which may appear to be homologous at first glance when shared by two taxa, can be attributed with greater probability to convergence when evaluated within the context of their morphofunctional complex. As an example, Schwartz's (1986) claim that an elongated calcaneus is a synapomorphy allying *Tarsius* with lorisoids was not corroborated by a careful morphofunctional analysis of the entire hindlimb (Jouffroy *et al.*, 1984).

A comprehensive evaluation of all features of an organ system is also essential in order to assess whether some components or traits of a system may be structurally and functionally correlated. If invariably correlated, these traits can be treated as a single, but complex, "character" during phylogenetic analysis. However, as demonstrated for some aspects of fetal membrane morphogenesis in Primates, extensive assessments may reveal that certain character correlations exist only in some eutherian taxa, but not others. Such an analysis warrants the continued independent use of these traits during phylogenetic reconstruction (see Fig. 11 for the differing patterns of allantoic vesicle and yolk sac associated with hemochorial placentation in primates, rodents, and lagomorphs). For this reason, it is erroneous to treat the "large complex of characters associated with haemochorial placentation" in primates as a single "character" in a multidisciplinary data matrix for phylogenetic analysis, as done by Andrews (1988) in his generally unbiased assessment of much of the available biological evidence.

In conclusion, developmental evidence from the fetal membranes and placenta, while confirming monophyly of the orders Primates and Chiroptera, provides no corroboration for hypotheses of superordinal relationships among "archontans." These ontogenetic data do not falsify an archontan hypothesis, because none of the orders traditionally included in Archonta (Primates, Dermoptera, Scandentia, and Chiroptera) can be shown to be cladistically related to any other eutherian order by analysis of the fetal membrane evidence. It seems highly improbable that a sister-group relationship exists among Primates, Rodentia, and Lagomorpha, contrary to the hypothesis of Miyamoto and Goodman (1986; see also Stanhope *et al.*, this volume). Rodents and lagomorphs share a suite of autapomorphous ontogenetic features from both the fetal membranes and dentition (Luckett, 1985), and, considered in conjunction with other shared, derived attributes from the cranial morphology and dentition of fossil and extant forms, these provide strong support for clustering the two orders in a monophyletic superorder Glires, as emphasized by Novacek (1990).

Because of the near-identical nature of the reconstructed primate and eutherian morphotypes for the entire developmental pattern of the fetal membranes and placenta, it is unlikely that additional data from this system will clarify the higher-level relationships of Primates. It is hoped that continued discovery of new fossil specimens from Asia and Africa, coupled with a renewed focus on obtaining additional molecular, neurobiological, cranial, and postcranial data from "key" taxa, including *Ptilocercus, Cynocephalus,* and a wider range of microchiropteran families, will lead to a further refinement and possible resolution of hypotheses concerning the superordinal affinities of Primates and other "archontans."

ACKNOWLEDGMENTS

Most of the developmental stages of Primates and other mammals examined for this study are housed at the Hubrecht Laboratory, Utrecht, The Netherlands, and have been studied with the aid of Drs. Gesineke Bangma, Else Boterenbrood, Siegfried de Laat, and Pieter Nieuwkoop. Additional specimens were examined at the former Carnegie Laboratories of Embryology, Davis, California, and in the private collections of Drs. A. Gopalakrishna, Nagpur, India; H.-J. Kuhn, Göttingen, Germany; the late H. W. Mossman, Madison, Wisconsin; D. Starck, Frankfurt, Germany; and the late W. A. Wimsatt, Ithaca, New York. All of these individuals and institutions are thanked for their help and encouragement. Special thanks are due to Dr. Nancy Hong for preparing the illustrations, and for her editorial and moral assistance throughout all phases of this study. Financial support was provided in part by a U. S. Senior Scientist Award from the Alexander von Humboldt Foundation, Bonn, Germany.

References

Amoroso, E. C. 1952. Placentation, in: A. S. Parkes (ed.), *Marshall's Physiology of Reproduction*, Vol. 2, pp. 127–311. Longmans, Green, London.

Andrews, P. 1988. A phylogenetic analysis of the Primates, in: M. J. Benton (ed.), *The Phylogeny and Classification of the Tetrapods*, Vol. 2, pp. 143–175. Oxford University Press (Clarendon), London.

Baker, R. J., Novacek, M. J., and Simmons, N. B. 1991. On the monophyly of bats. *Syst. Zool.* **40:**216–231.

Beard, K. C. 1990. Gliding behaviour and palaeoecology of the alleged primate family Paromomyidae (Mammalia, Dermoptera). *Nature* **345:**340–341.

Beard, K. C., Dagosto, M., Gebo, D. L., and Godinot, M. 1988. Interrelationships among primate higher taxa. *Nature* **331:**712–714.

Beard, K. C., Krishtalka, L., and Stucky, R. K. 1991. First skulls of the early Eocene primate *Shoshonius cooperi* and the anthropoid–tarsier dichotomy. *Nature* **349:**64–67.

Bock, W. J. 1977. Foundations and methods of evolutionary classification, in: M. K. Hecht, P. C. Goody, and B. M. Hecht (eds.), *Major Patterns in Vertebrate Evolution*, pp. 851–895. Plenum Press, New York.

Butler, H. 1967. The giant cell trophoblast of the Senegal galago (*Galago senegalensis senegalensis*) and its bearing on the evolution of the primate placenta. *J. Zool.* **152:**195–207.

Butler, H. 1982. The placenta and fetal membranes of Strepsirhini and Haplorhini, in: D. E. Haines (ed.), *The Lesser Bushbaby (Galago) as an Animal Model: Selected Topics*, pp. 183–197. CRC Press, Boca Raton, Fla.

Butler, H., and Adam, K. R. 1964. The structure of the allantoic placenta of the Senegal bush baby (*Galago senegalensis senegalensis*). *Folia Primatol.* **2:**22–49.

Campbell, C. B. G. 1980. The nervous system of the Tupaiidae: Its bearing on phyletic relationships, in: W. P. Luckett (ed.), *Comparative Biology and Evolutionary Relationships of Tree Shrews*, pp. 219–242. Plenum Press, New York.

Christensen, G. C. 1954. Angioarchitecture of the canine penis and the process of erection. *Am. J. Anat.* **95:**227–262.

Cronin, J. E., and Sarich, V. M. 1980. Tupaiid and Archonta phylogeny: The macromolecular evidence, in: W. P. Luckett (ed.), *Comparative Biology and Evolutionary Relationships of Tree Shrews*, pp. 293–312. Plenum Press, New York.

Czelusniak, J., Goodman, M., Koop, B. F., Tagle, D. A., Shoshani, J., Braunitzer, G., Kleinschmidt, T. K., De Jong, W. W., and Matsuda, G. 1990. Perspectives from amino acid and nucleotide sequences on cladistic relationships among higher taxa of Eutheria, in: H. H. Genoways (ed.), *Current Mammalogy*, Vol. 2, pp. 545–572. Plenum Press, New York.

Enders, A. C., and Wimsatt, W. A. 1968. Formation and structure of the hemodichorial chorioallantoic placenta of the bat (*Myotis lucifugus lucifugus*). *Am. J. Anat.* **122:**453–490.

Enders, A. C., Wimsatt, W. A., and King, B. F. 1976. Cytological development of yolk sac endoderm and protein-absorptive mesothelium in the little brown bat, *Myotis lucifugus. Am. J. Anat.* **146:**1–30.

Gardner, R. L. 1985. Clonal analysis of early mammalian development. *Philos. Trans. R. Soc. London Ser. B* **312:**163–178.

Gardner, R. L. 1989. Cell allocation and lineage in the early mouse embryo. *Ciba Found. Symp.* **144:**172–186.

Gérard, P. 1932. Études sur l'ovogénèse et l'ontogénèse chez les Lémuriens du genre *Galago. Arch. Biol.* **43:**93–151.

Gingerich, P. D. 1984. Primate evolution, in: T. W. Broadhead (ed.), *Mammals: Notes for a Short Course*, pp. 167–181. University of Tennessee, Knoxville.

Goodman, M. 1975. Protein sequence and immunological specificity: Their role in phylogenetic studies of Primates, in: W. P. Luckett and F. S. Szalay (eds.), *Phylogeny of the Primates*, pp. 219–248. Plenum Press, New York.

Goodman, M., Romero-Herrera, A. E., Dene, H., Czelusniak, J., and Tashian, R. E. 1982. Amino acid sequence evidence on the phylogeny of Primates and other eutherians, in: M. Goodman (ed.), *Macromolecular Sequences in Systematic and Evolutionary Biology*, pp. 115–191. Plenum Press, New York.

Gopalakrishna, A. 1958. Foetal membranes in some Indian Microchiroptera. *J. Morphol.* **102:**157–198.

Gopalakrishna, A., and Karim, K. B. 1979. Fetal membranes and placentation in Chiroptera. *J. Reprod. Fertil.* **56:**417–429.

Gopalakrishna, A., and Karim, K. B. 1980. Female genital anatomy and the morphogenesis of foetal membranes of Chiroptera and their bearing on the phylogenetic relationships of the group. *Natl. Acad. Sci. India Golden Jub. Comm. Vol.,* pp. 379–428.

Hill, J. P. 1932. The developmental history of the Primates. *Philos. Trans. R. Soc. London Ser. B* **22:**45–178.

Hill, J. P. 1965. On the placentation of *Tupaia. J. Zool.* **146:**278–304.

Hill, J. P., and Florian, J. 1963. The development of the primitive streak, head-process and annular zone in *Tarsius,* with comparative notes on *Loris. Bibl. Primatol.* **2:**1–90.

Hill, J. P., Ince, F. E., and Subba Rau, A. 1928. The development of the foetal membranes in *Loris,* with special reference to the mode of vascularisation of the chorion in the Lemuroidea and its phylogenetic significance. *Proc. Zool. Soc. London* **1928:**699–716.

Hubrecht, A. A. W. 1896. Die Keimblase von *Tarsius,* in: *Festschrift für Carl Gegenbaur,* Vol 2, pp. 149–178. Wilhelm Engelmann Verlag, Leipzig.

Hubrecht, A. A. W. 1899. Ueber die Entwickelung der Placenta von *Tarsius* und *Tupaja,* nebst Bemerkungen ueber deren Bedeutung als haematopoietische Organe, in: A. Sedgwick (ed.), *Proceedings of the 4th International Congress of Zoology,* pp. 343–412. C. J. Clay and Sons, London.

Hubrecht, A. A. W. 1902. Furchung und Keimblattbildung bei *Tarsius spectrum. Verh. K. Akad. Wet. Amsterdam* **2:**1–113.

Hubrecht, A. A. W. 1908. Early ontogenetic phenomena in mammals and their bearing on our interpretation of the phylogeny of the vertebrates. *Q. J. Microsc. Sci.* **53:**1–181.

Hubrecht, A. A. W. 1919. Früheste Entwicklungsstadien und Placentation von *Galeopithecus. Verh. K. Akad. Wet. Amsterdam* **16:**1–39.

Johnson, J. I., and Kirsch, J. A. W. 1991. Phylogeny through brain traits: Intraordinal relationships among mammals including Primates and Chiroptera. *Am. J. Phys. Anthropol. Suppl.* **12:**100.

Jones, J. K., Jr., and Genoways, H. H. 1970. Chiropteran systematics, in: B. H. Slaughter and D. W. Walton (eds.), *About Bats,* pp. 3–21. Southern Methodist University Press, Dallas.

Jouffroy, F.-K., Berge, C., and Niemitz, C. 1984. Comparative study of the lower extremity in the genus *Tarsius,* in: C. Niemitz (ed.), *Biology of Tarsiers,* pp. 167–190. G. Fischer Verlag, Stuttgart.

Kaas, J. H. 1993. Archontan affinities as reflected in the visual system, in: F. S. Szalay, M. J. Novacek, and M. C. McKenna (eds.), *Mammal Phylogeny.* Springer-Verlag, New York. (in press).

Karim, K. B. 1976. Embryology of some Indian Chiroptera. Thesis, Doctor of Science. Nagpur University, Department of Zoology.

Kaudern, W. 1911. Studien über die männlichen Geschlechtsorgane von Insectivoren und Lemuriden. *Zool. Jahrb.* **31:**1–106.

Kaufmann, P., Luckhardt, M., and Elger, W. 1985. The structure of the tupaia placenta. II. Ultrastructure. *Anat. Embryol.* **171:**211–221.

Kay, R. F., Thorington, R. W., Jr., and Houde, P. 1990. Eocene plesiadapiform shows affinities with flying lemurs not primates. *Nature* **345:**342–344.

King, B. F. 1984. The fine structure of the placenta and chorionic vesicles of the bush baby, *Galago crassicaudata. Am. J. Anat.* **169:**101–116.

King, B. F. 1986. Morphology of the placenta and fetal membranes, in: W. R. Dukelow and J. Erwin (eds.), *Comparative Primate Biology,* Vol. 3, pp. 311–331. Liss, New York.

Krause, D. W. 1991. Were paromomyids gliders? Maybe, maybe not. *J. Hum. Evol.* **21**:177–188.

Kuhn, H.-J., and Schwaier, A. 1973. Implantation, early placentation, and the chronology of embryogenesis in *Tupaia belangeri. Z. Anat. Entwicklungsgesch.* **142**:315–340.

Luckett, W. P. 1968. Morphogenesis of the placenta and fetal membranes of the tree shrews (family Tupaiidae). *Am. J. Anat.* **123**:385–428.

Luckett, W. P. 1974. The comparative development and evolution of the placenta in primates. *Contrib. Primatol.* **3**:142–234.

Luckett, W. P. 1975. Ontogeny of the fetal membranes and placenta: Their bearing on primate phylogeny, in: W. P. Luckett and F. S. Szalay (eds.), *Phylogeny of the Primates*, pp. 157–182. Plenum Press, New York.

Luckett, W. P. 1976. Cladistic relationships among primate higher categories: Evidence of the fetal membranes and placenta. *Folia Primatol.* **25**:245–276.

Luckett, W. P. 1977. Ontogeny of amniote fetal membranes and their application to phylogeny, in: M. K. Hecht, P. C. Goody, and B. M. Hecht (eds.), *Major Patterns in Vertebrate Evolution*, pp. 439–516. Plenum Press, New York.

Luckett, W. P. 1980a. Monophyletic or diphyletic origins of Anthropoidea and Hystricognathi: Evidence of the fetal membranes, in: R. L. Ciochon and A. B. Chiarelli (eds.), *Evolutionary Biology of the New World Monkeys and Continental Drift.* pp. 347–368. Plenum Press, New York.

Luckett, W. P. 1980b. The suggested evolutionary relationships and classification of tree shrews, in: W. P. Luckett (ed.), *Comparative Biology and Evolutionary Relationships of Tree Shrews*, pp. 3–31. Plenum Press, New York.

Luckett, W. P. 1980c. The use of reproductive and developmental features in assessing tupaiid affinities, in: W. P. Luckett (ed.), *Comparative Biology and Evolutionary Relationships of Tree Shrews*, pp. 245–266. Plenum Press, New York.

Luckett, W. P. 1980d. The use of fetal membrane data in assessing chiropteran phylogeny, in: D. E. Wilson and A. L. Gardner (eds.), *Proceedings of the Fifth International Bat Research Conference*, pp. 245–266. Texas Tech University Press, Lubbock.

Luckett, W. P. 1982. The uses and limitations of embryological data in assessing the phylogenetic relationships of *Tarsius* (Primates, Haplorhini). *Geobios Mém. Spéc.* **6**:289–304.

Luckett, W. P. 1983. Development of the fetal membranes and placenta in *Galeopithecus* (order Dermoptera). *Anat. Rec.* **205**:114A–115A.

Luckett, W. P. 1985. Superordinal and intraordinal affinities of rodents: Developmental evidence from the dentition and placentation, in: W. P. Luckett and J.-L. Hartenberger (eds.), *Evolutionary Relationships among Rodents*, pp. 227–276. Plenum Press, New York.

Luckett, W. P. 1993. Morphogenesis of the mammalian fetal membranes and placenta: Uses and limitations for phylogenetic reconstruction. *J. Exp. Zool.* (in press).

MacPhee, R. D. E., and Cartmill, M. 1986. Basicranial structures and primate systematics, in: D. Swindler (ed.), *Comparative Primate Biology*, Vol. 1, pp. 219–275. Liss, New York.

MacPhee, R. D. E., Cartmill, M., and Rose, K. D. 1989. Craniodental morphology and relationships of the supposed Eocene dermopteran *Plagiomene* (Mammalia). *J. Vertebr. Paleontol.* **9**:329–349.

Martin, R. D. 1975. The bearing of reproductive behavior and ontogeny on strepsirhine phylogeny, in: W. P. Luckett and F. S. Szalay (eds.), *Phylogeny of the Primates*, pp. 265–297. Plenum Press, New York.

Martin, R. D. 1990. *Primate Origins and Evolution.* Chapman & Hall, London.

Matthews, L. H. 1941. Notes on the genitalia and reproduction of some African bats. *Proc. Zool. Soc. London Ser. B* **111**:289–346.

Miyamoto, M. M., and Goodman, M. 1986. Biomolecular systematics of eutherian mammals: Phylogenetic patterns and classification. *Syst. Zool.* **35**:230–240.

Mossman, H. W. 1937. Comparative morphogenesis of the fetal membranes and accessory uterine structures. *Contrib. Embryol. Carnegie Inst.* **26**:129–246.

Mossman, H. W. 1987. *Vertebrate Fetal Membranes.* Rutgers University Press, New Brunswick, N.J.

Novacek, M. J. 1982. Information for molecular studies from anatomical and fossil evidence on higher eutherian phylogeny, in: M. Goodman (ed.), *Macromolecular Sequences in Systematic and Evolutionary Biology*, pp. 3–41. Plenum Press, New York.

Novacek, M. J. 1990. Morphology, paleontology, and the higher clades of mammals, in: H. H. Genoways (ed.), *Current Mammalogy*, Vol. 2, pp. 507–543. Plenum Press, New York.

Novacek, M. J., and Wyss, A. R. 1986. Higher-level relationships of the Recent eutherian orders: Morphological evidence. *Cladistics* **2:**257–287.

Pettigrew, J. D. 1986. Flying primates? Megabats have the advanced pathway from eye to midbrain. *Science* **231:**1304–1306.

Pettigrew, J. D. 1991. Wings or brains? Convergent evolution in the origins of bats. *Syst. Zool.* **40:**199–216.

Pettigrew, J. D., Jamieson, B. G. M., Robson, S. K., Hall, L. S., McAnally, K. I., and Cooper, H. M. 1989. Phylogenetic relations between microbats, megabats and primates (Mammalia: Chiroptera and Primates). *Philos. Trans. R. Soc. London Ser. B* **325:**489–559.

Prasad, M. R. N., Mossman, H. W., and Scott, G. L. 1979. Morphogenesis of the fetal membranes of an American mole, *Scalopus aquaticus*. *Am. J. Anat.* **155:**31–68.

Rasweiler, J. J., IV. 1979. Early embryonic development and implantation in bats. *J. Reprod. Fertil.* **56:**403–416.

Rasweiler, J. J., IV. 1990. Implantation, development of the fetal membranes, and placentation in the captive black mastiff bat, *Molossus ater*. *Am. J. Anat.* **187:**109–136.

Rasweiler, J. J., IV. 1991. Development of the discoidal hemochorial placenta in the black mastiff bat, *Molossus ater:* Evidence for a role of maternal endothelial cells in the control of the trophoblastic growth. *Am. J. Anat.* **191:**185–207.

Reng, R. 1977. Die Placenta von *Microcebus murinus* Miller. *Z. Säugetierkd.* **42:**201–214.

Rosenberger, A. L., Strasser, E., and Delson, E. 1985. Anterior dentition of *Notharctus* and the adapid–anthropoid hypothesis. *Folia Primatol.* **44:**15–39.

Sarich, V. M. 1993. Mammalian systematics: 25 years among their albumins and transferrins, in: F. S. Szalay, M. J. Novacek, and M. C. McKenna (eds.), *Mammalian Phylogeny*. Springer-Verlag, New York. (in press).

Schwartz, J. H. 1978. If *Tarsius* is not a prosimian, is it a haplorhine? in: D. J. Chivers and K. A. Joysey (eds.), *Recent Advances in Primatology*, Vol. 3, pp. 195–204. Academic Press, New York.

Schwartz, J. H. 1986. Primate systematics and a classification of the order, in: D. R. Swindler and J. Erwin (eds.), *Comparative Primate Biology*, Vol. 1, pp. 1–41. Liss, New York.

Schwartz, J. H., and Tattersall, I. 1985. Evolutionary relationships of living lemurs and lorises (Mammalia, Primates) and their potential affinities with European Eocene Adapidae. *Anthropol. Pap. Am. Mus. Nat. Hist.* **60:**1–100.

Schwartz, J. H., and Tattersall, I. 1987. Tarsiers, adapids and the integrity of Strepsirhini. *J. Hum. Evol.* **16:**23–40.

Schwartz, J. H., Tattersall, I., and Eldredge, N. 1978. Phylogeny and classification of the primates revisited. *Yearb. Phys. Anthropol.* **21:**95–133.

Shoshani, J. 1986. Mammalian phylogeny: Comparison of morphological and molecular results. *Mol. Biol. Evol.* **3:**222–242.

Simons, E. L., and Rasmussen, D. T. 1989. Cranial morphology of *Aegyptopithecus* and *Tarsius* and the question of the tarsier–anthropoidean clade. *Am. J. Phys. Anthropol.* **79:**1–23.

Smith, J. D. 1980. Chiropteran phylogenetics: Introduction, in: D. E. Wilson and A. L. Gardner (eds.), *Proceedings of the Fifth International Bat Research Conference*, pp. 233–244. Texas Tech University Press, Lubbock.

Smith, J. D., and Madkour, G. 1980. Penial morphology and the question of chiropteran monophyly, in: D. E. Wilson and A. L. Gardner (eds.), *Proceedings of the Fifth International Bat Research Conference*, pp. 347–365. Texas Tech University Press, Lubbock.

Smith, J. M., Burian, R., Kauffman, S., Alberch, P., Campbell, J., Goodwin, B., Lande, R., Raup, D., and Wolpert, L. 1985. Developmental constraints and evolution. *Q. Rev. Biol.* **60:**265–287.

Srivastava, S. C. 1952. Placentation in the mouse-tailed bat, *Rhinopoma kinneari*. *Proc. Zool. Soc. Bengal* **5:**105–131.

Starck, D. 1956. Primitiventwicklung und Plazentation der Primaten, in: H. Hofer, A. H. Schultz, and D. Starck (eds.), *Primatologia,* Vol. 1, pp. 723–886. Karger, Basel.

Stephens, R. J., and Easterbrook, N. 1968. Development of the cytoplasmic membranous organelle in the endodermal cells of the yolk sac of the bat *Tadarida brasiliensis cynocephala. J. Ultrastruct. Res.* **24:**239–248.

Strahl, H. 1899. Der Uterus gravidus von *Galago agisymbanus. Abh. Senckenb. Naturforsch. Ges.* **26:**155–199.

Strauss, F. 1978a. The ovoimplantation of *Microcebus murinus* Miller (Primates, Lemuroidea, Strepsirhini). *Am. J. Anat.* **152:**99–110.

Strauss, F. 1978b. Eine Neuuntersuchung der Implantation und Placentation bei *Microcebus murinus. Mitt. Naturforsch. Ges. Bern* **35:**107–119.

Szalay, F. S., Rosenberger, A. L., and Dagosto, M. 1987. Diagnosis and differentiation of the order Primates. *Yearb. Phys. Anthropol.* **30:**75–105.

Wheeler, Q. D. 1990. Ontogeny and character phylogeny. *Cladistics* **6:**225–268.

Wible, J. R., and Covert, H. H. 1987. Primates: Cladistic diagnosis and relationships. *J. Hum. Evol.* **16:**1–22.

Wible, J. R., and Novacek, M. J. 1988. Cranial evidence for the monophyletic origin of bats. *Am. Mus. Novit.* **2911:**1–19.

Wimsatt, W. A. 1954. The fetal membranes and placentation of the tropical American vampire bat *Desmodus rotundus murinus,* with notes on the histochemistry of the placenta. *Acta Anat.* **21:**285–341.

Wimsatt, W. A., and Kallen, F. C. 1952. Anatomy and histophysiology of the penis of a vespertilionid bat, *Myotis lucifugus lucifugus,* with particular reference to its vascular organization. *J. Morphol.* **90:**415–466.

Wislocki, G. B. 1929. On the placentation of primates, with a consideration of the phylogeny of the placenta. *Contrib. Embryol. Carnegie Inst.* **20:**51–80.

Cranioskeletal Morphology of Archontans, and Diagnoses of Chiroptera, Volitantia, and Archonta

6

FREDERICK S. SZALAY and
SPENCER G. LUCAS

> In order to emphasize the hypothesis that the orders Menotyphla, Dermoptera, Chiroptera and Primates have had a common origin, possibly from some upper Cretaceous family resembling in many characters the Tupaiidae, these four orders may be embraced in a single superorder, which may be named **Archonta** in allusion to the fact that Linnaeus included in the Primates the genera *Homo, Simia, Lemur* (including the Lemuroids and the "Flying Lemur"), *Vespertillio.*
>
> W. K. Gregory (1910, p. 322)

> Gregory's "Archonta" is almost surely an unnatural group.
>
> G. G. Simpson (1945, p. 173)

Introduction

The seemingly intractable problem of the earliest adaptations and the nature of relationships of the mixodectids, microsyopids, plesiadapiforms, tupaiids,

FREDERICK S. SZALAY • Department of Anthropology, Hunter College, City University of New York, New York, New York 10021. SPENCER G. LUCAS • New Mexico Museum of Natural History, Albuquerque, New Mexico 87104.
Primates and Their Relatives in Phylogenetic Perspective, edited by Ross D.E. MacPhee. Plenum Press, New York, 1993.

euprimates, colugos, and bats has been in the forefront of mammalian systematics for quite some time. The literature of the past two decades that has specifically focused on parts, or the whole, of this problem is briefly reviewed below, and a number of the specific character evolution hypotheses are examined. After a general neglect of the issues related to the Archonta, perhaps because of Simpson's (1945) influential views, there has been a resurgence of interest in the 1970s. Many contributions, on a considerably expanded data base, have grappled with the problems of adaptation and phylogenetic relationships. Beard (1989, 1993, this volume) in particular has clearly and firmly stated his several hypotheses of various character polarities, and of the taxonomic concepts of Primatomorpha and Eudermoptera, all based on carefully analyzed postcranial evidence. In these works he has rejected, more by implication than strong arguments, the Archonta. His hypotheses are important not only because of the greatly increased postcranial evidence that was analyzed in sophisticated functional-adaptive detail, but also because within their confines the recency of relationships of colugos, plesiadapiforms, euprimates, and bats, in that order, was seemingly supported by a number of apparently well-tested synapomorphies. Beard (this volume) has considered the tree shrews to have little relevance to the relationships of the taxa included in the Primatomorpha.

Testing of the synapomorphies suggested by Beard is a complex undertaking that involves not only additional fossils, or clearly applied methodology regarding the phyletic and cladistic relationships of characters and taxa, but also a reappraisal of basic osteological information on all of the groups in question. It also requires both functional and ecomorphological research. These issues are being considered in detail, within the confines of functional-adaptive analysis of colugo, scandentian, and euprimate skeletal morphology, and only a preliminary use of some of these results is possible at this time.

While the work on the analysis of selected living taxa is ongoing, some important, and so far the oldest, fossil materials relevant to the problems surrounding the concept of Archonta, the Early Paleocene skeletal remains of one individual animal of *Mixodectes* and of *Chriacus*, were described by Szalay and Lucas (1993). In light of the focused interest not only on archontan relationships, but also on the general lack of understanding of early placental ties from the earliest Late Cretaceous to the beginnings of the Eocene, new postcranial evidence is welcome. We refer to these two specimens in our discussion, although the complex pattern of associations and allocations that we described (Szalay and Lucas, 1993) cannot be repeated here because of space limitations. The designation "Nacimiento Specimen A" stands for the skeletal material assigned to the oxyclaenine dental morph, New Mexico Museum of Natural History No. (=NMMNH) 19995; and "Nacimiento Specimen B" refers to the skeletal material assigned to the clearly identifiable dental morph of *Mixodectes*, NMMNH 3088. Such specimens that cannot be assigned with confidence to either one of the two skeletons, but that most likely belong to one of the two, will be designated as "Nacimiento specimen." It is the analysis of these remains, described in detail by Szalay and Lucas (1993), that gave us a focused perspective on some of the numerous issues that remain unresolved concerning intra- and intertaxonal relationships of the Archonta.

Table I is a list of abbreviations, designating homologies, that are used in the text and figures.

For the two known living species of Dermoptera we use the distinctive generic designations *Galeopithecus volans* and *Galeopterus variegatus* (see Szalay, 1969, p. 241) rather than a general reference to *Cynocephalus*. References in the literature to colugo morphology have often been lumped under the latter term, when in fact either of the two species has been used. Dental, cranial, and postcranial differences support the generic distinction of these taxa.

We dedicate this paper to W. D. Matthew and G. G. Simpson, whose extraordinary monographs, both of 1937, different from one another but highly complementary, still rank as exemplars and major achievements in Paleogene mammalian paleontology.

Cranioskeletal Evidence and Archontan Phylogeny

At present there exists the equivalent of a gaping hole in the body of understanding of placental evolution, from 80 to 55 MYBP. This considerable degree of ignorance certainly includes the origin and diversification of the placentals we consider archontans, as well as those groups that were derived from the lineage that gave rise to *Protungulatum* and its closely clusterable descendants (i.e., the paraphyletic Condylarthra). As concluded by W. D. Matthew in 1925 (edited and annotated by G. G. Simpson in 1943), the differentiation of the various archontan and other eutherian groups was very likely a Cretaceous event (see also Beard, 1989; Kielan-Jaworowska and Dashzeveg, 1989; Kielan-Jaworowska and Nessov, 1990). Consequently, a comparative analysis of the group-specific attributes of these various archaic groups with the diagnostic apomorphies (the *derived combinations* of new traits of the ancestor of a monophyletic group; Szalay *et al.* 1987) of the other respective taxa is, in our view, the most theory-faithful and therefore valuable method of understanding archontan, or other group origins and branchings, and therefore their monophyletic (either paraphyletic or holophyletic) status. Because of the reality of both phyletic and cladistic evolutionary processes, tested taxonomic properties of any one group may be considered as hypotheses of ancestral states for others. To arrive at any level of confidence, these hypotheses of taxonomic properties, however, should be tested through functional-adaptive analysis as suggested by Bock (1981), Szalay and Bock (1991), and others.

The metatherian Deltatheroida have recently been authoritatively examined by Kielan-Jaworowska and Nessov (1990). They have convincingly shown that major central Asiatic radiation not to be eutherian. Yet, from the same beds where metatherian deltatheroidans occur (putatively as old as Albian to

Table I. Abbreviations Used in the Text and Figures[a]

Bones, joints, and joint facets

ACu	astragalocuboid
AFi	astragalofibular
AN	astragalonavicular
ANJ	astragalonavicular joint
ANl	lateral astragalonavicular
As	astragalus
ASN	astragalosustentacular-navicular
ATi	astragalotibial
ATia	anterior astragalotibial
ATid	distal astragalotibial
ATil	lateral astragalotibial
ATim	medial astragalotibial
ATip	posterior astragalotibial
Ca	calcaneus
CaA	calcaneoastragalar
CaAa	auxiliary calcaneoastragalar
CaAd	distal calcaneoastragalar
CaCu	calcaneocuboid
CaCud	distal calcaneocuboid
CaCul	lateral calcaneocuboid
CaCum	medial calcaneocuboid
CaCup	proximal calcaneocuboid
CaFi	calcaneofibular
CaN	calcaneonavicular
CaTi	calcaneotibial
CCJ	calcaneocuboid joint
CLAJ	(chiropteran) continuous lower ankle joint
EMJ	entocuneiform-mesocuneiform joint
EMt1	entocuneiform-first metatarsal
EMt1J	entocuneiform-first metatarsal joint
EMt2	entocuneiform-second metatarsal
En	entocuneiform
EnMc	entocuneiform-mesocuneiform
EnPh	entocuneiform-prehallux
HRJ	humeroradial joint
Hu	humerus
HUJ	humeroulnar joint
HuRac	capitular humeroradial
HuRal	lateral humeroradial
HuRam	medial humeroradial
HuUl	humeroulnar
HuUlm	medial humeroulnar
HuUlr	radial humeroulnar
LAJ	lower ankle joint
NaEn	naviculoentocuneiform

(continued)

Table I. (*Continued*)

Ra	radius
RaCa	radiocarpal
RaLu	radiolunate
RaSc	radioscaphoid
RaUld	distal radioulnar
RaUlp	proximal radioulnar
RCJ	radiocarpal joint
RL	radiolunate
RSCe	radioscaphocentrale
RSCL	radioscaphocentralunate
RUJd	distal radioulnar joint
RUJp	proximal radioulnar joint
SCL	scaphocentralunate
Su	sustentacular
Sud	distal sustentacular
Sus	superior sustentacular
UAJ	upper ankle joint
UCJ	ulnocarpal joint
UWJ	upper wrist joint

Topographical bony details, ligaments, tendons, muscles, and anatomical directions

ac	astragalar canal
ampt	astralgar medial plantar tubercle
at	anterior plantar tubercle
capit	calcaneal pit for cuboid process
eplp	plantar process of entocuneiform
fdbt	common tendon of flexor digitorum brevis
fft	flexor fibularis (= flexor hallucis longus) tendon
fhd	femoral head
ftt	flexor tibialis (= flexor digitorum longus) tendon
ge	groove for extensors
gf	groove for flexors
gfft	groove for flexor fibularis tendon
gpbt	groove for peroneus brevis tendon
gtpl	groove for tendon of peroneus longus
hbgr	humeral bicipital groove
hbrfl	humeral brachioradialis flange
hcap	humeral capitulum
hcfo	humeral coronoid fossa
hdpcr	humeral deltopectoral crest
hdtrc	humeral deltotriceps crest
hecf	humeral entepicondylar foramen
hepf	humeral epitrochlear fossa
hlfcap	humeral lateral flange of capitulum
hmcr	humeral medial crest
hmec	humeral medial epicondyle
hmtub	humeral medial tuberosity
hofo	humeral olecranon fossa
hrfo	humeral radial fossa
hsucr	humeral supinator crest

(*continued*)

Table I. (*Continued*)

htr	humeral trochlea
hzcon	humeral zona conoidea
pp	peroneal process of calcaneus
rbbrt	radial biceps brachialis tuberosity
rcefo	radial central fossa
rcpr	radial central process
ret	retinaculum
rlcr	radial lateral crest
rmcr	radial medial crest
sup	sustentacular process
ulco	ulnar coronoid process
ulgapl	ulnar groove for abductor pollicis longus
ulol	ulnar olecranon process
ulscfo	ulnar subcoronoid fossa
ulscs	ulnar subcutaneous surface
ulst	ulnar styloid process
ultrn	ulnar trochlear notch
utfdp	ungual tubercle for flexor digitorum profundus

[a]Specific joints are abbreviated by the combination of the first letters in capitals of the names of those units that contribute to the joint, and the letter J for joint. The combination adjectives that designate two articulating bones usually stand for particular articular facets. Abbreviations entirely in lowercase designate landmarks on specific bones, anatomical directions, ligaments, tendons, or muscles. These abbreviations are listed under two separate headings in order to facilitate retrieval of information. Most of the characters abbreviated in this table and used in the figures and in the text are specific homology designations rather than only topographical descriptive terms.

Santonian Cretaceous in the Kyzyl Kum desert of Uzbekistan), eutherians that may be antecedent to the North American Maastrichtian *Protungulatum* were present in west-central Asia. To confound the process of assessing the temporal and geographical context, Aptian Cretaceous eutherians are known (*Prokennalestes;* see Kielan-Jaworowska and Dashzeveg, 1989) even earlier, from Mongolia. Furthermore, the latest Cretaceous dispersal of mammals into North America was only the first of three major faunal events that drastically altered the animals we encounter in the fossil record of North America. The next dispersal event was the Torrejonian invasion, probably from Asia (Lucas, 1984; Sloan, 1969), and the third was the Clarkforkian–Wasatchian one most immediately from Asia but perhaps ultimately from more southern latitudes of Africa and India (see particularly Sloan, 1969; Bown and Rose, 1990; Krause and Maas, 1990).

It is our view that while the skeletal evidence from the fossil record is increasing, its integration with an evolutionary (functional and ecological) osteology of living small mammals, with the exception of a number of primate studies, has not been pursued as much as this subject deserves (see discussion in Szalay, 1993). This lack of in-depth literature on the living species (many of which are becoming increasingly rare or on the brink of extinction) makes proper analysis of the fossil record exceedingly difficult, and any tests/ conclusions more tentative than they could be. Other groups in addition to the living archontans are clearly closely relevant to the issues raised in the multidimensional Archonta problem. The still-to-be-described Cretaceous and Early Tertiary groups such as various arctocyonids, mixodectids, microsypids, and of course plesiadapiforms play a large role in this ongoing inquiry, as Beard's (1989) study has shown. Apatemyids, however, contra Gunnell (1989), are unlikely to be part of relevant comparisons. The association of the dentition of *Mixodectes* with diagnostic postcranial elements described by Szalay and Lucas (1993) is, however, significant.

Before we embark on a discussion of comparative osteology, we will briefly review the literature as it relates to our discussion of the craniodental evidence bearing on the Archonta, and then comment on the phylogenetic status of the Chiroptera, Volitantia, and Archonta, respectively.

Brief History of Some Recent Research on the Archonta

The history of the concept Archonta has its beginnings with Gregory's (1910) monumental study in evolutionary mammalogy, but clearly the bulk of the work on which it was based extends back to the golden age of post-Darwinian comparative morphology in the last half of the 19th century. Leche's (1886) monograph in which he concluded that colugos are the closest relatives of bats is a major and meticulous study, despite Winge's (1923; 1941, pp. 277–296; the original Danish version written between 1887 and 1918) often exaggerated and sometimes erroneous contradictions of its factual and interpreted details. Winge's seemingly crushing rebuttal of the Volitantia hypothesis of Illiger (1811) was accompanied by his idiosyncratic consideration of the galeopithecids as basal insectivorans.

Simpson (1937) advocated the close ties of the mixodectids and microsyopids (which he, like many others before him, considered primatelike), and was intrigued by the possibility of the relationship of these with the Plagiomenidae. Simpson (1937, p. 131), however, also noted that the " . . . inconclusive evidence of possible relationship of the plagiomenids to the Dermoptera has not been significantly altered since Matthew (1918) wrote. The chain of evidence thus tending to link the mixodectids with the Dermoptera is so weak at every point as not to merit serious consideration at present." In his 1945 mammal classification he rejected the concept of the Archonta, as seen in the introductory quote above. The checkered history of taxonomic allocations of these relevant fossil groups was later reviewed in detail by Szalay (1969). He rejected the possible dermopteran affinities of mixodectids, considered microsyopids possibly allied to early Primates (*sensu lato*) on dental evidence only, and regarded tupaiids as unrelated to any of these but to some "Insectivora." The unresolved issues raised in the past, then abandoned and raised again, are thus particularly relevant to the understanding of the nature of the frustrating taxonomic tasks that were faced then, and are confronted today, by students trying to make sense of the evolutionary history of early eutherian groups.

Many of the vascillating opinions that ensued about the archontans (although rarely raised as specific issues of that taxon until McKenna's 1975 return to the formal usage of the concept) were based on dental and basicranial information only, and this clearly limited the tests of the various hypotheses. D. E. Russell's (1959, 1964) analyses of the skull of *Plesiadapis* and McKenna's (1966) study of the first known skull of a microsyopid revealed a problem that was further complicated by some obvious differences from the living archontan groups (see Szalay, 1969; Gingerich, 1976). While the dental record has become much richer, there has been only a slow growth in information and interpretation of cranial and postcranial attributes of these animals.

Szalay and Decker (1974, Fig. 2) in their study of some primitive eutherian, plesiadapiform, and euprimate tarsals have presented a phylogenetic hypothesis in which they have unequivocally shown the primates, dermopterans, and chiropterans converging toward a common stem in the Cretaceous, in exclusion of the macroscelidids. McKenna (1975) made use of an emended Archonta of Gregory (1910) as part of his concept of Tokotheria without, however, noting any attributes for this clade. He considered the tupaiids to be the stem of the Archonta, a stem that he believed was "partly terrestrial . . . with arboreal capabilities" (p. 38).

Szalay *et al.* (1975) in their study of the postcranial remains of *Plesiadapis* reaffirmed not only the clear arboreal adaptations of the genus but also its sufficiently detailed similarity in various advanced eutherian features to Paleogene euprimates. Szalay (1977) and Szalay and Drawhorn (1980) discussed some tarsal-based attributes for the archontan common ancestry, predicted on the approach of comparative joint analysis and some of the results developed in Szalay and Decker (1974). While they discussed tupaiids, plesiadapiforms, and euprimates, they did not consider chiropteran morphology but merely assumed that Leche's (1886) study, and his and Gregory's (1910) views of colugo relationships with bats, were correct. With these assumptions, they examined the then-available and relatively well-represented relevant Paleogene tarsal samples and generated a complex of diagnostic features that in their view linked the archontans (bats, again, having been omitted). Novacek (1980), who also examined aspects of cranioskeletal morphology as it related to tupaiid affinities, criticized the methodological perspective employed in Szalay (1977) and Szalay and Drawhorn (1980) who used a combination of diagnostic characters, of which some also occur independently in other groups.

Cartmill and MacPhee (1980) in their analysis of the tupaiid cranial skeleton and carotid circulation provided a comprehensive discussion of some of the taxonomic problems of the cranial anatomy of the archontans. Their verdict was: "With respect to cranial anatomy, the archontan morphotype is therefore indistinguishable from the eutherian morphotype" (p. 127). Furthermore, they also remained unpersuaded by the diagnostic traits of the tarsus proposed by Szalay (1977).

Novacek (1982) through his studies of placental ordinal relationships, particularly those of bats and insectivorans, has come to advocate strongly the examination of Dermoptera–Chiroptera ties, the Volitantia of Illiger 1811, for which at the time he saw no viable alternative. He nevertheless viewed the published evidence on the tarsus in relation to the Archonta as ambiguous, ". . . forcing one to the conclusion that tupaiids, primates, and insectivorans represent three distinct ordinal-level taxa of uncertain relationships among themselves and with other Eutheria" (pp. 23–24). Novacek's (1986) study on leptictid cranial morphology has reaffirmed his skeptical outlook on the "Archonta," but continued to consider the concept of Volitantia probable.

Shortly thereafter, Novacek and Wyss (1986) in their brief examination of the archontans in the context of eutherian relationships acknowledged a previous overstatement concerning the relative merits of tarsal characters for the diagnosis of the Archonta. They cited 17 characters for the support of the Volitantia, but only 2 for Archonta itself. The first trait—"pendulous penis suspended by reduced sheath between genital pouch and abdomen" (p. 265)—while it cannot be tested against the fossil record, is a uniquely possessed feature of some significance. The second character, as stated by them in isolation from its closely related functional components of the putative protoarchontan tarsus, was "sustentacular facet of astragalus in distinct medial contact with distal astragalar facet" (p. 260), on which we comment below. It appears then that among most taxonomists actively pursuing research on higher-level eutherian relationships it was only a single neontological trait that was considered the only unequivocal support for the Archonta. The osteological trait as stated by Novacek and Wyss (1986) is present in several ordinal groups of eutherians other than archontans where it is not unequivocally present.

A detailed discussion of the literature of the archontan attributes, with many original observations on cranial morphology, was published by Wible and Covert (1987). They advocated tupaiid–euprimate special ties, yet they exercised extreme caution in endorsing the Archonta concept (seemingly based on the same traits as those cited by Novacek and Wyss, 1986), which they considered as essentially a *faute de mieux* arrangement, although they saw merit in a dermopteran–chiropteran clade within the Archonta. Wible and Novacek (1988), in their examination of bat monophyly, not only came out strongly against the somewhat quixotic views of bat diphyly (marshaling 26 characters for the holophyly of bats), but also provided strong support (7 characters) for the Volitantia Illiger, 1811, gave a weaker endorsement for the Euprimates–Scandentia clade (4 characters), and listed the same two traits to support the Archonta as those given in Novacek and Wyss (1986). Recently, the study of the basicranial evidence bearing on primate origins by MacPhee *et al.* (1988) has provided new data analysis but no positive conclusions as to primate or colugo basicranial affinities. Gunnell (1989) has recently presented an account of a number of Paleogene genera that he considered Plesiadapiformes, and in particular attempted a review of the postcranial literature, but in our view without a full appreciation of what has been established in that literature. His inclusion of the Apatemyidae is presumably based on some dental features. Gunnell placed the Mixodectidae in the Insectivora?, and the Microsyopidae in the Primates? (*sensu lato*).

Beard (1989) in his extensive, functional-adaptive morphological analysis of the plesiadapiform, dermopteran, tupaiine, and euprimate postcranial skeletons introduced much new evidence from Paleogene skeletal morphology, and in our view generated insightful and welcome testable issues regarding the Archonta. He presented his taxic conclusions based on extensive character analysis in the form of the new, allegedly holophyletic, concepts of Primatomorpha and Eudermoptera. He grouped the Plesiadapiformes (without Microsyopidae or Mixodectidae) with the Dermoptera, within which he united the Paromomyidae and Galeopithecidae as the Eudermoptera, and considered the Primates as the sister of his expanded concept of Dermoptera. The Chiroptera and the Scandentia were, respectively, omitted from the Primatomorpha, in fact leaving the possibility of the Archonta open. In the following year, Beard (1990) published a short paper about not only gliding, but finger-gliding plesiadapiform paromomyids that he considered to be the sister group of the living Dermoptera. In addition, in the same issue of *Nature*, Kay *et al.* (1990) reported on a cranium of *Ignacius grabullianus* and argued that it had not a petrosal but an entotympanic bulla. These studies, by virtue of the wide dissemination of *Nature* and the new views they presented, had the singular effect of reviving interest in the Archonta problem, and at the same time brought into focus the importance of fossils, their careful scrutiny as to the reality or artifactual nature of some of their characteristics, and the functional-adaptive appraisal of the postcranial skeleton in Paleogene mammals. What immediately followed in the semipopular accounts, characteristically widespread today, were quick and superficial recounting of the issues (Martin, 1990; Shipman, 1990; Zimmer, 1991).

Beard (1989, pp. 586–593) in his monographic study used the following characters to corroborate the Primatomorpha and Eudermoptera, taxa that he considered holophyletic clades, and two other unnamed groups, referring to the traits as synapomorphies of the groups. The numbers given these characters are ours, in order to facilitate the discussion to follow.

Unnamed holophylum "Plesiadapidae–Paromomyidae–Micromomyidae–Galeopithecidae" (p. 586):

1. Lunate positioned distal to scaphoid (rather than ulnar to it); triquetrum (cuneiform as used in this chapter) articulates with both lunate and scaphoid on its radial aspect
2. Strong development of accessory synovial articulation between posterior side of sustentaculum tali and plantar side of medial astragalar pillar

Unnamed holophylum "Plesiadapidae–Paromomyidae–Galeopithecidae" (p. 586):

3. Ossification of the auditory bulla
4. Extreme reduction or loss of the internal carotid artery

Eudermoptera (Paromomyidae and Galeopithecidae) (pp. 586–587):

5. Triquetrum (cuneiform) becomes strongly narrowed ulnarly
6. Ulnocarpal articulation restricted to radial and palmar aspects of ulnar styloid process
7. Intermediate manual and pedal phalanges longer than corresponding proximal phalanges

Primatomorpha (pp. 592–593):

8. Lesser tuberosity on proximal humerus projects strongly medially
9. Capitulum almost globular; central fossa on radius nearly circular in outline and well excavated; groove-like zona conoidea separates capitulum from trochlea
10. Acetabulum craniocaudally elongated; acetabular margin strongly buttressed cranially, poorly buttressed dorsally
11. Area for insertion of M. quadratus femoris extensive on posterior aspect of femoral shaft
12. Groove for tendon of M. flexor fibularis located laterally on posterior astragalus
13. Calcaneocuboid articulation with plantar excavation on calcaneus and corresponding proximally projecting process on cuboid

Beard's confident assertion about paromomyid ties, and gliding and finger-gliding (we believe that these are two significantly *different issues, with discrete implications*) were immediately challenged by Krause (1991), who has provided an exhaustive account of the caveats surrounding phalangeal proportions and allocations in Beard's original work.

Discussion

It should be obvious that the adaptive modifications of a segment of a lineage establish the phylogenetic constraint, the genomic foundation for the following generations that again are shaped by the ecology. Thus, a functional-adaptive perspective as a primary handle on characters in macro-taxonomy, and the methods that guide its foundations in order to construct the phylogenetic patterns, need not be particularly justified here (see Szalay and Bock, 1991). One major problem in mammalian phylogenetics (and this is clearly not limited to the contentions surrounding the archontans), in addition to the ever-important need to improve the fossil record, the various theocratic views notwithstanding (e.g., Patterson, 1981; Forey, 1982), is the analysis of morphological characters beyond their taxic distribution. Yet the infusion of "process" (or its omission) into "data" in what is considered objective taxonomy, is often an instant prescription for scientific conflict. As virtually every debate reflects, a full scrutiny of the extant osteology and the fossil record [McKenna's (1987, p. 75) comments notwithstanding concerning the "excellent results" supplied by distribution analyses alone] is often blocked by the lack of any, or only few attempts to consider the implications of eco-morphological studies on living animals for phylogenetic analysis. One must obtain what there is a distribution of, and there is no "objective" way to draw limits for characters. In general, there is no consensus concerning what are acceptable taxonomic characters. Yet the results of functional-adaptive analyses of character complexes (*sensu* Bock, 1981) potentially permit taxonomists to gain at least a partially causal understanding of morphological differences in morphoclines, and allow the necessary connections between seemingly distantly related homologies. A prevalent assumption of many taxonomists, such as that "we can see it all objectively in taxonomic practice without the burden of any process," is not an exaggerated aphorism. Such an approach, we believe, results not only in rather puzzling atomization of closely interdependent characters, but also in the inability to peel back the constraints of distinctive and successive adaptive solutions. It is these constraints that allow an increasingly precise understanding of the phyletic and cladistic ties of the taxonomic properties, and subsequently the taxa themselves.

Most disagreements are rooted in problems of homology versus homoplasy, and consequently in the differences between the methods that are employed to ascertain these. As we maintain that these problems require partial resolution via functional-adaptive analysis, the issues of the putative common ancestors of Archonta, Plesiadapiformes, microsyopids, mixodectids, and tupaiids partly represent problems related to the range and manner of particular solutions for arboreality and gliding that are difficult to assess. For the relationship of various Paleogene groups to the Dermoptera it is the separation of finger-gliding versus obligate slow claw-climbing arboreality that are some of the major postcranial issues of contention, and character transformation series must eventually reflect this.

While we provide lists of attributes that we believe corroborate the holophyly of the Volitantia and the monophyly of Archonta, we cannot hope to analyze in any detail the totality of the cranioskeletal evidence for the holophyly of the Volitantia, a concept that found its recent champion in Novacek. Nevertheless, we present some additional evidence here that is strongly suggestive of transformation series of several characters from the condition seen in living colugos to the protobats. Unlike Beard (1989, 1990, this volume) implied, his analyses do not supply evidence that bats and colugos acquired their interdigital patagia (or elbow and carpal similarities) independently. Even if Beard was to be correct in his assessment of finger-gliding in Paromomyini, and we believe that he may well be correct, such an established hypothesis in light of the synapomorphies linking Galeopithecidae–Chiroptera would make the concept of Dermoptera as he defined it even more likely to be paraphyletic.

We briefly discuss below the three major areas of hard anatomy, the dentition, cranial structure, and postcranial morphology where we place emphasis on our conclusions concerning the hitherto enigmatic Mixodectidae and where there is broader relevance to various problems related to the Archonta.

Dentition

There has been substantial growth in the fossil record of teeth and even whole dentitions all of which have considerable bearing (with an understanding of their feeding-related phyletic constraints) on the affinities of the various poorly known Paleogene families. One recent source for a review of this new information is that of Gunnell (1989), although as we stated above we do not agree with many of his stated character- and taxon-related hypotheses of relationships, but which we cannot pursue further in this chapter.

Because much of the evidence we discuss is postcranial, we will not unduly stress here the important dental evidence. The special similarity of the primitive plesiadapiform dental pattern to that of Eocene euprimates suggests, as it has for a long time, a common stem for these groups—as far back as the Cretaceous. At the present the fossil record of the dentition helps little with the understanding of the internal relationship of the archontans, although we believe that such issues as the affinities of the Plagiomenidae need reexamination (see particularly MacPhee *et al.*, 1989). It can be defended, however, that tupaiids, mixodectids, and microsyopids are clearly not in possession of, or unlikely to have had, the synapomorphies suspected to be at the roots of the monophyletic Primates (*sensu* Szalay and Delson, 1979; Szalay *et al.*, 1987). We will not linger on fine points of classifications here, but clearly the issues of paraphyly cannot be resolved as yet.

The obvious must be emphasized here: none of the probably gliding plesiadapiforms (lineages of *Phenacolemur* and *Tinimomys;* see Beard, 1989) displays dental morphology that is even remotely similar to that of colugos. Given the nature of the mammalian record, however, the dental evidence is destined to grow in importance. We are struck, as were workers in the 1930s who had far less information available to them, by the potential significance of the similarities in dilambdodonty (*Icaronycteris,* mixodectids, plagiomenids, and colugos), of the construction of the hypocone (*Icaronycteris, Ptilocercus,* mixodectids), and the twinning of the entoconid and hypoconulid (tupaiids, mixodectids, microsyopids, plagiomenids, colugos, and the chiropteran morphotype, the latter *fide* Slaughter, 1970). In general, the special dental similarities of mixodectids, plagiomenids, colugos, and the probable chiropteran ancestry are of as yet unclear significance (with regard to the plagiomenids we do not agree with MacPhee *et al.*, 1989).

Because of the difficult and unresolved issue of mixodectid–plagiomenid affinities, some comments on the recent description and allocation of new taxa to the plagiomenids are relevant here. The recent referral of the omomyid primate *Ekgmowechashala* by McKenna (1990) to the Plagiomenidae is, to us, irreconcilable in light of the published morphological evidence. This omomyid primate, which has a hypoconulid displaced toward the hypoconid, a well-developed euprimate-like (as in *Washakius* or *Rooneyia*) hypocone, and conules on the buccal half of the molars, was allocated with the newly described genera *Tarka* and *Tarkadectes* to the Plagiomenidae. Furthermore, the new taxa were assigned to the Ekgmowechashalinae itself. One of these (*Tarka*) has clearly preserved upper molars with no hypocone, relatively huge lingually displaced para- and metaconules, and large stylar cusps that would befit a mixodectid, plagiomenid, and galeopithecid, but entirely unlike the omomyid to which it was compared. The two types of upper molar patterns and their occlusal mechanics simply do not support the assumption of that paper that the buccal cusps of the omomyid are stylar cusps, rather than what they certainly appear to be: a small mesial accessory cuspule and the paracone and metacone.

Cranial Anatomy

We present no new cranial evidence here, but some remarks on recently described and interpreted cranial morphology of relevant taxa are appropriate here. Publications of the past decade—those of Cartmill and MacPhee (1980), Novacek (1980, 1986, 1987), MacPhee (1981), MacPhee and Cartmill (1986), Zeller (1985, 1986), Szalay *et al.* (1987), Wible and Covert (1987), Wible and Novacek (1988), and MacPhee *et al.* (1989)—have presented descriptions and syntheses of importance for the evaluation of archontan basicranial evidence. The recently described skull of *Ignacius* by Kay *et al.* (1990) raised the issue of the composition of the bulla in paromomyids in particular and in the Plesiadapiformes in general. These authors stated that the bulla is of an entotympanic construction. The specimen is slightly but critically crushed, and shows symmetrical fractures or sutures in the dorsal tympanic space, interpreted by Kay *et al.* (1990) as a petrosal–entotympanic suture. The disappearance of this gap posteriorly, where the putative entotympanic is smoothly continuous with the petrosal, raises issues of (empirical) identification and the nature of tests of this proposed homology. Is the alleged suture merely a separation of the thin dorsal bone in the tympanic cavity? And if it is not, are we witnessing a stage prior to the ontogenetic fusion of the entotympanic and petrosal, a condition that may be antecedent to the one diagnostic of euprimates? The putative entotympanic of *Ignacius*, as well as the details of the morphology of the basicranium, are, of course, fundamentally distinct from that of the Recent Galeopithecidae (see especially Hunt and Korth, 1980; MacPhee *et al.*, 1989). In particular, in colugos the entotympanic abuts against the petrosal immediately lateral to the basioccipital–basisphenoid. There is no fusion of any sort, as there is above and lateral to the basioccipital in *Ignacius*. Both in *Ignacius* and in *Plesiadapis*, as in the primitive euprimate adapids, the central position of the bubbled promontorium and the wide and high medial portion of the hypotympanic sinus (Szalay *et al.*, 1987) are shared with adapids and therefore suggest the possibility of a derivation from a common primate stem. The reduced intrabullar arterial circulation in plesiadapiforms, a trait shared with colugos as noted by Kay *et al.* (1990), may represent either a synapomorphy, or independently attained derived reductions, or possibly even a primitive eutherian retention from some unknown Cretaceous ancestry. (The last, however, is not very likely.)

Importantly, in spite of claims about the homology of the bulla of *Ignacius,* Kay *et al.* (1990) did not show any significant structural similarities to the living colugos beyond the reduced intrabullar arterial circulation. Independent reduction of the carotid circulation in plesiadapiforms and galeopithecids certainly cannot be ruled out in light of its (apparently) unreduced presence on the petrosal of *Tinimomys* reported by Gunnell (1989, p. 87). As the association of the petrosal allocated to *Tinimomys* with teeth is circumstantial—they are not part of an intact specimen—the ear region may not be that of *Tinimomys* or a plesiadapiform.

Beard's (1989) characters 3 and 4 with regard to the reduction of the carotid circulation (see above) or the ossification of the bulla, therefore, are problematical. In particular, the absence of an ossified bulla in *Tinimomys* does not account for an "ossified bulla synapomorphy" of a "Eudermoptera + Plesiadapidae" clade.

Postcranial Anatomy

Given a correlation of the functional-adaptive aspect of some of the shared similarities among some archontans specifically with arboreal slow-climbing (as convincingly developed by Beard, 1989), but only a few of these associations with the mechanics of dermopteran gliding (discussed below, and Szalay, in preparation), the conceptual subdivision, and the (methodological) manner of evaluation of such similarities is of particular concern, as we voiced throughout and elsewhere (e.g., Szalay and Bock, 1991). Furthermore, there are some differences in several of the homologous elements of the two living genera of colugos that well illustrate the issues that make *both* horizontal (cladistic) and vertical (phyletic) comparisons of characters a complex, necessary and valid matter (Bock, 1977, 1981). Many traits, which in a purely distribution-based analysis might be considered as synapomorphies of one group, may, after proper character analysis, appear to be the ancestral stage (derived at that point) to more inclusive groupings. The nature of similarities of the elbow is instructive in this regard.

The distal humerus, ulna, and proximal radius of *Mixodectes* (Figs. 1–4) described by us (Szalay and Lucas, 1993) are especially similar to their homologues in plesiadapiforms, *Ptilocercus,* and primitive euprimates, nevertheless with some taxon-specific differences being present in all of these (see Szalay and Lucas, 1993). While the rounded radial central humeroradial (HuRac) facet is shared, among eutherians, with colugos, the lack of the well-developed

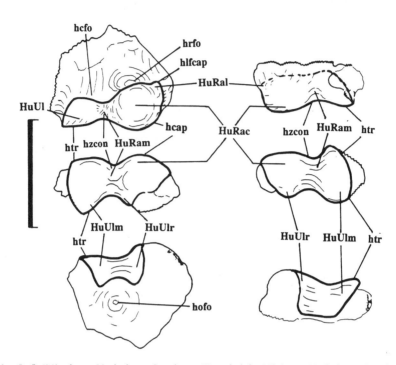

Fig. 1.   Left (*Mixodectes;* Nacimiento Specimen B) and right (*Chriacus;* Nacimiento Specimen A) distal humeral fragments of the two species from the Torrejonian Early Paleocene of New Mexico (see Szalay and Lucas, 1993) in anterior (top row), distal (middle row), and posterior (bottom row) views. For abbreviations see Table I. Scale represents 1 cm.

radial biceps brachialis tuberosity (rbbrt) and a modest radial extension, the radial medial humeroradial (HuRam) facet medially in the latter represents a significant (probably *primitive* archontan) difference from dermopterans. While plesiadapiforms, *Ptilocercus,* early euprimates, and *Mixodectes* share special similarities in this area, dermopterans differ from these significantly! They have an expanded HuRam facet both on the radius and on the humerus. Our analysis strongly suggests that a unique combination of similarities between the distal humeri and proximal and distal radii, the fused distal radii and ulnae, as well as the unique special similarity of what is probably a scaphocentralunate, of colugos and bats, indicate a direction for the probable transformation. This becomes particularly convincing in light of the mechanics and adaptations in several living mammalian gliders that are unlike the colugos in the form-function of their adaptive strategies. The galeopithecid condition of the elbow and upper wrist joint are far more likely closely representative of the antecedents of the protobat than what is known in any other mammal. The evidence from the carpus in colugos and bats

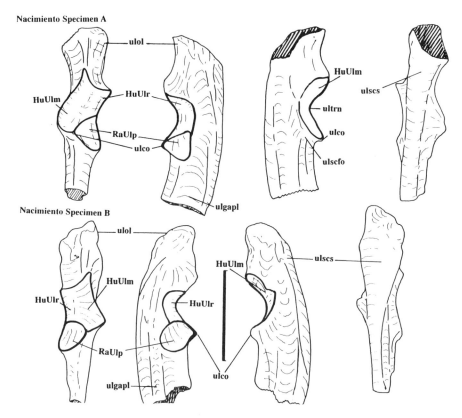

Fig. 2. Left ulna of *Chriacus* (above) and right ulna of *Mixodectes* (below) from the Torrejonian Early Paleocene of New Mexico; from left to right: anterior, lateral, medial, and posterior views. For abbreviations see Table I. Scale represents 1 cm.

(Szalay, in preparation) contradicts Beard's (1989) character 1. Character 6 of Beard is clearly a trait shared by colugos and bats to a greater extent than with paromomyids. The dominant role of the radius in the distal fusion of the radius and ulna is clearly a colugo–bat similarity (see below), and not that of the Eudermoptera.

Characters 8–11 of Beard may be valid synapomorphies stemming from the last common ancestor of the plesiadapiforms, euprimates, and primitive tree shrews (e.g., *Ptilocercus*), a poorly understood group internally, from which a lineage led to colugos (and thence another to the putative proto-chiropteran).

The close similarity of both the calcaneus and astragalus of Nacimiento Specimen B, *Mixodectes* (Fig. 5), to those of plesiadapiforms and colugos is highly significant, as we discuss below. Nacimiento Specimen A, an early oxy-claenine, *Chriacus* (Fig. 6), however, also holds particular interest for this study because of its taxon-specific arboreal adaptation. While *Chriacus* sheds only

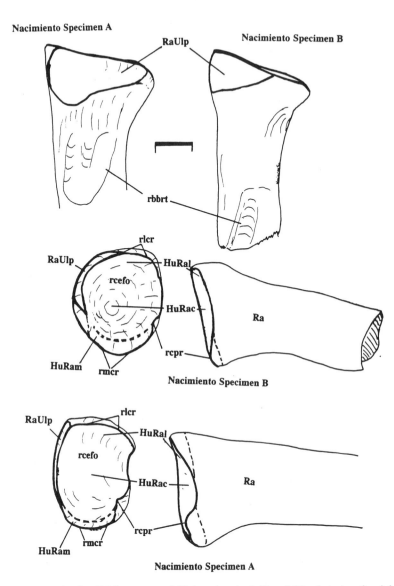

Fig. 3.   Left proximal radial fragments of *Chriacus* (on the left) and *Mixodectes* (on the right); top row shows posterior views; middle and bottom rows show the distal (on the left) and anterior (on the right) views. For abbreviations see Table I. Scale represents 2 mm.

minimal light on the problem of the Archonta, in its contrasting morphology it does reaffirm the uniqueness of the diagnostic archontan postcranial complex. It is only the astragalocalcaneal complex that unequivocally corroborates arboreality for the oxyclaenine, but it does so very strongly. Furthermore, the adaptively squirrellike conformation of the ankle complex of *Chriacus* makes

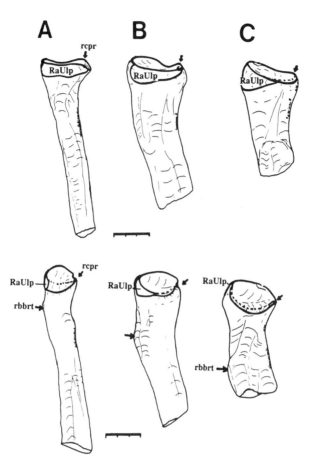

Fig. 4.   Comparisons of the left proximal radii of the Puercan *Anisonchus* (A), and the Torrejonian *Chriacus* (B) and *Mixodectes* (C). Top: posterior views; bottom, medial (i.e., proximally ulnar) views. For abbreviations see Table I. Scales represents 3 mm.

it unlikely that this oxyclaenine might be particularly closely related to the plesiadapiform complex. We are, however, not rejecting possible close ties between some primitive condylarths and stem plesiadapiforms or stem archontans. The known postcranial elements described by us (Szalay and Lucas, 1993) leave little doubt that *Protungulatum* and similar early condylarths retain postcranial traits more primitive than this oxyclaenine (see Szalay and Decker, 1974, Szalay and Drawhorn, 1980, Szalay, 1984, 1985, and Szalay and Dagosto, 1988, on the tarsal evidence; Szalay and Dagosto, 1980, on the humerus; Rose, 1987, 1990, on a wide variety of Eocene skeletal morphs). Nevertheless, elements in addition to the calcaneus and astragalus from the tarsus (see Szalay and Lucas, 1993) give important information on the conformation of the cuboid and cuneiforms in an early, relatively small arcto-

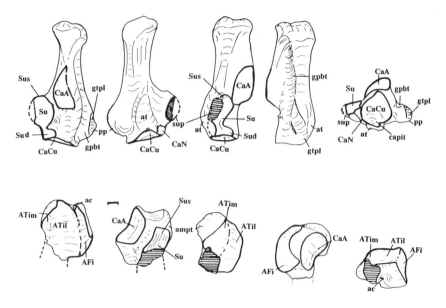

Nacimiento Specimen B

Fig. 5.   Left calcaneus (above) and astragalus (below) of *Mixodectes* from the Torrejonian Early Paleocene of New Mexico. From left to right: dorsal (extensad), ventral (flexad), medial, lateral, and distal views. Hatching represents broken surfaces, and broken lines round out the sustentaculum on the calcaneus and indicate the position of the neck distal to the broken surfaces of the body on the astragalus. For abbreviations see Table I. Scale represents 1 mm.

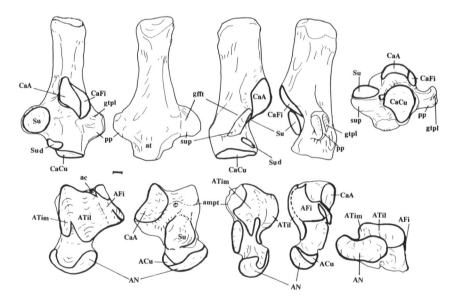

Nacimiento Specimen A

Fig. 6.   Reconstructions of left calcaneus (top) and astragalus (bottom) of *Chriacus* from the Torrejonian Early Paleocene. From left to right: dorsal (extensad), ventral (flexad), medial, lateral, and distal views. For abbreviations see Table I. Scale represents 1 mm.

206

cyonid. The perhaps clade-specific similarities between plesiadapiforms and the Nacimiento *Chriacus* cuboids, and particularly of the similarities of the entocuneiforms (see Fig. 7), may mean more in terms of condylarth–archontan relationships than what is possible to seriously entertain at present. This statement is made in light of our current understanding of the morphology of the earliest known skeletal remains of Cretaceous and Paleocene leptictids and Eocene erinaceomorphs.

In Szalay and Lucas (1993) we have noted the special similarity of the tarsals of *Mixodectes* to those of described plesiadapiforms (Szalay *et al.*, 1975; Beard, 1989). What is particularly significant is the deep osseous and fibrous groove (judged by its extreme similarity to this condition in the living colugos) on the lateral side of the calcaneus, along with the alignment of the anterior tubercle of the calcaneus. The functional conclusions were discussed in Szalay and Lucas (1993).

Our conclusion about the inferred mechanics of the tarsus in *Mixodectes* and its correlated biological roles is closely relevant to any phylogenetic assessment of the Archonta. The calcaneal grooves for the tendons of the flexor fibularis (gfft), peroneus longus (gtpl), and peroneus brevis (gpbt) are present

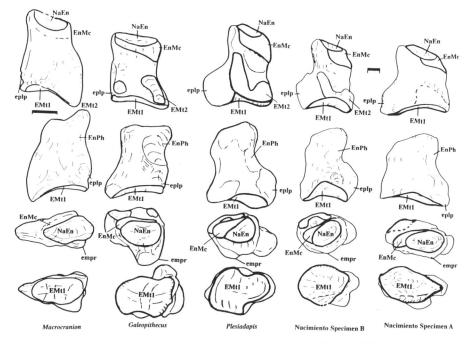

Fig. 7.   Comparisons of right entocuneiforms, from left to right, of the Middle Eocene *Macrocranion*, the colugo *Galeopithecus*, the Late Paleocene *Plesiadapis tricuspidens*, the Early Paleocene *Mixodectes* (Nacimiento Specimen B), and an Early Paleocene *Chriacus* (Nacimiento Specimen A). From top to bottom: medial, lateral, proximal, and distal views. For abbreviations see Table I. Scale for *Macrocranion* and the scale for the others is 1 mm.

in both *Chriacus* and *Mixodectes*—but with significant differences. In *Mixodectes* the anterior tubercle of the calcaneus is directly distal and flexad of the calcaneocuboid joint (CCJ). Its alignment, or rather that of the calcaneocuboid ligament flexad which braces the joint with the foot distal to the calcaneus, is identical to that of both living colugos. This is significant because it may be related to the tensile forces generated in underbranch hanging. Furthermore, the gfft is less sharply defined than in *Chriacus*. The functional-adaptively, and perhaps phylogenetically, most telling comparison concerns the inferred path of the tendon of the peroneus brevis (as it is reflected by the gpbt). This prominent groove (bony–fibrous tunnel) represents the osteological expression of the peroneal retinaculum which firmly restricts both the peroneus brevis and peroneus longus before the peroneus longus is shunted in a groove of its own. This connection between the bony morphology and the strapping of the tendons was confirmed on a foot of *Galeopterus* by us. While the gpbt on the dorsal surface of the peroneal process is wide and bordered laterally by a sharp crest, the homologous groove in *Chriacus* is strikingly different. In *Mixodectes*, as in *Galeopterus* (but not as much in *Galeopithecus*), the groove is both sharply defined and narrowly extended onto the entire distal two-thirds of the calcaneus. Like the orientation of the anterior tubercle, the great depth of the gpbt also strongly suggests the alignment of tensile forces along the long axis of the calcaneus and the entire foot distal to it in the mixodectid.

It is important to note that insertion of a massive tendon of the peroneus brevis on the proximal end of Mt5, on its lateral tubercle, allows for powerful abduction, eversion (pronation), and flexion of the foot, which is therefore mechanically suited for suspension. In fact, in living colugos (and also in what we infer to be the protochiropteran condition as well; see below) it is the two lateral pedal rays that are the stoutest and longest. This is then the manner of form–function relationship of the pes in colugos, which is in turn strongly correlated (almost certainly causally) with their characteristic manner of frequent underbranch suspension (Macdonald, 1984), and not with any particular aspect of their gliding behavior. Yet the nongliding behavior of colugos is undoubtedly connected to their aerial activities, but this important aspect of their ecological morphology has not been studied at all. Another telling similarity of the calcaneus to colugos is the presence of a calcaneal (and undoubtedly navicular as well) calcaneonavicular (CaN) facet, a character rare in mammals.

The calcaneal calcaneocuboid (CaCu) facet in *Mixodectes*, like in plesiadapiforms, dermopterans, *Ptilocercus,* and euprimates, is nearly circular. Like these taxa (except for the tupaiids, as pointed out by Beard, 1989), the calcaneus has its deepest point flexad and approximately on the midline of the facet (the calcaneal pit, capit), aligned in a spoutlike manner with the calcaneal anterior tubercle. The mobility-related aspect of this shift has been explored in detail by Szalay and Decker (1974).

The astragalus of *Mixodectes* lacks the head and neck, but the entire body is well preserved. It is diagnostically archontan and dermopteran in having, in combination, a long, posteriorly and medially narrow ATil facet, a high AFi facet, a well-offset astragalar gfft, and particularly a strongly Sus facet above the astragalar medial plantar tubercle (ampt). The CaA facet, as in *Chriacus*, is sellar, with a ridge on the medial side of the facet. This character is somewhat different in colugos which have only a very slightly concave astragalar CaA facet. The body clearly indicates one of the more striking signs of the habitually inverted protoarchontan foot. The medially and posteriorly narrow ATil facet, an area that is enlarged compared with other contemporary eutherians, broadens distally on the body of the astragalus, yet remains relatively long and narrow. This is a combination that is uniquely shared by *Ptilocercus*, probably the stem archontan and the stem plesiadapiform, and to a lesser degree by colugos. It must be appended that one Nacimiento specimen described by Szalay and Lucas (1993), a terminal phalanx (Fig. 8), which belonged to either *Chriacus* or *Mixodectes*, is highly suggestive of slow claw-climbing. As Fig. 8 suggests, it is reminiscent of the homologous element in plesiadapids.

We simply want to emphasize here that (1) the greater similarity of the molar dentition of mixodectids (a fluctuating concept in the past literature; e.g., Simpson, 1937; Szalay, 1969; Russell *et al.*, 1973; Gunnell, 1989) to that of colugos (with the notable exception of plagiomenids which are dentally very colugolike, some recent skepticisms as to its significance notwithstanding) and (2) the probably hanging-related adaptations reflected along the extremely similar form–function solutions of the pes of mixodectids to those of colugos, together suggest more than convergence. We believe that in spite of the many shared primitive archontan characters between plesiadapiforms and mixodectids, this primarily Paleocene family, more than any other group of archaic archontans ("primates"), may hold special ties to the colugos. As developed elsewhere (Szalay and Walsh, in preparation), the same may be true for the Microsyopidae. Such allocations, of course, will not solve the problems of paraphyly that *any* classificatory arrangement beyond and within the Archonta creates.

The condition we described for *Mixodectes* and a similar one we postulate for the relatively primitive UAJ condition of archontans are different from that of the most ancient and most primitive of euprimates (early adapiforms and omomyids). It must be remembered, however, that the origin of euprimates was closely related to what has been called grasp-leaping behavior (Szalay and Delson, 1979; Szalay and Dagosto, 1980; Dagosto, 1988). In contrast to the slow-climbing habits of the plesiadapiforms, an inference massively documented by Beard (1989), the euprimate shift to powerful grasping and nails (Gregory, 1920; Szalay and Delson, 1979; Dagosto, 1985, 1986, 1988; Gebo, 1985, 1986, 1987; Szalay and Dagosto, 1988) clearly signals the context within which the UAJ form-function of the protoeuprimate was altered.

In the protoeuprimate lineage the arc of the astragalar component of the UAJ increased, facilitating greater flexion–extension of the foot, and load

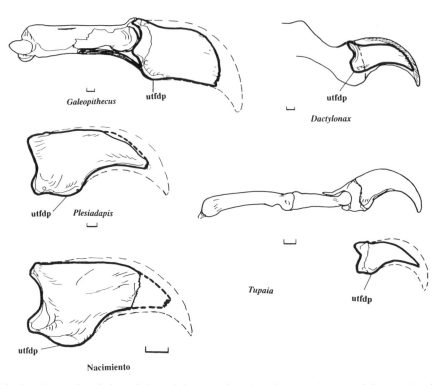

Fig. 8.   Comparison in lateral view of the ungual phalanx from locality 313 of the New Mexico Natural History Museum where a single individual of *Chriacus* and *Mixodectes* occur with the European *Plesiadapis tricuspidens*, a colugo (*Galeopithecus*), a striped possum (*Dactylonax*), and a tree shrew (*Tupaia* sp.). The ungual phalanx of the extant forms are shown in relation to the more proximal segments, and the tendon of the digital flexor is shown in the colugo. For abbreviations see Table I. Scales represent 1 mm.

transfer between the astragalus and tibia was simply widened in the extended position of the foot that occurred when landing after a leap. The area of the astragalus under load at that point, then, was shifted to the posterior part of the ATil facet, and therefore its relatively greater width in euprimates makes adaptive sense—and causally accounts for the derived differences from a more primitive archontan condition.

It is important to rectify a minor point of difference here. Beard (1989) in his seminal study of plesiadapiform postcranial morphology allocated previously published material (Szalay and Drawhorn, 1980, Fig. 10IA) to *Phenacolemur simonsi*. In our view the tarsals, particularly the calcaneus, (AMNH 89518), the larger of the two dermopterans (*Cynodontomys*) illustrated by Szalay and Drawhorn (1980), are distinct from *Phenacolemur* and from any other plesiadapiform, contra Beard's view. Most critically, the far greater flexad and posterior extension of the calcaneal pit (capit; see illustrations of

Mixodectes for this structure) on the Sand Quarry specimen, as well as the relatively lesser flexad–extensad depth of the CaCu facet are some of the traits that distinguish it from *Phenacolemur.*

Beard's (1989) character that states the presence of the well-developed Sus facets on the proximal tarsals cannot be considered as anything but an archontan diagnostic feature—in combination with other functional-adaptive attributes of the foot (see below). Its absence in early euprimates is probably a derived archontan trait, as is the tibiocalcaneal facet of tupaiids (see below).

The entocuneiforms (Fig. 7), when known, are of special significance for this phylogenetic discussion because the colugos stand out in their unique morphology, and because hallucial function may be ascertained from their configuration. Plesiadapiforms, *Mixodectes* (Nacimiento Specimen B), and *Chriacus* (Nacimiento Specimen A) resemble one another more than either of these resembles colugos, as discussed in detail in Szalay and Lucas (1993). This is not a particular obstacle, however, for the special relationship of any of these with colugos. We suspect that these similarities represent either a general eutherian, or Archonta–Condylarthra relationship. The lack of a hypertrophied eplp, in both colugos and early erinaceomorphs, may be independently similar in these taxa, as the reduction of this flange may be related to the deemphasis of the hallux. In colugos there is a derived emphasis of the lateral pedal digits, and a virtual loss of hallucial grasping is quite likely.

As we discussed (Szalay and Lucas, 1993), we cannot recognize specific unequivocally gliding-related tarsal adaptations in either rodent or phalangeriform gliders. So it appears that the similarity between many of the Paleogene archontans to selected areas of colugo anatomy may be due to a common phylogenetic constraint derived from the apomorphy of their common ancestor. It is, at present, still very difficult to isolate in the tarsus the grasping- and hanging-related functions from those that may be involved in some hitherto unrecognized way with the mechanics linked to colugo-specific gliding. Nevertheless, the total tarsal similarity of *Mixodectes* to that of colugos resulted in an assessment that we believe to be both adaptively convincing (arboreal, and often underbranch hanging) as well as strongly indicative of phylogenetic ties. We repeat that some of these similarities are not casually related to the gliding behavior itself but to the highly characteristic suspensory and slow climbing behavior of colugos. A simplistic translation of this interconnected *adaptive* and *phylogenetic* assessment (based on both phyletic and cladistic analysis of traits) into a purely cladistic frame of reference is, however, still not warranted, we believe, in light of the total available evidence. Parallel (and presumably gliding-related) solutions of a number of plesiadapiform lineages, all sharing the common inheritance of the early slow climbing arboreal radiation of the Archonta, for a gliding and suspensory mode of life cannot be ruled out. In fact, unequivocally shared and derived similarities (*within* the context of the Archonta) between even such probable gliders as *Tinimomys* or *Phenacolemur* (Beard, 1989, 1993) and the living colugos are still not documented, largely, however, because of lack of complete skeletons. We do, however, recognize synapomorphies (which are also unique within the Archonta) of colugos and bats (see also Novacek, 1982, 1986), as we state below. The grasp-leaping, fast-moving adaptations of the euprimates must be looked upon as a step removed, if not from a primitive (but specifically archontan) condition, then from some hitherto unidentified and similar eutherian base.

While we cannot adequately comment in greater depth on Beard's (1989) major study without much new ecological morphology efforts on various living mammals and the subsequent new comparative analyses that should follow, his adaptive conclusions concerning *gliding* in *Paromomys* and *Tinimomys* appear sound to us. Yet the critical issue of finger-gliding remains unresolved. This of course leaves the previously noted possibility open that gliding has evolved independently in the archaic paraphyletic assemblage, the Plesiadapiformes. To put it differently, we do not doubt the validity of (archontan, or perhaps even plesiadapiform) homologies established between Plesiadapiformes and Dermoptera, although many of these are also shared by *Ptilocercus* (Szalay, in preparation). The concept of homology, of course, is a different one from that of synapomorphy.

Nevertheless, because of the highly unique nature of living colugos and their (putatively) homologous similarity in some significant areas (as discussed above and below) with bats, we cannot simply ignore the fact that the patterns of characters shared among such groups as Plesiadapiformes, Mixodectidae, Primates, Dermoptera, Chiroptera, or Scandentia may well represent both the ancestral condition to the Archonta and any of the possible subdivisions within it. Furthermore, extremely derived as the bats may be, their impressive list of unique similarities to colugos (e.g., Wible and Novacek, 1988) and the few traits we report here, appear to place them within a monophyletic Archonta. To support our stand, we will now attempt to give a tentative list of biologically meaningful (in some cases even causally understood) morphological attributes for a number of relevant taxa that we maintain were the diagnostic apomorphies in combination with the various antecedent conditions attained already in the lineage before the ancestor of that taxon. In other words, in the various groups included in the Archonta, traits of the protoarchontan may either persist or occur in the modified designation of that group's ancestral apomorphies. Such a procedure dictates no obligate holophyletic classificatory practice, yet allows a genuinely phylogenetic diagnosis of monophyletic taxa that takes into account phyletics as well as cladistics (Szalay *et al.,* 1987; Szalay, 1993).

The characters we list below are restricted to the skeletal system. These features not only allow the connection of the living taxa with the extinct ones, hence the testing of the most complete phylogenetic hypotheses possible (i.e., functional, adaptive, and historical), but many of these are also available for the phylogenetic diagnoses below only in fossils.

Regarding the Chiroptera

As stated before, there has never been a successful challenge to the holophyletic concept of the Chiroptera, and we will not discuss this in light of the ample literature in this area during the past decade. Recently, Wible and Novacek (1988), Baker *et al.* (1991), and Simmons *et al.* (1991) have given a detailed rebuttal of the contested nature of bat holophyly raised by Pettigrew (1986) and Pettigrew *et al.* (1989). In particular, the neurological details on which Pettigrew based his hypothesis are discussed and refuted in detail by Kaas and Preuss (1993) who are specialists in comparative neuroanatomy. While we strongly support the phyletic unity of bats, the nature of character transformations are serious problems for phyletics within and outside the chiropterans. Questions about the Archonta are closely related to our understanding of bat character phylogenies. We would like to note here, therefore, because it is relevant, that in spite of our support for the reaffirmation of chiropteran holophyly by Baker *et al.* (1991) and Simmons *et al.* (1991), we strongly disagree with their theoretical perspective on character analysis.

Simmons *et al.* (1991, p. 240) state that ". . . a hypothesis of relationships is essential before one can hope to identify instances of either reversal or convergence. Neither reversal or convergence should be ruled out a priori before a thorough phylogenetic analysis has been conducted." Such a stance places primary importance on taxic analysis (i.e., phylogenetic analysis in the cladist's sense) as the determinant of character polarities. The latter approach is probably the primary motivation for the increasing virtue attributed to ever-larger data bases consisting of increasingly nondescript (simple) morphological attributes. The result is often the runaway atomization of areas of anatomy which need be understood as integrated complex functional units. Because *a posteriori* weighting is the practice followed by taxic methods (not only by cladists, but see also Mayr and Ashlock, 1991; see discussion in Szalay and Bock, 1991), the congruence of impressively large numbers of characters in purely cladistic schemes, without regard to what these really may be, is supposed to be the "test" of such clustered traits. An approach detailed under a discussion of null-group comparisons of homologies in Szalay and Bock (1991) mandates independently tested (causally understood) apomorphies, synapomorphies, character transformation hypotheses (taxonomic properties) against which the taxon phylogenies themselves should be tested.

The first well-known fossil bat, *Icaronycteris,* is probably temporally and phyletically far away from the origins of the order. Yet, as fully appreciated by Novacek (1987), such a relatively early fossil is especially worth close scrutiny. The osteological traits we cite for the protochiropteran, while preliminary, do add support to the concept of Volitantia, as a suite of characters most similar to the primitive combination displayed by galeopithecids (see below). Jepsen's (1970) excellent stereophotos show the distal ulna joining the distal radius extremely closely, and it is probably fused. The proximal ulna clearly shows the separation of the olecranon ossification (process), against the posterior side of the distal humerus, and the area where it was attached to the end of the intact part of the ulna shows up as being heavily pitted, characteristic of surfaces with epiphysial contact. The olecranon is not the "ulnar patella" which is clearly visible on the lateral side in *Pteropus* in Fig. 11. In fact, the olecranon is as thin and reduced as the condition in *Pteropus*. The small bone in Jepsen (1970) labeled as a sesamoid in his Fig. 11 is most likely the prehallux. The large size and articular configuration of what we call the scaphocentralunate, judged from its articular contacts, are as one finds in colugos and living pteropids.

The following brief list (without an expansion on the complex modifications and correlated changes of the entire pectoral girdle) merely lists or alludes to a virtually unassailable combination of diagnostic osteological traits present in the protochiropteran.

1. The entire musculoskeletal system of the pectoral limb can be subdivided into a number of character complexes which we will not do in any detail at this time.
2. The first phalanx of the hallucial digit of the foot is approximately 1.5 times longer than the equivalent units lateral to it (adaptation to accommodate habitual hanging when not flying; it is probably the result of phyletic compensation for the heritage of hypertrophied lateral pedal cheiridia.
3. Continuous (coalesced) CaA and Su facets on calcaneus and astragalus in the lower ankle joint, referred to here as chiropteran (calcaneal or astragalar) lower ankle joint (CLAJ) facet.
4. Pubic spine for insertion of psoas minor.
5. For a number of important cranial characters, see Novacek and Wyss (1986), Novacek (1987), and Wible and Novacek (1988).

Regarding the Volitantia

Whatever the phyletic affinities of mixodectids (and their position within the Archonta proper is probably secure), the elbow remains of *Mixodectes* described and discussed in Szalay and Lucas (1993) point to an animal with a highly mobile and freely rotating radius, which in turn postdicts pronation and supination equal to what we see in the earliest euprimates, plesiadapiforms, *Ptilocercus*, and didelphids—but nevertheless quite distinct from colugos. In light of the terrestrial mechanics of the tarsus of what may be considered a protoeutherian condition (see Szalay, 1984, 1985), the similarities of the elbow of archontans to didelphids are probably convergent. The mixodectid distal humerus, however, while comparable to those of advanced plesiadapiforms (described in Beard, 1989), is more similar to those of *Ptilocercus* and early Euprimates.

We, however, do not doubt that the shared primitive archontan, plesiadapiform, and mixodectid conditions for the elbow are antecedent to those of the protodermopteran. Nevertheless, the osteological similarity of this area in the living Dermoptera to the Chiroptera is probably one of the most significant ones phylogenetically, along with other osteological characters, because eventual testability against new fossils is a realistic prospect. In the living colugos the manner of reduction of the ulna, the fusion at some point in ontogeny with the distal radius, the increased importance of the proximal radius in articulating with the humerus (Figs. 9–13), and a scapho-centralunate are all traits that are not generally related merely to gliding. We suspect that in some hitherto unknown way they were a response to the demands of finger-gliding. These traits are group specific. The detailed ecological morphology of the colugos is not known. Neither marsupial nor rodent gliders, however, opted for such locomotor or form–function strategy.

The similarity of the dermopteran and protochiropteran elbow and antibrachium patterns, functionally explicable, suggests a chiropteran phylogenetic heritage from a precisely colugolike (not eudermopteran) ancestry. These unique and testable (through functional examination) special similarities of the dermopteran elbow and lower arm (specifically the distal humerus and aspects of the proximal and distal ulna) are with bats (Figs. 9–13), but these were unexamined in the creation of the concepts Primatomorpha and Eudermoptera. These similarities (see also Thewissen and Babcock, 1991, this volume; Beard, this volume) should not be dismissed as convergent, as their uniqueness, within the context of the highly probable causal connection between the finger-gliding of colugos and the hand-flying of bats, suggests otherwise.

It should be noted here concerning the chiropteran elbow that references in the literature to trochlear characters of the humerus can be misleading. If one considers the area of the distal humerus that articulates with the ulnar trochlear, or semilunar, notch in most mammals, the designation "trochlea" on the distal humerus has a precise meaning of homology that pertains to interacting functional components between the humerus and ulna. Using "trochlea" that way, as we do, means then that the last common ancestor of bats had no appreciable trochlea that was homologous to that of other mammals. The medial margin of the distal articular area of bats is in contact with the radius, and this area is the homologue of the zona conoidea of other mammals. As the figures (Figs. 9–12) show, the dermopteran elbow is characterized by the restriction of anterior humeroulnar contact together with an expansion of the radial contact with the zona conoidea, and therefore a great increase in the HuRam facets. In fact, of the two valid genera of living colugos, *Galeopterus* displays the extreme of this condition. In chiropterans (and for us the megabats are acceptably representative exemplars postcranially and cranially of a quasi-primitive condition, but not in terms of cheek tooth morphology), there is of course a remnant of the proximal ulna. It is fragmented into two pieces and has all but lost its restrictive function related to axial rotation and compression-related stability in quadrupeds, for the obvious

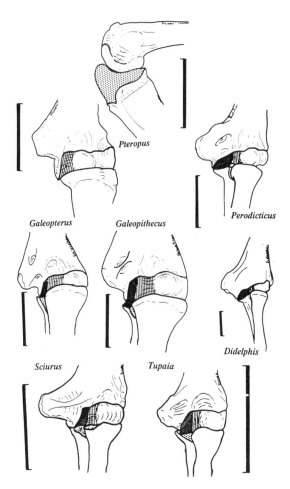

Fig. 9. Comparison of the elbow in anterior view (except in *Pteropus* which is also shown in medial view) in a sciurid, a tupaiid, a didelphid, a highly derived (somewhat opossum-like) primate *Perodicticus*, the two extant colugos, and a pteropid bat. The humeral trochlea proper which articulates with the ulna is shown dark (finely stippled), the zona conoidea (the part of the distal humerus medial to the capitulum which articulates with the radius) is hatched, and the stippled area (where shown) on the ulna and radius represents the visible radioulnar (and annular ligament contact on the radius in *Pteropus*) articulation. Note the diminished contact of the ulna with the trochlea in Dermoptera, and the complete loss of ulno-trochlear articulation in bats. Scales are 1 cm.

reasons of flight-related mobility. It still (flexibly) anchors the lower arm to the upper one. The distal end of the radius in these two orders is no less specially similar (Fig. 13). The distal end of the ulna is fused to the radius, and the widened upper wrist joint (UWJ, composed of the RCJ and UCJ), the troughlike distal combined radial–ulnar facet, and the deeply sculpted grooves for the extensor tendons of colugos are astonishingly similar to bats.

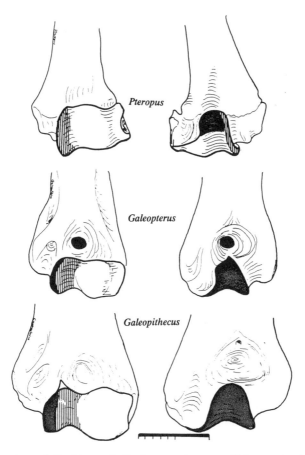

Fig. 10.   Comparison of distal humeri of colugos and a pteropid in anterior (on the left) and posterior (on the right) views. Hatched area represents the contact of the radius with the zona conoidea, in particular its extension onto the posterior side in bats which is correlated with the raised posterior rim of their radial head; finely stippled (dark) areas represent the anterior and posterior contact areas of the ulna with the humerus. Note the extreme reduction of this articulation in colugos and its further restriction to the posterior surface in bats. Scale represents 1 cm.

The orientation (torsion) of the distal radius (+ ulna) compared with the orientation of the proximal end is identical in colugos and the megabats. There are no such similarities to the chiropteran condition in any of the five living gliding groups of therians (two suprageneric rodent taxa, Petauristinae and Anomaluridae, and three genera of phalangeriform syndactylans), or the plesiadapiforms probably correctly identified as gliders by Beard (1989, 1990; see Thorington, 1984, for an analysis of the sciurid gliders). All of these have robust distal ends on the ulnae.

Coupled with the special similarity of the radius (+ ulna), a preliminary note must be added concerning the carpals. As noted, it appears most likely

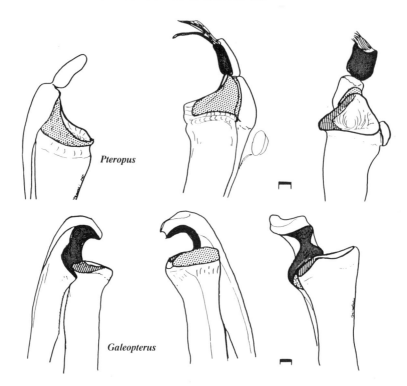

Fig. 11.   Comparison of left proximal radius and ulna in articulation in a colugo and a bat. Dark, finely stippled, areas show the ulnar surfaces in articulation with the humerus, hatching shows radial contact areas with the humerus (with the zona conoidea in colugos and additionally with the posterior surface in bats; see Fig. 10), and stippled areas show the radioulnar (and annular ligament against the radius) articular contacts. Note extreme reduction of the olecranon process in the colugo and its practical absence in the bat. The "ulnar patella" of bats is visible on the two drawings to the right on the top. From left to right: medial, lateral, and anterior (flexad) views. Scales are 1 mm.

that a scaphocentralunate (SCL), a fusion of the scaphoid, centrale, and lunate, is shared by the colugos and bats. In addition, *Icaronycteris*, the oldest known bat skeleton (Jepsen, 1970), has deep ungual phalanges, as seen in colugos.

It is important to note here that Krause (1991) cast doubt *only* on Beard's evidence for the association of the intermediate phalanx of the hand, which was the basis for attributing the extension of the patagium onto the hand as an interdigital membrane. This issue, given the scarcity of fossils and their often probabilistic assignment, clearly cannot be decided at present. In his conclusion, Krause (p. 187) also stated: ". . . there is no convincing evidence *from these elements* that paromomyids were either gliders or dermopterans." The remaining evidence, the long bones, the innominate, the few carpals, and the good comparative tarsal samples, were not discussed. Krause must have been

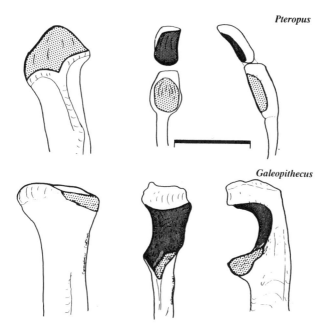

Pteropus

Galeopithecus

Fig. 12.   Comparison of posterior view of left proximal radii (on the left), and anterior (in the middle) and lateral (on the right) views of left proximal ulnae of a colugo and a bat. Dark, finely stippled, areas show ulnar surfaces in articulation with the humerus, and stippled surfaces show areas of contact between radius and the ulna and radius and the annular ligaments. Note extreme reduction of the olecranon process in the colugo and its practical absence in the bat. Scale represents 1 cm.

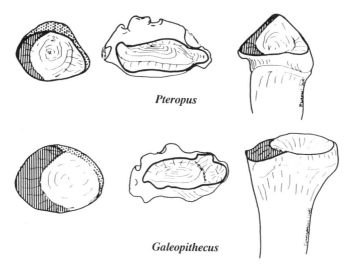

Pteropus

Galeopithecus

Fig. 13.   Comparison of left radii (and part of the fused ulnae distally) of a colugo and a bat. From left to right: proximal view of radius, distal view of radius-ulna, and anterior (flexad) view of radius. Stippled areas represent contact between radius and ulna and radius and annular ligaments; hatched areas are humeroradial articular surfaces. Note that the extension to the right (middle row) on the upper wrist joint articular surfaces represents the fusion of the distal ulna with the radius.

aware that he did not refute Beard's account of gliding within the Plesiadapiformes.

Our interpretations suggest that dermopterans became hand-modulated gliders, or finger-gliders. We designate it so because independent movement of the fingers may have significant effect on their glide control. The hand-flying bats became volant (only once), and we believe, as Leche (1886), Novacek (1982, 1986, 1987), and Thewissen and Babcock (1991) did, that they came from an ancestry that was dermopteran, i.e., a close cladistic relative of galeopithecids.

The following are a diagnostic combination of skeletal attributes of the last common ancestor of colugos and bats, the protovolitantian:

1. (Skeletal evidence for) interdigital patagium (i.e., this taxonomic property is testable in fossils)
2. Fusion of distal ulna to distal radius that is transversely widened; deep grooves for carpal extensors on dorsal surface of distal radius; reduction of the ulna from anterior humeral contact; the latter is extreme in bats (in order to increase freedom of radius to rotate)
3. Fusion of scaphoid, centrale, and lunate into scaphocentralunate
4. Flattening of ribs
5. Proximal displacement of the areas of insertion for the pectoral and deltoid muscles, a coalesced single proximal humeral torus [see also Novacek, (1987); this trait occurs independently in a number of mammalian clades also]
6. Elongation of the fourth and fifth pedal rays (to accommodate habitual, but not obligate, underbranch hanging)
7. Both proximally and distally deep ungual phalanges, compressed mediolaterally

Regarding the Archonta

Throughout much of this chapter we either discussed or alluded to the monophyly of the Archonta. In spite of our confidence in the holophyly of the Volitantia, we are unsure of the specific phyletic or cladistic ties of plesiadapiforms, mixodectids, microsyopids, tupaiids, and euprimates. While tupaiine tupaiids, from their fully arboreal and grasping roots (see above on *Ptilocercus*, and particularly Le Gros Clark, 1926; Szalay, in preparation), have become scamperers, the euprimates originated with a lineage of rapidly jumping and pedal grasping forms, what have been called grasp-leapers (Szalay and Delson, 1979; Dagosto, 1988). While we consider support for the monophyly of the Archonta substantial, we certainly cannot be secure in its holophyly in light of its obvious ancient origins in the Cretaceous, and therefore the possible hitherto perhaps unimagined and untested connections to other eutherians. We, however, cannot endorse any meaningful close relationship of apatemyids with the Archonta, contra Gunnell (1989). The well-known skeleton of *Heterohyus* documented by Koenigswald (1990) suggests no postcranial traits that may be considered archontan. The single osteological trait listed by Novacek and Wyss (1986) for the support of the Archonta is isolated from the remainder of the relevant tarsal character complex. Yet it is the totality of the latter that makes the isolated trait relevant together with the remainder of the complex. Furthermore, the continuity of the astragalar sustentacular facet with the astragalonavicular one is far from unique. It is conspicuously present in tenrecs and hyraxes, among others. Alone it cannot suffice as a diagnostic trait.

We list below those diagnostic attributes that we believe were present in combination in the putative protoarchontan, a fully arboreal probably Late Cretaceous eutherian with a probable habitus distinct from the earliest terrestrial placentals (Szalay, 1984):

1. Entotympanic auditory bulla (both rostral and caudal entotympanics).
2. Elbow characterized by rounded humeral capitulum, shallow humeral olecranon fossa, a rounded proximal radius with a bowl-shaped radial central fossa, an anteriorly "bent" olecranon process on the ulna, and a concavo-convex proximal radioulnar articulation, all of these together indicating not only greatly increased pronation–supination of the forearm, but also its habitual flexion.
3. On distal femur patellar groove shallow, as in living slow arborealists, and the robust third and lateral trochanters are near the proximal end of the bone.
4. Fully eutherian upper ankle joint in which the arc (in flexion and extension of the pes) of the path of the crus is very small but its linear path is elongated and characteristic, and the fibular side of the astragalus is relatively (unusually) high compared with other eutherians.
5. The astragalar lateral astragalotibial contact (ATil facet) is posteriorly and medially narrow (the body of the astragalus is posteriorly hypertrophied compared with other early eutherians, thus "squeezing" the astragalar canal laterally) whereas it is distally wide both medially and laterally, indicating a habitually flexed and highly inverted foot that can be further turned by the rotation of the calcaneus (see below).
6. Path of strongly restrained tendon of flexor fibularis forms a well-defined and delineated groove on the posterior end of the astragalus; the offset astragalar ATil facet assures considerable inversion during flexion of the foot on the crus.
7. Lower ankle joint characterized by extreme mobility (a derived eutherian compensation for the mortise–tenon mediolateral restriction of the UAJ), particularly an ability for inversion and supination by lateral rotation of the calcaneus on its long axis (with consequences on the foot distal to it), manifested by a helical calcaneal CaA facet and a long astragalar sustentacular contact that almost reached the AN facet.

8. Facets for calcaneocuboid articulation (CaCu facets on both calcaneus and cuboid) as deep flexad–extensad as mediolaterally, and with a slight concavo-convex articulation which allows free rotation of the calcaneus while it pushes the forefoot distally and into a supinated position.

Summary

The postcranial remains of the Early Paleocene mammal *Mixodectes* (described in Szalay and Lucas, 1993) from the San Juan Basin of New Mexico enabled us to compare mixodectids with arctocyonids, various lipotyphlans, creodonts, carnivorans, rodents, plesiadapiforms, tupaiids, and the family Galeopithecidae (Dermoptera). The postcranials assigned to the mixodectids show special archontan-level similarities with the plesiadapiforms, *Ptilocercus*, and the protoeuprimates, and share some advanced archontan characters with the slow climbing and pedal-hanging related morphology of the living colugos. The taxonomic concepts Primatomorpha and Eudermoptera are examined, along with those of the Dermoptera, Chiroptera, Volitantia, and Archonta, particularly in light of the recent cranial and postcranial characters proposed in the literature. Through our phyletic (vertical) and cladistic (horizontal) comparative analysis of a number of osteological complexes, we marshal what we consider strong support for the concepts of Archonta Gregory 1910 (without the macroscelidids, but with the inclusion of a number of fossil groups) and Volitantia Illiger, 1811. We strongly support much of Beard's (1989) functional-adaptive analysis of plesiadapiform character clines, but disagree with the putative Eudermoptera or the Primatomorpha, and find no convincing support for them as either holophyletic or paraphyletic groups. In addition to several archontan apomorphies present in the last common ancestor of the Volitantia, we find strong support for this concept in the nature of the similarities (and phyletically significant differences as well) in the elbow complex, and in Leche's (1886) original discovery of the unique presence of an interdigital membrane of the hand in bats and colugos. We present what we believe to be strong evidence for the monophyletic (but not necessarily holophyletic) reality of the Archonta: apomorphous osteological traits that were probably present in the last common ancestor of the Scandentia, Plesiadapiformes, Microsyopidae, Mixodectidae, Euprimates, and Volitantia. Within this assemblage it appears that at present only one holophyletic grouping, the Volitantia, can be supported; other exclusively vertical relationships are not warranted based on the evidence we are familiar with.

ACKNOWLEDGMENTS

We thank our friend and colleague Henry Galliano for his gift of skeletons of *Galeopterus*, bats, and other mammals to Szalay, and Ross D. E. Mac-

Phee for his generous invitation to contribute to this volume. We acknowledge with pleasure the positive impact of the contributions of Christopher Beard to the study of the "archaic archontans", and for his partaking in many open exchanges of stimulating ideas. We are grateful to several museum curators who allowed access to the collections in their care.

References

Baker, R. J., Novacek, M. J., and Simmons, N. B. 1991. On the monophyly of bats. *Syst. Zool.* **40:** 216–231.

Beard, K. C. 1989. Postcranial anatomy, locomotor adaptations, and paleoecology of early Cenozoic Plesiadapidae, Paromomyidae, and Micromomyidae (Eutheria, Dermoptera). Ph.D. dissertation, The Johns Hopkins University School of Medicine, Baltimore.

Beard, K. C. 1990. Gliding behaviour and palaeoecology of the alleged primate family Paromomyidae (Mammalia, Dermoptera). *Nature* **345:** 340–341.

Beard, K. C. 1993. Phylogenetic systematics of the Primatamorpha, with special reference to Dermoptera, in: F. S. Szalay, M. J. Novacek, and M. C. McKenna (eds.), *Mammalian Phylogeny: Placentals*, pp. 129–150. Springer-Verlag, Berlin.

Bock, W. J. 1977. Adaptation and the comparative method, in: M. Hecht P. C. Goody, and B. M. Hecht (eds.), *Major Patterns in Vertebrate Evolution*, pp. 57–82. Plenum Press, New York.

Bock, W. J. 1981. Functional-adaptive analysis in evolutionary classification. *Am. Zool.* **21:**5–20.

Bown, T. M., and Rose, K. D. (eds.). 1990. Dawn of the Age of Mammals in the northern part of the Rocky Mountain Interior, North America. *Geol. Soc. Am. Spec. Pap.* **243.**

Cartmill, M., and MacPhee, R. D. E. 1980. Tupaiid affinities: The evidence of the carotid arteries and cranial skeleton, in: W. P. Luckett (ed.), *Comparative Biology and Evolutionary Relationships of Tree Shrews*, pp. 95–132. Plenum Press, New York.

Dagosto, M. 1985. The distal tibia of primates with special reference to the Omomyidae. *Int. J. Primatol.* **6:**45–75.

Dagosto, M. 1986. The joints of the tarsus in the strepsirhine primates. Ph.D. dissertation, City University of New York.

Dagosto, M. 1988. Implications of postcranial evidence for the origin of euprimates. *J. Hum. Evol.* **17:**35–56.

Forey, P. L. 1982. Neontological analysis versus paleontological stories. *Syst. Assoc. Spec. Pap.* **21:**119–157.

Gebo, D. L. 1985. The nature of the primate grasping foot. *Am. J. Phys. Antrhopol.* **67:**269–378.

Gebo, D. L. 1986. The anatomy of the prosimian foot and its application to the primate fossil record. Ph.D. dissertation, Duke University.

Gebo, D. L. 1987. Functional anatomy of the tarsier foot. *Am. J. Phys. Anthropol.* **73:**9–31.

Gingerich, P. D. 1976. Cranial anatomy and evolution of early Tertiary Plesiadapidae (Mammalia, Primates). *Univ. Mich. Pap. Paleontol.* **15:**1–141.

Gregory, W. K. 1910. The orders of mammals. *Bull. Am. Mus. Nat. Hist.* **27:**1–524.

Gregory, W. K. 1920. On the structure and relations of *Notharctus*, an American Eocene primate. *Mem. Am. Mus. Nat. Hist.* **3:**49–243.

Gunnell, G. F. 1989. Evolutionary history of Microsyopoidae (Mammalia, ?Primates) and the relationship between Plesiadapiformes and Primates. *Univ. Mich. Pap. Paleontol.* **27:**1–157.

Hunt, R. M., Jr., and Korth, W. W. 1980. The auditory region of Dermoptera: Morphology and function relative to other living mammals. *J. Morphol.* **164:**167–211.

Illiger, C. 1811. Prodromus systematis mammalium et mvium additis terminis zoographicis utriudque classis. C. Salfeld, Berlin.

Jepsen, G. L. 1970. Bat origins and evolution, in: W. A. Wimsatt (ed.), *Biology of Bats*, Vol. 1, pp. 1–64. Academic Press, New York.

Kaas, J. H., and Preuss, T. M. 1993. Archontan affinities as reflected in the visual system, in: F. S. Szalay, M. J. Novacek, and M. C. McKenna, (eds.), *Mammalian Phylogeny: Placentals,* pp. 115–128. Springer-Verlag, Berlin.

Kay, R. F., Thorington, R. W., Jr., and Houde, P. 1990. Eocene plesiadapiform shows affinities with flying lemurs not primates. *Nature* **345:**342–344.

Kielan-Jaworowska, Z., and Dashzeveg, D. 1989. Eutherian mammals from the early Cretaceous of Mongolia. *Zool. Scr.* **18:**347–355.

Kielan-Jaworowska, Z., and Nessov, L. A. 1990. On the metatherian nature of the Deltatheroida, a sister group of the Marsupialia. *Lethaia* **23:**1–10.

Koenigswald, W. von. 1990. Die Paläobiologie der Apatemyiden (Insectivora, s.l.) und die Ausdeutung der Skelettfunde von *Heterohyus nanus* aus dem Mitteleozän von Messel bei Darmstadt. *Palaeontog. Abt. A* **210:**41–77.

Krause, D. W. 1991. Were paromomyids gliders? Maybe, maybe not. *J. Hum. Evol.* **21:**177–188.

Krause, D. W., and Maas, M. C. 1990. The biogeographic origins of late Paleocene–early Eocene mammalian immigrants to the Western Interior of North America. *Geol. Soc. Am. Spec. Pap.* **243:**71–105.

Leche, W. 1886. Uber die säugethiergattung *Galeopithecus:* eine morphologische untersuchung. *Kongl. Svenska Vet. Akad. Handl.* **21:**1–92.

Le Gros Clarke, W. E. 1926. On the anatomy of the pen-tailed tree-shrew (*Ptilocercus lowii*). *Proc. Zool. Soc. London* **1926:**1179–1309.

Lucas, S. G. 1984. Taxonomic status of *Titanoides simpsoni* Simons, 1960 and some observations on *Pantolambda* Cope, 1882 (Mammalia, Pantodonta). *N.M.J. Sci.* **24:**46–52.

Macdonald, D. (ed.). 1984. *The Encyclopedia of Mammals.* Facts and File Publications, New York.

McKenna, M. C. 1966. Paleontology and the origin of the Primates. *Folia Primatol.* **41:**1–25.

McKenna, M. C. 1975. Toward a phylogenetic classification of the Mammalia, in: W. P. Luckett and F. S. Szalay, (eds.), *Phylogeny of the Primates: A Multidisciplinary Approach,* pp. 21–46. Plenum Press, New York.

McKenna, M. C. 1987. Molecular and morphological analysis of high-level mammalian interrelationships, in: C. Patterson (ed.), *Molecules and Morphology in Evolution: Conflict or Compromise?* pp. 55–93. Cambridge University Press, London.

McKenna, M. C. 1990. Plagiomenids (Mammalia: ?Dermoptera) from the Oligocene of Oregon, Montana, and South Dakota, and middle Eocene of northwestern Wyoming. *Geol. Soc. Am. Spec. Pap.* **243:**211–234.

MacPhee, R. D. E. 1981. Auditory regions of primates and eutherian insectivores: Morphology, ontogeny and character analysis. *Contrib. Primatol.* **18:**1–282.

MacPhee, R. D. E., and Cartmill, M. 1986. Basicranial structures and primate systematics, in: D. R. Swindler and J. Erwin (eds.), *Comparative Primate Biology,* Vol. 1, pp. 219–275. Liss, New York.

MacPhee, R. D. E., Novacek, M. J., and Storch, G. 1988. Basicranial morphology of early Tertiary erinaceomorphs and the origin of Primates. *Am. Mus. Novit.* **2921:**1–42.

MacPhee, R. D. E., Cartmill, M., and Rose, K. D. 1989. Craniodental morphology and relationships of the supposed Eocene dermopteran *Plagiomene* (Mammalia). *J. Vertebr. Paleontol.* **9:**329–349.

Martin, R. D. 1990. Some relatives take a dive. *Nature* **345:**291–292.

Matthew, W. D. 1918. A revision of the lower Eocene Wasatch and Wind River faunas. Part V. Insectivora (continued), Glires, Edentata. *Bull. Am. Mus. Nat. Hist.* **38:**565–657.

Matthew, W. D. 1937. Paleocene faunas of the San Juan Basin, New Mexico. *Trans. Am. Philos. Soc.* **30:**1–510.

Mayr, E., and Ashlock, P. D. 1991. Principles of systematic zoology, second edition. McGraw Hill, New York.

Novacek, M. J. 1980. Cranioskeletal features in tupaiids and selected eutherians as phylogenetic evidence, in: W. P. Luckett (ed.), *Comparative Biology and Evolutionary Relationships of Tree Shrews,* pp. 35–93. Plenum Press, New York.

Novacek, M. J. 1982. Information for molecular studies from anatomical and fossil evidence on higher eutherian phylogeny, in: M. Goodman (ed.), *Macromolecular Sequences in Systematic and Evolutionary Biology,* pp. 3–41. Plenum Press, New York.

Novacek, M. J. 1986. The skull of leptictid insectivorans and the higher-level classification of eutherian mammals. *Bull. Am. Mus. Nat. Hist.* **183**:1–111.

Novacek, M. J. 1987. Auditory features and affinities of the Eocene bats *Icaronycteris* and *Palaeochiropteryx* (Microchiroptera, *incertae sedis*). *Am. Mus. Novit.* **2877**:1–18.

Novacek, M. J., and Wyss, A. R. 1986. Higher-level relationships of the Recent eutherian orders: Morphological evidence. *Cladistics* **2**:257–287.

Patterson, C. 1981. Significance of fossils in determining evolutionary relationships. *Ann. Rev. Ecol. Syst.* **12**:195–223.

Pettigrew, J. P. 1986. Flying primates? Megabats have the advanced pathway from eye to midbrain. *Science* **231**:1304–1306.

Pettigrew, J. D., Jamieson, B. G. M., Robson, S. K., Hall, L. S., McNally, K. I., and Cooper, H. M. 1989. Phylogenetic relations between microbats, megabats, and primates (Mammalia: Chiroptera and Primates). *Philos. Trans. R. Soc. London Ser. B* **325**:489–559.

Rose, K. D. 1987. Climbing adaptations in the early Eocene mammal *Chriacus* and the origin of the Artiodactyla. *Science* **236**:314–316.

Rose, K. D. 1990. Postcranial skeletal remains and adaptations in early Eocene mammals from the Willwood Formation, Bighorn Basin, Wyoming. *Geol. Soc. Am. Spec. Pap.* **243**:107–133.

Russell, D. E. 1959. Le crâne de *Plesiadapis*. *Bull. Soc. Geol. Fr. Ser.* 7 **1**:312–314.

Russell, D. E. 1964. Les mammifères paleocènes d'Europe. *Mem. Mus. Natl. Hist. Nat. N.S.* 13, **1**:1–324.

Russell, D. E., Louis, P., and Savage, D. E. 1973. Chiroptera and Dermoptera of the French early Eocene. *Univ. Calif. Publ. Geol. Sci.* **95**:1–57.

Shipman, P. 1990. Primate origins up in the air again. *New Sci.* **126**:57–60.

Simmons, N. B., Novacek, M. J., and Baker, R. J. 1991. Approaches, methods and the future of the chiropteran monophyly controversy: A reply to J. D. Pettigrew. *Syst. Zool.* **40**:239–243.

Simpson, G. G. 1937. The Fort Union of the Crazy Mountain Field, Montana, and its mammalian faunas. *Bull. U.S. Natl. Mus.* **169**:1–287.

Simpson, G. G. 1945. The principles of classification and a classification of mammals. *Bull. Am. Mus. Nat. Hist.* **85**:1–350.

Slaughter, B. H. 1970. Evolutionary trends of chiropteran dentition, in: B. H. Slaughter and D. W. Walton (eds.), *About Bats*, pp. 51–83. Southern Methodist University Press, Dallas.

Sloan, R. E. 1969. Cretaceous and Paleocene terrestrial communities of western North America. *North Am. Paleontol. Conv. Proc.* 1(E):427–453.

Szalay, F. S. 1969. Mixodectidae, Microsyopidae, and the insectivore–primate transition. *Bull. Am. Mus. Nat. Hist.* **140**:193–330.

Szalay, F. S. 1977. Phylogenetic relationships and a classification of the eutherian Mammalia, in: M. K. Hecht, P. C. Goody, and B. M. Hecht (eds), *Major Patterns in Vertebrate Evolution*, pp. 315–374. Plenum Press, New York.

Szalay, F. S. 1984. Arboreality: Is it homologous in metatherian and eutherian mammals? *Evol. Biol.* **18**:215–258.

Szalay, F. S. 1985. Rodent and lagomorph morphotype adaptations, origins, and relationships: Some postcranial attributes analyzed, in: W. P. Luckett and J.-L. Hartenberger (eds.), *Evolutionary Relationships among Rodents—A Multidisciplinary Analysis*, pp. 83–132. Plenum Press, New York.

Szalay, F. S. 1993. *Evolutionary History of the Marsupials and an Analysis of Osteological Characters.* Cambridge University Press, New York.

Szalay, F. S., and Bock, W. J. 1991. Evolutionary theory and systematics: Relationships between process and pattern. *Z. Zool. Syst. Evolutionsforsch.* **29**:1–39.

Szalay, F. S., and Dagosto, M. 1980. Locomotor adaptations as reflected on the humerus of Paleogene primates. *Folia Primatol.* **34**:1–45.

Szalay, F. S., and Dagosto, M. 1988. Evolution of hallucial grasping in primates. *J. Hum. Evol.* **17**:1–33.

Szalay, F. S., and Decker, R. L. 1974. Origin, evolution and function of the tarsus in Late

Cretaceous Eutheria and Paleocene primates, in: F. A. Jenkins, Jr. (ed.), *Primate Locomotion*, pp. 223–259. Academic Press, New York.

Szalay, F. S., and Delson, E. 1979. *Evolutionary History of the Primates*. Academic Press, New York.

Szalay, F. S., and Drawhorn, G. 1980. Evolution and diversification of the Archonta in an arboreal milieu, in: W. P. Luckett (ed.), *Comparative Biology and Evolutionary Relationships of Tree Shrews*, pp. 133–169. Plenum Press, New York.

Szalay, F. S., and Lucas, S. G. 1993. The postcranial morphology of *Mixodectes* and *Chriacus* from the Paleocene, with comments on the phylogeny of the Archonta. *Bull. N.M. Mus. Nat. Hist.*

Szalay, F. S., Tattersall, I., and Decker, R. L. 1975, Phylogenetic relationships of *Plesiadapis*— Postcranial evidence. *Contrib. Primatol.* **5:**136–166.

Szalay, F. S., Rosenberger, A. L., and Dagosto, M. 1987. Diagnosis and differentiation of the order Primates. *Yearb. Phys. Anthropol.* **30:**75–105.

Thewissen, J. G. M., and Babcock, S. K. 1991. Distinctive cranial and cervical innervation of wing muscles: New evidence for bat monophyly. *Science* **251:**931–936.

Wible, J. R., and Covert, H. H. 1987. Primates: Cladistic diagnosis and relationships. *J. Hum. Evol.* **16:**1–22.

Wible, J. R., and Novacek, M. J. 1988. Cranial evidence for the monophyletic origin of bats. *Am. Mus. Novit.* **2911:**1–19.

Winge, H. 1923. *Pattedyr-Slaegter*. H. Hagerrups Forlag, Copenhagen.

Winge, H. 1941. *The Interrelationships of Mammalian Genera* (translation of Winge, 1923). C. A. Reitzels Forlag, Copenhagen.

Zeller, U. A. 1985. Die Ontogenese und Morphologie der Fenestra rotunda und des Aquaductus cochleae von *Tupaia* und anderen Säugern. *Gegenbaurs morph. Jahrb.* **131:**179–204.

Zeller, U. A. 1986. Ontogeny and cranial morphology of the tympanic region of the Tupaiidae, with special reference to *Ptilocercus*. *Folia primatol.* **47:**61–80.

Zimmer, C. 1991. Family affairs. *Discover* **12:**64–65.

A Molecular Examination of Archontan and Chiropteran Monophyly

7

RONALD M. ADKINS
and RODNEY L. HONEYCUTT

Introduction

Comparative morphological and paleontological data have provided valuable information on higher-level eutherian mammal relationships, and several phylogenetic hypotheses are well supported by these comparative data (Wyss *et al.*, 1987; Novacek *et al.*, 1988). Nevertheless, many aspects of eutherian mammal relationships have not been resolved, and considerable conflicts between branching patterns diagnosed by different types of characters exist (Goodman *et al.*, 1982; Miyamoto and Goodman, 1986; Novacek and Wyss, 1986; Shoshani, 1986; McKenna, 1987; Wyss *et al.*, 1987; Novacek *et al.*, 1988; Novacek, 1992; Stanhope *et al.*, this volume).

The majority of molecular systematic studies pertaining to the higher relationships among eutherian mammals have utilized two primary sources of data, amino acid sequences (Miyamoto and Goodman, 1986; Shoshani, 1986) and immunological/serological distances (Kirsch, 1977; Cronin and Sarich, 1980; Sarich, 1985). Immunological distance data, derived primarily from

RONALD M. ADKINS and RODNEY L. HONEYCUTT • Department of Wildlife and Fisheries, Texas A&M University, College Station, Texas 77843.
Primates and Their Relatives in Phylogenetic Perspective, edited by Ross D.E. MacPhee. Plenum Press, New York, 1993.

albumin and transferrins, have proven useful in deriving phylogenetic hypotheses for mammals, especially those pertaining to primates and their relatives. In many cases the apparent information content of trees derived from immunological distances has been extremely good (Prager and Wilson, in press). Immunological distance data, however, have been criticized because they do not provide information in the form of discrete character states and there are no mechanisms for adjudicating homology. In addition, immunological data from one study usually cannot be incorporated directly into another done separately or under different conditions, and concern has been shown for the analytical difficulties associated with immunological data being nonmetric (Farris, 1983).

Amino acid sequence data circumvent many of the criticisms directed toward immunological distance data, and the bulk of recent (within the past 10 years) molecular systematic studies of eutherian mammals have consisted of phylogenies derived from amino acid sequences of globins, α-crystallin, and pancreatic ribonuclease (Beintema and Lenstra, 1982; Beintema, 1985; deJong, 1985; Goodman et al., 1985; Miyamoto and Goodman, 1986; Shoshani, 1986). In many cases, relationships suggested by morphology have been supported by amino acid sequence data, yet in still other cases incongruence has been observed (Tagle et al., 1986). For instance, the latest synthesis of amino acid sequence data revealed a phylogeny of Eutheria that supported some common concepts of mammalian interordinal relationships, but the overall topology contradicted at least a few conclusions supported by nearly every other study of eutherian relationships (Miyamoto and Goodman, 1986). This critique of amino acid sequence data does not mean that such information is not useful for determining mammalian relationships. It simply means that amino acid sequence data have their limitations, and these limitations are probably related to the small number of potential synapomorphies supporting particular clades, selective constraints on how proteins evolve, and artifacts that might result from augmentation procedures whereby hypothetical mRNA sequences are used to construct minimal-length networks between potential codon sequences (Mayr, 1986; Wyss et al., 1987; Allard, 1990).

Over the past 5 years the primary emphasis in mammalian molecular systematics has been a phylogeny reconstruction using nucleotide sequence data from both mitochondrial and nuclear DNA. Mitochondrial DNA (mtDNA) sequence variation has proven useful in several systematic studies of primates (Brown et al., 1982; Hixon and Brown, 1986; Hayasaka et al., 1988; Ruvolo et al., 1991; Disotell et al., 1992). In addition, mtDNA nucleotide sequences have been used successfully to address hypotheses pertaining to the higher-level relationships of eutherian mammals (Miyamoto and Boyle, 1989; Adkins and Honeycutt, 1991; Mindell et al., 1991; Kraus and Miyamoto, 1991; Irwin et al., 1991; Allard et al., 1991b; Allard and Honeycutt, 1991). By far the most detailed comparative studies using nuclear DNA sequences have focused on the relationships among primate lineages (Hayasaka et al., 1988; Koop et al., 1989). With few exceptions (Bailey et al., 1992), most other studies of mammalian relationships using nuclear DNA sequences have been less robust than the primate comparisons (Easteal, 1988, 1990; Gouy and Li, 1989; Li *et al.*, 1990; Graur *et al.*, 1991).

The increased emphasis on nucleotide sequence data over immunological distances and amino acid sequences stems, in part, from the fact that with the advent of the polymerase chain reaction it has become much easier to obtain nucleotide sequence data from large numbers of diverse taxa. Aside from the technical reasons as to why nucleotide sequences are being used in systematic studies, these data have several advantages over protein sequence data. First, nucleotide sequence data allow for more detailed examination of the process of gene evolution (Li *et al.*, 1985). At the DNA level the extent of sequence divergence can be evaluated not only in coding regions (both RNA and protein) but also in noncoding regions, such as introns, nontranscribed regions, untranslated regions, and pseudogenes. Second, in genes encoding proteins, sequence divergence can be partitioned into different types of substitutions such as synonymous (not inducing an amino acid replacement) versus nonsynonymous (inducing an amino acid replacement) substitutions, transition (purine to purine or pyrimidine to pyrimidine) versus transversion (purine to pyrimidine or vice versa) substitutions, and substitutions at the first, second, and third position of a codon. Third, nucleotide sequences provide information on thousands of independently evolving characters, and these sequences coupled with the fact that different regions of the genome vary in rate of evolution provide considerably more characters than either morphology or amino acid sequences.

This chapter represents a detailed appraisal of nucleotide sequence variation in a mitochondrial gene, cytochrome c oxidase subunit II (COII). The COII gene has promise in resolving phylogenetic relationships not only within Primates but also among several orders of mammals (Ruvolo *et al.*, 1991; Adkins and Honeycutt, 1991; Disotell *et al.*, 1992). This gene codes for a protein, and changes occur by single base substitutions or to a lesser extent by the insertion/deletion of entire codons at the 3′ end of the gene. In a protein-coding gene, single base changes offer more direct ways for evaluating issues such as differences in the rate of change relative to codon position and the frequency and information content associated with transitional versus transversional differences.

We address three specific topics in this chapter. First, we evaluate relationships among the major lineages of primates and compare those results with other molecular data on nucleotide sequence variation. Second, we evaluate the monophyly of the superorder Archonta. Third, we address the recent controversy concerning the diphyletic origin of bats.

Methods and Materials

Molecular Techniques

mtDNA was isolated by CsCl gradient centrifugation using the technique of Honeycutt *et al.* (1987). The COII gene was amplified by the polymerase chain reaction (Saiki *et al.*, 1988) using Taq polymerase (Perkin–Elmer Cetus) and the reaction profiles and primers published previously (Adkins and Honeycutt, 1991). Sequences were determined by the dideoxynucleotide chain termination method (Sanger *et al.*, 1977). For sequencing, either single-stranded DNA was produced via the polymerase chain reaction (Allard *et al.*, 1991a) or double-stranded amplifications were ligated into the pCR™ plasmid (Invitrogen Corporation) and sequenced via the technique of Kraft *et al.* (1988). Because Taq polymerase has an inherent error rate (Saiki *et al.*, 1988; Tindall and Kunkel, 1988; Keohavong and Thilly, 1989), at least two separate clones were sequenced for each taxon. When sequence discrepancies were found between two clones, a third clone was sequenced and the base present in two of the clones was accepted as the correct base. Alignment was done by eye, and no insertions/deletions were observed within the coding region.

Analytical Approaches

Hypotheses of relationships were based on the identification of synapomorphies as defined by Hennig (1966), and the topology requiring the least number of changes was accepted as the "best" topology. All parsimony analyses were done using the PAUP 3.ON program (Swofford, 1990) or Hennig86 (Farris, 1988). The data were analyzed at two different levels. As an initial approach, all substitutions were considered when constructing a most-parsimonious tree. Alternatively, transversions alone were used to diagnose relationships. The rationale for this two-tiered approach is based on the high transition/transversion ratios (Brown *et al.*, 1982; Miyamoto and Boyle, 1989) commonly observed for mtDNA sequences which indicate the relatively more conservative nature of transversions.

Specimens Examined

We utilized previously published sequences of *Homo sapiens* (Anderson *et al.*, 1981), *Bos taurus* (Anderson *et al.*, 1982), *Mus* sp. (unspecified species; Bibb *et al.*, 1981), *Rattus norvegicus* (Gadaleta *et al.*, 1989), *Pan paniscus, Gorilla gorilla, Hylobates syndactylus, Macaca fascicularis, M. mulatta* (Ruvolo *et al.*, 1991), *Dasypus novemcinctus, Galago senegalensis, Cynocephalus variegatus, Tupaia glis, Phyllostomus hastatus, Rousettus leschenaulti* (Adkins and Honeycutt, 1991), and *Capra hircus* (Janecek *et al.*, in press). New sequences were obtained from *Lagothrix lagotricha, Tarsius bancanus, Lemur catta, Elephantulus rufescens, Macrotus californicus, Rhinolophus darlingi,* and *Georychus capensis.*

Results and Discussion

Relationships within Primates

An analysis of sequence variation in the COII gene was performed using data from *Homo, Gorilla, Pan, Hylobates, Macaca fascicularis, M. mulatta, Galago,*

Tarsius, Lagothrix, and *Lemur* (Fig. 1). *Bos* was utilized as the outgroup. In addition, we performed analyses on the nearly 0.9 kb of mitochondrial sequence representing three tRNAs and portions of NADH-dehydrogenase subunits 4 and 5 (Brown *et al.,* 1982; Hayasaka *et al.,* 1988). These published sequences were realigned with *Bos,* and the new alignment was found to be nearly identical to the previously published one. The principal difference between our COII gene data set and that of Hayasaka *et al.* (1988) is that the latter authors collected data from *Saimiri sciureus* rather than from *Lagothrix.*

The phylogenetic analysis of transversions alone in both the COII gene data and the 0.9 kb previously reported produced very similar results (Fig. 2). On the basis of both data sets, *Tarsius* is placed sister to all other primates. According to one popular conception of primate intraordinal relationships, tarsiers and Anthropoidea are sister groups (Pocock, 1918; MacPhee and Cartmill, 1986) within the suborder Haplorhini. This idea suggests that *Tarsius* is phylogenetically closer to simians than are Lorisoidea or Lemuroidea. We compared the phylogeny in Fig. 2 with the popular view using the criteria of Templeton (1983) [as implemented in DNApars of PHYLIP 3.3 (Felsenstein, 1990)], and we found both hypotheses to have equal support under these criteria. From a parsimony standpoint, the placement of the tarsier as part of the Haplorhini to the exclusion of the prosimians adds five steps to the COII tree and six to the tree based on the data of Hayasaka *et al.* (1988).

For the COII data set, the analysis based on all substitutions differs from one based on transversions alone (Fig. 2) by placing *Hylobates* sister to a hominoid-macaque clade. However, that tree differs from one placing *Hylobates* sister to *Homo/Pan/Gorilla* by only one step. Interestingly, when *Lagothrix* is removed, the analysis of all substitutions results in the placement of *Hylobates* sister to the hominoids. One possible explanation for the apparent problem with *Lagothrix* may be that New World monkeys are chronologically rather divergent primates and that such a long undivided branch could have a destabilizing effect on the analyses because of numerous homoplastic transitions. Perhaps the inclusion of more than one lineage from this diverse group would stabilize the results (Felsenstein, 1978; Swofford and Olsen, 1990).

The placement of *Tarsius* is rather interesting considering the controversy surrounding its phylogenetic affinities. The results are even more curious if one inserts published sequences of *Mus* and *Rattus* into both mitochondrial data sets: the rodents are placed between *Tarsius* and the remaining primates. If *Tarsius* is forced to lie with the other primates, an analysis of transversions alone in the COII data set produces two trees that are two steps longer than the shortest tree (394 steps) and in which *Tarsius* is sister either to all other primates or to the strepsirhines. If the results are constrained to contain a haplorhine clade, a tree five steps longer than the shortest one results that places *Tarsius* sister to the other haplorhines.

Although a large number of traits (Martin, 1990) can be quoted in support of Haplorhini, the highly autapomorphic nature of tarsiers (MacPhee and Cartmill, 1986) has always complicated resolution of their affinities. We cannot comment in detail on the morphological data, but we find interesting

```
DASYPUS       ATGGCCTTACCCATTCAACTAGGATTCCAAGACGCAACATCACCAATCAT
HOMO          .....ACATG.AGCG...G....TC.A........T..T..C..T.....
PAN           .....ACATG.AGCG...G....TC.A........T..T..C..T.....
GORILLA       .....ACATG.AGCG...G....TC.A........T..T..T..T.....
HYLOBATES     .....ACATG.AGC....A....CC.A........T.....C..T.....
M. FASC.      ......CAC..AG........A.CC.G........C....C..C.....
M. MULATTA    ......CAC..AG........A.CC.G........C.....T..TG.T..
LAGOTHRIX     ......CAT..AGCC........T..A...A....T.....C........
TARSIUS       .....ACATT..T.........T..T.....T..T..C..C..T..T..
LEMUR         .......AT..AG.....T.......T.....T..TG.T..T..C..T..
GALAGO        .....TCACG.AG.A...TAC..C..T..........G..G.C..T..T..
CYNOCEPHALUS  .....ACAC..AC.A..GT.....C.T..G..T..TT.C..C..C.....
TUPAIA        .....A.AC..GC.A..G.................TT........T..
ELEPHANTULUS  .....A.AC..AT.........C..T........TT....C..C..T..
PHYLLOSTOMUS  .....A......T.C..T..........T.....C..C.....
MACROTUS      .....A.AT..AT.C...........T.........C..G.....T..
RHINOLOPHUS   .....GCAC..TT.........C.....G..T..C.........C.....
ROUSETTUS     .....A.AT...T.C..........C.........C..T.....
BOS           .....A.AT.....A.................T.....
CAPRA         .....A.AC...A........T..T.................C..T..
GEORYCHUS     .....A.AC..TCAC.....T..C..T.....T..C.....T.......
MUS           .......AC..AT.C.....T..TC.A........C.....C..T..T..
RATTUS        .....T.AC..AT.......T..C..A........C........T.....

DASYPUS       AGAAGAACTCCTACATTTCCATGACCACACACTAATAATTGTGTTCCTAA
HOMO          ......G..TA.CACC..T.....T...G.C..C.....CA.T.....T.
PAN           .........TA.TATC..T........TG.C.......A.C..T..C
GORILLA       .........AA.CACC..T.....T..TG.T..C.....CA.T.....C.
HYLOBATES     .........AA.CTC......C.....TG.C..C.....CA.T.....T.
M. FASC.      ......GT.AA.TACC.....C.....TG.TT.C...GCCA.A.C.....
M. MULATTA    ...G..GT.GA.TAC........TG.TT.T...GCCA.A.CT..T.
LAGOTHRIX     ...G.....TA.CGC...........TG.C..T......A.T........
TARSIUS       .......T.A..T..C..T.........T.........C...T..
LEMUR         .........TT..T.C..T..C.........TT........A.A.....G.
GALAGO        .......T.AT..T.C..........T..T.....C..A...A...
CYNOCEPHALUS  .......A...T.C..T........TG.C.......C..A...T...
TUPAIA        .......T.A..T..C.....................C.......C.
ELEPHANTULUS  ........AACC..C.............TT.....C...C.ATTCT.
PHYLLOSTOMUS  .........A.........T..C.....T..C......C..TA...
MACROTUS      .........TT...C..T.........C..T.........T....
RHINOLOPHUS   .........G..G..C........T..T..C........T......
ROUSETTUS     G........T..C..C..T......T..G.........C.....T.
BOS           .........A..T..C..T..........G.........C...T...
CAPRA         .........A.........T..C..T.............T......
GEORYCHUS     .........AT..AGC..T................C..A...TT...
MUS           ......G..AA..A..........T...........T......
RATTUS        .........TAC.A.C..T..........C...........A.....C.
```

Fig. 1. Aligned nucleotide sequences of the cytochrome oxidase subunit II gene. Periods indicate a nucleotide that is identical to the base present at that position in the armadillo, *Dasypus novemcinctus*. M. Fasc., *Macaca fascicularis;* M. mulatta, *Macaca mulatta.* The 12 additional nucleotides at the 3′ end of the sequence for *Lagothrix* were not used in phylogenetic analyses.

```
DASYPUS        TTAGCTCTTTAGTCCTTTACATTATTACCCTGATATTAACAACAAAACTA
HOMO           .CT...TCC.......G..TGCCC..TT...A.C.C.C............
PAN            .CT...TCC.......A...GCCC..TT...A.C.C.C............
GORILLA        .CT...TC.......G...GCCC..TT...A.C.C.C........G....
HYLOBATES      .C....TCC.......A.TGCCC.CTT...A.C.C.C............
M. FASC.       .C....TCC....G..A...GC.C.GCT.TCA.C.C.C...........
M. MULATTA     .C....TCC....AT.A..TGCCC.GCT.TCA.CGC.C...........
LAGOTHRIX      .C.....AC....AT.A..T.....CT........C....T......T..
TARSIUS        .......AC....T.....T..........A...C.....C...T..
LEMUR          ....T...C............CT....T...C.C..T..TG....T
GALAGO         .C.....AC....A..C..T.....T....T...C..T....TG....C
CYNOCEPHALUS   .......CC........C....C....T....A..G.....C..C..G..T
TUPAIA         ......GC...G..A......CT.T..C..G.....C..G......
ELEPHANTULUS   .C........G.....C..TG.A..CT.AGCT........C..C...
PHYLLOSTOMUS   .C........G.....A...T.A..CT..TCT...C.......CGC...
MACROTUS       .......CC....AT.G..TG....CT.AGCT...........TT.G
RHINOLOPHUS    .......C.....G..G..T.....CT....TC..C.....C..T.GC..T
ROUSETTUS      .......C.T..G.....TC.A..CT.AG.T..GC........GC..G
BOS            .......A.....A........T.A..A...C....G.....G..G
CAPRA          ......AC.G..A....T.....T.A..A.................
GEORYCHUS      ......AC....A..A..T.A...T.A..T...........C...T..
MUS            .......C........C..T..C..CT.G..A..............
RATTUS         .C.....CC....A....T......T.A..A.......C..........

DASYPUS        ACCCATACAAGCACAATAGACGCACAAGAAGTAGAAACCGTATGAACAAT
HOMO           ..TA....T.A..TCTC......T..G...A...........C.....T..
PAN            ..TA....T..T.TTTC......C..G...A...........C.....T..
GORILLA        ..TAGC..C.A..TCTC......C......A.....G...A.C.....C..
HYLOBATES      ...A.C..T.AT.TT.CG.....C.....GA...........C.......
M. FASC.       ...A....T.A..TC.C......C......A...........A.C.....T..
M. MULATTA     ...A.C..T....TC.C......C......A..........TA.C.....T..
LAGOTHRIX      .....C..C..T..C...A.T..T......A.T....TA..C......C..
TARSIUS        ..T........T..........C.............T.....G..C..
LEMUR          .TA.....A...C.........C..........A..........
GALAGO         .....C..............T..........G..A........C..
CYNOCEPHALUS   .TG..C.......T...A.T..C...AT.....G...A......C..
TUPAIA         ..T.....G....T........C..G..G..C..G..AA.T......
ELEPHANTULUS   .....C..T.....T........T......G..AA.C......C..
PHYLLOSTOMUS   .....C......T........T..C...........AA.C......
MACROTUS       ..G........T..T........C.........AA.T......
RHINOLOPHUS    ..A..C..T..T................G....G..AA.C.....T..
ROUSETTUS      .....C..C..................G........A.C.....T..
BOS            ...............G.....T..........G..AA.C.....C..
CAPRA          .....C..C.....C...............G..C......C..
GEORYCHUS      ..A..C.......T.....T..........A...........
MUS            ..A..................T.......T......A.T.....T..
RATTUS         ..A..C..............C...........AA.T.......
```

Fig. 1. Continued.

the paper by Schwartz and Tattersall (1987) in which some of these data are interpreted as convergences brought about by functional necessity and other characters are put forward to support a prosimian clade containing tarsiers (see Luckett, this volume).

Karyological data have proven useful in most primatological studies, yet tarsiers have a karyotype bearing such drastic rearrangements that it is impossible to draw any phylogenetic conclusions. Nevertheless, Dutrillaux and

```
DASYPUS        CTTGCCCGCAGTTATCCTAATCCTAATTGCACTACCCTCACTACGAATTT
HOMO           .C.......CA.C......G....C..C..C..C..A..C.....C..CC
PAN            .C.......CA.C......G....T.....C.....A..C..G..T..CC
GORILLA        .C.A.....TA.....T..G....G..C..C..C..A..T........CC
HYLOBATES      .C.A..T..TA....T...G.T........C..C..A..C..C..C..CC
M. FASC.       ........G....................T..C..A..C....CG..CC
M. MULATTA     ...A.....A.....T............T..C..A..T.....C..CC
LAGOTHRIX      .C.A..T...A........TA.G.....C..C..A..CT....C..CC
TARSIUS        ...A..A..TA.......T..........T..C.....C....T...C
LEMUR          .C.A..T....CA.....C..T..T.....T..T..A...T....C...C
GALAGO         TC....A..T........C.....C..C..C..C..A..T..T..G..CC
CYNOCEPHALUS   ...A.....C........G.CTTC......C..C..A.....G..G...C
TUPAIA         TC.T..A..CA.....T............C.....A...G.....CC
ELEPHANTULUS   .C.C.....TA....T..C........C..C.........T.......C
PHYLLOSTOMUS   TC.T..A..CA.C.........AC......C..C..A..T.....C..C.
MACROTUS       .C.C..A...A.........TAC......C..G..T..CT....C...C
RHINOLOPHUS    .C.C..A..CA.C.....T..TA......T..T..A...........C
ROUSETTUS      TC.A..A...A.A........A........C..T..A..C..C.....C
BOS            TC......CA.C...T....T........T..T..T..TT.......C
CAPRA          ...A..A..CA....TT....TA.G.....T..C..A..TT.......C
GEORYCHUS      T..A..A..TA....T..T..TA......C..T..A..T........C.
MUS            TC.A..A..T..A.....T...A......T..C.....T.....C...C
RATTUS         TC.C..A..T..C..T..T..T..T..T.....C..T.....C.....C

DASYPUS        TATATATAATAGACGAAATCAACAACCCACTCCTCACCATTAAAGCTATA
HOMO           .T..C....C......GG.....G.T..CTC...T.....C...T.A..T
PAN            .T..C....C......GG.....G....CTC.T.T..T.....T.A..C
GORILLA        .C.....G.C...T........TG....CTC...T..T..C..T.A..C
HYLOBATES      .T...T...C...........G....CTC.T.T..T..C..G..A..C
M. FASC.       ....CC...C...T...G.....G.....TC.T.T.....C..AT.A..T
M. MULATTA     ....CC...C...T...G.....G.....TC.T...........T.A..C
LAGOTHRIX      .G..C..G.CG......T.T.....A..CTA.T.A..AC.......A..T
TARSIUS        .C..CT..........T..T.C..TC...A..TG...GA.C...
LEMUR          .G..................TGCT.CA..CTC.T.A...C....A.C...
GALAGO         .T..C........T......G.G.C...TTC.T.A...C.A...A.AG..
CYNOCEPHALUS   .T..C....C................C...T.A.....C.......G
TUPAIA         .G..C........T.....T........TCT..A..AG.....A....
ELEPHANTULUS   ....C........T...........GC.A.A..TG....A.A...
PHYLLOSTOMUS   ....C................T.....CTC.A.G..T......A.C...
MACROTUS       ................G..T..T..T..CGCTA.A.........A.C...
RHINOLOPHUS    ....C......T.G..T.......GTC..A...G..GA.C...
ROUSETTUS      .C..C......T.....T..T..T...TAT..A..G.A...A.....
BOS            ....C........T.......T......TCT..T..AG.A...A.C...
CAPRA          ....C..........G...........TC......AG.A...A....G
GEORYCHUS      ..........................T..A..T..AG.A...A.C..G
MUS            ........................CG.AT.A...G.....A.C...
RATTUS         ....C..............G..T..T......G.T..A..AG.A...A.....
```

Fig. 1. Continued.

Rumpler (1988) favored the hypotheses that either tarsiers did not share a common ancestry with strepsirhines and simians, or they branched from these groups after a relatively short time. There is some fossil evidence that may support a short common ancestry of tarsiers and other primates. If Beard *et al.* (1991) are correct in thinking that *Tarsius* is the closest living relative of the Eocene omomyid *Shoshonius*, any common ancestry between tarsiers and other

```
DASYPUS        GGTCACCAGTGATACTGAAGCTATGAATACACAGATTATGAAGACCTAAA
HOMO           ..C.....A..G.......C...C..G.....C..C..C.GC.GA....T
PAN            ..C..T..A....T...C...C........C..C..C.GC.GG....T
GORILLA        .......A........C...C........C..C....GT.GAT.G.T
HYLOBATES      ..C..T...........GC...C.....T.....C..C.GT.GG....T
M. FASC.       ..A.......G.....C...C...........C..C.G..G.....T
M. MULATTA     ..A...........C...C..........C.GG.G.....T
LAGOTHRIX      ..C.....A........C...TT....C....T....T..GC
TARSIUS        ..A...........T.............T..G..C.....G........
LEMUR          .......A........C...................T..TG
GALAGO         .......A........C..G........C..C..CA.G...TG
CYNOCEPHALUS   .......A...............T.....C..ACG..T.....
TUPAIA         ..C...........T.....C..G..T....C.......G....C
ELEPHANTULUS   ..C..T..A....T.....T.....G..T..T..C..CAC...T..C.G
PHYLLOSTOMUS   ..C.......A................T..C...TC...A...TG
MACROTUS       ..C.....A...........T.........G..C..CC.T....T..TC
RHINOLOPHUS    ..A.....A................C..C...........G
ROUSETTUS      .......A........C.....T.....C........A.G.G
BOS            ..A..T...............G..T............G...T...G
CAPRA          ..A..T..A................T.....C........T...G
GEORYCHUS      ..G..T............C...C....T....C..C.....AT....
MUS            ..G.....A........C....T..T..C..........TG
RATTUS         ..A.....A................T..T..C..........TG

DASYPUS        TTTTGACTCCTATATAGTCCCAACATCAGACCTCAAACCAGGAGAACTAC
HOMO           C..CA.......C...C.T..CC...T.TT...AG.......C..C..G.
PAN            C..CA.......C...C....CC...T.TTT..AG.......T..T....
GORILLA        C...A.......C....CC...T.TT...AG.......T..C....
HYLOBATES      C..CA.T..T..C..C.A...C...T.TTT..AG....G..G..C..T.
M. FASC.       ...CA.T..A..C...C.A..CC.GCTGTT.T.A..C.........C..T.
M. MULATTA     ...CA.......C...C.A..CC..CT.TT...A..C.........C..C.
LAGOTHRIX      C..C....AA....TACA..C....ACTTT..TG.......T...T.T.
TARSIUS        ...........C..A.T..T...G....T...............T.
LEMUR          .........G..C...ACT..TT.C.......T....T.......T.
GALAGO         ...C....T..C...AC...C...C.T...T.AG.........C..T.
CYNOCEPHALUS   C..C.....G..C...AC...T..CCTC.....A...........C.
TUPAIA         C..C.....A.....GA.T..C..TCTC.....A..G..T.....CG.C.
ELEPHANTULUS   C..C.....A.....A....T..CAAT..A..TCC........G...G.
PHYLLOSTOMUS   C.....T..A..C...A....T..G..C...T.A...T......G...C.
MACROTUS       .........C.....A..C.....C..T.A..G....G..G..T.
RHINOLOPHUS    ...C.....A.....A....T.....C....A..G......C....C.
ROUSETTUS      C........T..C..............CA.....A.........T...
BOS            C..C........C...A.T..........AT.A..G.....G..G....
CAPRA          C..C..T........A.T..........AT.A.....T.......
GEORYCHUS      C.....T.............T......A.G..T..T..C.......T...
MUS            C.....T..A......A.........AAC.....A.....T..T.....
RATTUS         C..........C...A.......CAAT.....A........T.....T.
```

Fig. 1. Continued.

extant primates must have occurred earlier. This means that either the primate lineage is much older than previously thought, or else that tarsiers did indeed share only a brief common path of evolution with other primates.

Much of the molecular data are less definitive than might be supposed, but tend to support the unity of the Haplorhini. Using immunodiffusion, Dene *et al.* (1976) found *Tarsius* to belong at the base of the haplorhine clade,

```
DASYPUS        GACTATTAGAAGTCGATAATCGACTTGTTCTACCAATAGAGCTATCAATT
HOMO           ....CC.T..C..T..C......G.A..A..C..G..T..AGCCC.C...
PAN            ....CC.T..C..T.....C......G..C..C...G.T..AGCCC.CG..
GORILLA        ....CC.T..C.....C..C...G.A..C..T...G.T..AGCCC.CG.C
HYLOBATES      ....CC.T........C..C...G.A....T....T..AGCCC.TG.C
M. FASC.       ....CC.......T..C......G.A..C..T....T..AGCC..CG.A
M. MULATTA     ....CC.......T..C......G.A....T....T..AGCTC.TG.A
LAGOTHRIX      ....CC.T.....A..C......ACAAC...T........AGC.GAT...
TARSIUS        .....C.......T..C.....TG.A.....C..T.....AT..C.....
LEMUR          .C...C.T........C......G.C..A..T..C.C..A...G..G..
GALAGO         ...........A..C..C...G.C..CT....C.C...AA....C..C
CYNOCEPHALUS   ....T..G.....T.....C...ACC..A..C......AACCC.G..C
TUPAIA         ....CC.T.....A......G.G..C..T..C.....AA.CC..G..
ELEPHANTULUS   .T...C.......T..C.........CC.CT....G.....AG.CC.....
PHYLLOSTOMUS   .C..........G.....G.A...A.C..T.....AA..A.GG.C
MACROTUS       .TT.GC.......A.....C...G.G..C..T..TG....AA..A.....
RHINOLOPHUS    .T...C.....A..C....G.G..A.....C.....AA.GA.T..C
ROUSETTUS      .T...C.G.....G..C....G.A..AT.........AT..A.T..C
BOS            .................G...A........AA..A...C
CAPRA          ....GC....G..A.....C...G....A.....C.....AA.GA.....
GEORYCHUS      .....C.T.....A......G.C...T.........AAC.C.T...
MUS            ....GC.......T.....C...G.C.....G........A..TC....C
RATTUS         .T...........T.......GG.A..CT..........A..TC.....

DASYPUS        CGTATACTAATTTCATCCGAAGATGTACTACACTCATGAGCCGTACCATC
HOMO           ......A.....A....AC....C..CT.G..........T..C..CA.
PAN            ......A.....A....AC....T...............T..T..CA.
GORILLA        ......A.....A....C......CT.............T..T..CA.
HYLOBATES      ......A.....A....AC....C..C...........A.T..T..C..
M. FASC.       ......A.....A....C......C..C...........A.TA.T..CA.
M. MULATTA     ......A.....A....TC....C..CT...........A..A.C..CA.
LAGOTHRIX      ......T.....C......C..C.C.............T..T.....
TARSIUS        ..A...T....C......A.......T..T..T.......T......C..
LEMUR          ..A...T....C..T..A...........A.T..C..T..
GALAGO         ......T.....A.C..A...........T.........A.A..G...G.
CYNOCEPHALUS   ..C...T....C.........C...........C.....T..G..C..
TUPAIA         ..A..G..........A.........G..C.........T..C.....
ELEPHANTULUS   ..A.....C..C.....A...........C.C.......TA.C...G.
PHYLLOSTOMUS   ..C.....T....T..........C....T...........T.....
MACROTUS       ..A..G.....C.....G.........T..T..T.......C..C..
RHINOLOPHUS    ..........C.........C..C..G.....G........T..
ROUSETTUS      ..A.......C.....T....C...............C..C..
BOS            ..A..GT..G.C..C..T.....C...T...........T..G..C..
CAPRA          ..A...T....C..T.......C..T...........A..T..C..
GEORYCHUS      ..A.......C.....T.....C..CT..........A.A..T..T..
MUS            ......T...........T....C..C..C...........A..C..C..
RATTUS         ..........C...........C..C..G...........A.C..T..
```

Fig. 1. Continued.

but Sarich and Cronin (1976), using microcomplement fixation, found no special affinity for *Tarsius* and placed it as part of a polytomy with prosimians, simians, tree shrews, and flying lemurs. Similarly, deJong and Goodman (1988) utilizing amino acid sequence data of the α-crystallin protein concluded that *Tarsius* held a position at the base of the haplorhine radiation. However, this placement required the assumption that primates are mono-

```
DASYPUS        CTTAGGACTAAAAACAGACGCAATCCCAGGTCGCCTAAACCAAGCCACAT
HOMO           A.....CT..........T.....T..C..A..T.........A....T.
PAN            A.....C...................T..C..A..T.........A....T.
GORILLA        A.....C...................C..A.....G.....A......
HYLOBATES      .C....T..T.....G..T..T.....C..A...........A......
M. FASC.       AC....C....AG.....T..CG.A..CGGA..T.....T...A.CGTT.
M. MULATTA     AC...C...........CG.A..T..A..T.....T...A..GTT.
LAGOTHRIX      AC.C..TG.C........T.......C..A...........T..
TARSIUS        TC....CT.......T..T..C.....C..A..T.....T.....T..TC
LEMUR          .C.....G............T.......A............T....
GALAGO         AC....TA.T........T..T........C..........A..CC
CYNOCEPHALUS   AC....T.....G..T........A..T..G.......A.A..CC
TUPAIA         ......C...................G..A..........A..CC
ELEPHANTULUS   AC.C..CA.G.....T.....C.......C..A.....T.......TC
PHYLLOSTOMUS   .C.....G...................A..T......A.A..CC
MACROTUS       .C....CT....G.....T..T.....C.....G.........A.A..CC
RHINOLOPHUS    .C.G..C...........T..T........A..T..G......A.A..CC
ROUSETTUS      A..G..C.............T........A..T..C..T...A.A..CC
BOS            TC......................C..T........A.A..CC
CAPRA          TC.....T.................T........TT....T..A.A..CC
GEORYCHUS      AA...........T..T.....T.....A..A.....T.....A..TC
MUS            .C......T.....T..T..C......C..A.....T.....A...G
RATTUS         AC....GT.......C..........C..C...........T...G

DASYPUS        TAATAGCAACACGACCTGGCCTATATTACGGTCAATGCTCAGAAATCTGT
HOMO           .C.CC..T........G..GG....C.................T.........
PAN            .C.CC..C........A..AG....C.....C.............
GORILLA        .C.CC..C........A..AG....C.....C..G.........
HYLOBATES      .C.CC..C.......C..A..AG....C.....C.........
M. FASC.       .C.CG..C........A...G.C........A.........
M. MULATTA     .C.CG..C........A...G.C........A..........G......
LAGOTHRIX      ..GCCT.T.T..............G.T...T..A..............C
TARSIUS        .T...T....T.....A......C.....T..C.........
LEMUR          .......CT.C..T..A..TG.T..C.....A...............T..C
GALAGO         .T..GA.CT.T...GTA..AA.T..C.......C.......C..G....C
CYNOCEPHALUS   .GTC.................G.TC..T..A..G..........C
TUPAIA         .T..CT.T........A..T..G.T.....A..G....T.........
ELEPHANTULUS   ..TCCT.......C..A..T..T.TC..T..C...........
PHYLLOSTOMUS   .CC..T.C..T..G..C..T.....C.....C..G.....C.........C
MACROTUS       .TC..T...TT........T..T.....T..C..G.....C.........
RHINOLOPHUS    .CT.GT.T.AC........T..T..C.....C.............T..C
ROUSETTUS      .TT..T.T........A..T.........T.........C.....T..
BOS            .T...T.GT.C..T..A...T.....................T..C
CAPRA          .T..GT.G..T..T..A..T....TC.....C.............C
GEORYCHUS      ...C.T.TT....C..A..TT...T......A...........T..C
MUS            ...C.T...AC.....A..GT...TC..T..C.......T.....T...
RATTUS         .C.C.T...AC.....A..T....TC..T..C.........T.....T..C
```

Fig. 1. Continued.

phyletic. The sequence for *Tarsius* was actually most similar to those of the guinea pig and springhare, differing only by a substitution at position 146—the only position providing a putative synapomorphy shared with other primates. Hemoglobin amino acid sequences have provided more positive information. Based on the amino acid sequences of the alpha- and beta-globin proteins, Beard and Goodman (1976) inferred seven nucleotide replacements to support a branch containing *Tarsius* and the simians.

```
DASYPUS       GGGTCCAACCACAGCTTCATACCTATTGTACTTGAACTAGTCCCCCTAAA
HOMO          ..AG.A........T.....G..C..C..C..A...T..A.T........
PAN           ..AG.A........T..T.....C..C..C..A...T..A....T.....
GORILLA       ..AG.........T..T..G..C.....T..A..G...A..........
HYLOBATES     ...G.......T.....T.....A.....T..A.....A.T...T....
M. FASC.      ..AG.A..T............A.....CGCA......A....A.....
M. MULATTA    ..CG.A..T..T.........A.....TGCA......A...A.......
LAGOTHRIX     .....A........T.....C..C..C..A...T.TA..TA.T.CC.
TARSIUS       .....T..T..T........G..C..C..C..A.......A..A.....
LEMUR         ..TG.A..T.....T..T.....A.............A..T.....
GALAGO        ..CG....T......A...G.....C...T.A...T......T...
CYNOCEPHALUS  ..C..A..T..T..........C..C...T.G.....C..T..A..T..
TUPAIA        ..C..T..T...........G..A..C..........T..A.....
ELEPHANTULUS  ..C..T..............A.....T..A...A.C...A.....
PHYLLOSTOMUS  ..C.................C..C...A....TA.....
MACROTUS      ..C..A.....T..T..T..............G.G..A......G.
RHINOLOPHUS   ..A..A..T...........G.....C..C..........T..A..T..
ROUSETTUS     ..A.........T..T.......A.....C..C.........AT....
BOS           .....A........T.......C.....C....GT....A....
CAPRA         ..A..A.....T..T.......A..C..T..C..G.....T........
GEORYCHUS     .......T.....T.......A.....A.....A....T..A....
MUS           ..A..T.....T.....T..G..C.....C..A...A.G..T..A.....
RATTUS        ..C..A..T.............C.........A...A....G..T.....

DASYPUS       ACACTTCGAAGACTGATCTGCCTCAATACTTTAA
HOMO          .AT...T...ATAG.GC.C.TA.TT.CC..A..G
PAN           .AT...T...ATAG..C.C.TA.TC.CT..A..G
GORILLA       .AT...T...ATAG.GC.C.TA.TCGCC..A...
HYLOBATES     .AT......GATAG.GC...TA.TT.CC..A...
M. FASC.      .AT...T...ATAG..C...TA.TT.CT..G...
M. MULATTA    .AT...T...ATGG..C...TACTT.CT..G...
LAGOTHRIX     .G.........TA...G.CT.A.ACT..TACATCGTATCACTGTAA
TARSIUS       ...T......A.T......A.A.....A.....
LEMUR         ...T......G...CTATTA..C.....A...
GALAGO        .T.T..T.....G...CTCCTTAA..CC.....
CYNOCEPHALUS  ...T......AC......AA..CT..C.TCA...
TUPAIA        G.....T...A.T...A.AA.AA.......C...
ELEPHANTULUS  .T.T.....CA........T..A.CC..TAA..T
PHYLLOSTOMUS  CTGT.....GA.A......A.A..T.....G...
MACROTUS      .T.T......A.A.......TA..T...T.A...
RHINOLOPHUS   ......T...A.A.....AT........G.A...
ROUSETTUS     .T.T......A.A.....CT.A..C.....G...
BOS           GT....T...A.A........G......T.A...
CAPRA         .T.T..T...A.A.......A........A...
GEORYCHUS     .TCG......A.T...A.CA.A..T...ACA...
MUS           .T.T......A..........T......A.....
RATTUS        .T.T......A.......A..T..T...A.....
```

Fig. 1. Continued.

By far the most compelling case for the monophyly of Haplorhini, based on molecular data, is that of Koop *et al.* (1989). For each of the major prosimian and simian primate lineages, they sequenced several kilobases from the beta-globin gene family, including epsilon, gamma, nu, delta, and beta. They found that *Tarsius* consistently grouped at the base of the haplorhine radiation. Analyzing each locus separately, they found most-parsimonious solu-

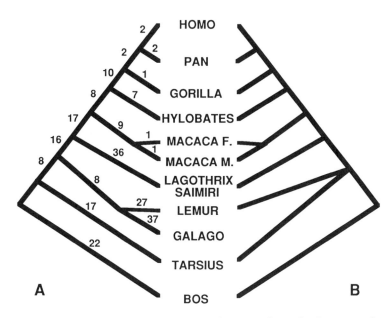

Fig. 2.   Phylogenetic hypotheses derived from parsimony analyses of only transversions in the nucleotide sequence of (A) the COII gene and (B) the 0.9-kb fragment of Brown *et al.* (1982) and Hayasaka *et al.* (1988). Macaca M., *Macaca mulatta;* Macaca F., *Macaca fascicularis.* The tree based on COII gene sequences was 322 steps long and had a consistency index of 0.528, excluding autapomorphies. Minimum possible branch lengths are shown on the tree based on COII gene sequence.

tions ranging from 911 to 1779 steps. To join *Tarsius* to the stem of the prosimians required at least 28 extra steps, and to place *Tarsius* at the base of Primates required at least 35 more steps. Further, the results of Koop *et al.* (1989) are corroborated by the DNA–DNA hybridization data of Bonner *et al.* (1980) who also found tarsiers to group more closely with simians than did lemurs or lorises. In fact, the same relative distances between the four major lineages corresponded remarkably well between the two studies.

Archontan and Chiropteran Monophyly

The superorder Archonta contains the orders Chiroptera (bats), Scandentia (tree shrews), Dermoptera (flying lemurs), and Primates, and there are several recent debates relative to the monophyly of both the superorder Archonta (Adkins and Honeycutt, 1991; Bailey *et al.*, 1992) and the order Chiroptera (Pettigrew, 1986; Pettigrew *et al.*, 1989). To evaluate both of these controversies, we examined variation in the COII gene in the following taxa: (1) Dermoptera (*Cynocephalus*), (2) Scandentia (*Tupaia*), (3) Chiroptera (Megachiroptera—*Rousettus;* Microchiroptera—*Phyllostomus, Rhinolophus, Macrotus*),

(4) Primates (Haplorhini—*Homo;* Strepsirhini—*Galago*), (5) Macroscelidea (*Elephantulus*), (6) Artiodactyla (*Bos* and *Capra*), (7) Rodentia (Sciurognathi—*Mus;* Hystricognathi—*Georychus*). We used the order Xenarthra (represented by *Dasypus*) as our outgroup because of its apparently divergent position relative to other eutherians (McKenna, 1975; Novacek and Wyss, 1986).

As with the primate data set, we separately analyzed all substitutions and transversions alone. An exact search considering all nucleotide substitutions among these 14 taxa was computationally unfeasible with our facilities. Therefore, using PAUP 3.ON we performed 1000 heuristic analyses involving random addition of taxa and the tree bisection–reconnection search strategy. Four equally parsimonious topologies resulted, of which a consensus is shown in Fig. 3A. These same topologies were produced by the combined commands mhennig*/bb to the HENNIG86 program. The analysis of transversions alone using HENNIG86 (commands ie-/bb) produced only a single most-parsimonious topology (Fig. 3B). To evaluate the phylogenetic structure (Hillis, 1991) present in the data at the levels of divergence being examined, we used PAUP 3.ON to sample randomly one million topologies and count either all substitutions or only transversions. The distribution of tree lengths when considering only transversions ($g_1 = -0.40$) was more left-skewed than when all substitutions ($g_1 = -0.25$) were counted. This indicates that there is most likely greater phylogenetic structure to the transversions than to all substitutions taken together. On this basis, the analysis of only transversions is assumed to be more reliable, and our attention will be focused mainly on these results.

A surprising result of our analyses is that Macroscelidea, Scandentia, and Dermoptera consistently form a clade. The relationships of these taxa have been extraordinarily difficult to determine. These three orders were once allied in the suborder Menotyphla within Insectivora by Haeckel (1866), principally because of their common possession of a cecum. This grouping led to the concept of "lower" insectivores (Lipotyphla) and "higher" insectivores (Menotyphla). Recently, Menotyphla has fallen into extreme disfavor (Cartmill and MacPhee, 1980; Cronin and Sarich, 1980; Luckett, 1980; Novacek, 1989; Simons *et al.*, 1991), and it is now generally felt that the similarities between tree shrews and elephant shrews in particular are the result of convergence or the retention of primitive characters. Our results indicate that there may actually be a close relationship between tree shrews, elephant shrews, and flying lemurs, and we suggest that this problem should be examined with more molecular data.

A sister-group relationship has also been proposed between Macroscelidea and Glires (Rodentia and Lagomorpha) (Novacek and Wyss, 1986). Glires is considered by some students to be the best supported hypothesis of mammalian superordinal relationships (Gregory, 1910; Hartenberger, 1985; Luckett, 1985; Novacek and Wyss, 1986) on the basis of morphological data, although even this hypothesis has its detractors (Martinez, 1985; Szalay, 1985). Unfortunately, the association of Macroscelidea with Glires is much more

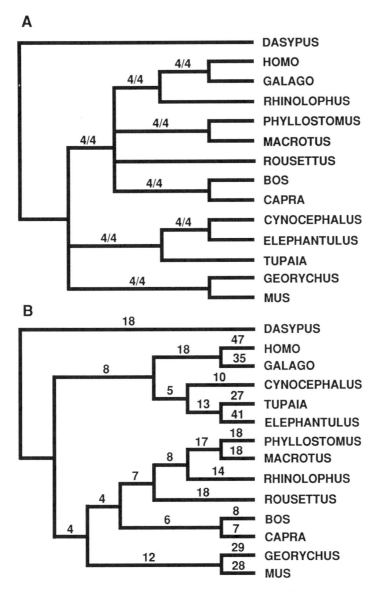

Fig. 3. Phylogenetic hypotheses derived from parsimony analysis of COII gene sequences. (A) Majority-rule consensus of four trees based on all substitutions. The length of the trees was 1310, and the consistency index was 0.424, excluding autapomorphies. The number of trees in which a particular internode was present is shown. (B) Tree based on transversions alone in the COII gene. Tree length is 553 steps, and the consistency index is 0.324, excluding autapomorphies. Minimum possible branch lengths are shown.

tentative. Of 68 characters studied by Novacek and Wyss (1986), only two synapomorphies supported a Macroscelidea–Glires clade. By contrast, many superficially similar traits of the postcranial skeleton in lagomorphs and macroscelideans have been interpreted as probable convergences (Szalay, 1985), casting some doubt on the relationship of Glires to Macroscelidea. In addition, recently discovered fossils (Simmons *et al.*, 1991) indicate that a relationship of macroscelideans with either lagomorphs or scandentians is unlikely. To impose a clade containing *Elephantulus*, *Mus* and *Georychus* on the COII gene data produces three trees that are 6 steps longer (559 steps) than the shortest one based on transversions alone, as judged by 1000 heuristic searches, using random addition of taxa (on the concept of Glires, see Stanhope *et al.*, this volume).

The relationships of Scandentia and Dermoptera are the subject of considerable debate. For a number of years the tree shrews were viewed as closely related to Primates, and perhaps representative of the primitive morphotype of Primates (Le Gros Clark, 1971). However, this idea was based on the retention of primitive skeletal traits and similarities in the trends of morphological modification in primates and tree shrews. Nevertheless, more rigorous, cladistic treatments by Wible and Covert (1987) and Novacek (1989) have supported a sister-group relationship between primates and tree shrews. Our data (Adkins and Honeycutt, 1991; this chapter) indicate that Scandentia is no more closely related to Primates than is Dermoptera, and instead place *Cynocephalus* as the sister group of a Scandentia–Macroscelidea clade (Fig. 3B). In contrast, Cronin and Sarich (1980), using microcomplement fixation, found that Dermoptera grouped more closely with Primates than with Scandentia. Beard (1990) and Kay *et al.* (1990) have described well-preserved fossils of paromomyids indicating that they are relatives of flying lemurs rather than of primates as previously believed. As a whole, these studies support a close relationship among primates, flying lemurs, and tree shrews but do not provide a clear resolution of their specific branching arrangement.

In recent years the interordinal relationship and monophyly of Chiroptera have come under increasing attention. Novacek and Wyss (1986) have suggested a sister-group relationship between Dermoptera and Chiroptera in addition to a Primates–Scandentia clade within superorder Archonta. The entire superorder is supported by the unique, derived presence of a pendulous penis and the structure of the sustentacular facet of the astragalus (although the facet is absent in microchiropterans, some of which have a sheathed penis similar to the primitive condition) (Wible and Covert, 1987). Dermoptera and Chiroptera were joined on the basis of characters principally relating to the structure of the auditory region and the modified anatomy of the forelimbs for gliding/flying. This arrangement was not supported by Cronin and Sarich (1980) who placed Chiroptera at the base of the eutherian radiation as part of a polytomy. When our analyses of the COII gene are constrained to contain the clade [[(*Homo*, *Galago*), *Tupaia*], (*Cynocephalus*, {[(*Phyllostomus*, *Macrotus*), *Rhinolophus*], *Rousettus*})] as suggested by Novacek and Wyss (1986), three trees are produced that are 20 steps longer (573 steps) than the shortest tree based on transversions alone and no constraints. Therefore, on the basis of our data, the hypothesis of Archonta is very unlikely. Chiroptera, it would appear, does not belong within Archonta.

A rather unusual hypothesis that has been put forward in the last few years is that Chiroptera is not monophyletic, with suborder Megachiroptera sister to Primates and Microchiroptera being distantly related to all orders within Archonta (Pettigrew, 1986; Pettigrew *et al.*, 1989). Pettigrew *et al.* (1989) utilized data from neural anatomy, amino acid sequences of beta-hemoglobin, and the ratio of the metacarpal length to the length of the first phalanx of the third and fourth digits (metacarpophalangeal index) to support this hypothesis. Detailed criticisms of the coding and interpretation of these data have been presented elsewhere (Baker *et al.*, 1991; Simmons *et al.*, 1991). Although the neural pathway data do provide a number of striking similarities between Megachiroptera and Primates, the hemoglobin data were rather inconclusive and the metacarpophalangeal index served only to distinguish between Megachiroptera and Microchiroptera, thus providing no information on the relationships of bats to other orders. The primary evidence against chiropteran diphyly is based on morphological data with a number of cranial and neural traits supporting chiropteran monophyly (Wible and Novacek, 1988; Thewissen and Babcock, 1991). Furthermore, molecular data from the mitochondrial genes cytochrome oxidase subunits I and II (Adkins and Honeycutt, 1991; Mindell *et al.*, 1991) and 12 S rRNA (Ammerman and Hillis, 1990) and the nuclear epsilon-globin gene (Bailey *et al.*, 1992) indicate monophyly of bats.

We expanded our original study of chiropteran monophyly (Adkins and Honeycutt, 1991) by including two additional taxa of bats, a representative of the order Macroscelidea, and an additional artiodactyl and rodent. An analysis of transversions supports the monophyly of Chiroptera but not the monophyly of Archonta (Fig. 3B). The results derived from transversions only are somewhat complicated when all substitutions are used. Each of the four resultant topologies derived from all substitutions (Fig. 3A) shows the orders Primates, Artiodactyla, and Rodentia to be monophyletic. Each topology, however, splits Chiroptera, placing *Rhinolophus* (microchiropteran family Rhinolophidae) with the primates, grouping together the phyllostomid bats (*Phyllostomus, Macrotus*), and shuffling the placement of the phyllostomids and *Rousettus* (a megachiropteran). These results are suggestive of rather old divergence times for bat lineages, including families within the Microchiroptera. The oldest unambiguous microchiropteran, *Icaronycteris index*, is at least 50 million years old (Novacek, 1987). This indicates that the period of common ancestry between the suborders, and perhaps even among some families, of bats was in the early Eocene or before. Indeed, the family Rhinolophidae is considered very divergent relative to Phyllostomidae (Hill and Smith, 1984). If the three lineages of bats in our data set are really as divergent as it appears they may be, evidence of their common ancestry may be concealed by numerous homoplasic transitions, and this may explain why weighting for more conservative transversions provides a considerably different result.

Conclusions

It is clear from the results of this study that not all systematic problems have been resolved with the use of nucleotide sequence data from one mitochondrial gene. This effort, however, has provided discrete character data that can be compared directly with other phylogenetic hypotheses produced by similar methods of analysis of character-state data, and we propose that the larger the amount of such empirical data, the more likely one is to approximate congruence in resultant phylogenetic hypotheses.

In the past, molecular systematic studies of mammalian relationships have been an afterthought, and in too many cases "bits and pieces" of molecular data on an extremely small subset of mammalian species have been used to address questions of higher-level relationships. Although this "exemplar" approach is an excellent starting point for understanding the patterns and processes of gene evolution, one must consider resultant phylogenies as first approximations or outlines awaiting the addition of more taxa and genes. This is especially true for situations like the chiropterans where one is dealing with old and divergent lineages presumably associated with short internodes. In this case there is no choice but to sample widely among the major divergent lineages of bats using a battery of both molecular and nonmolecular characters in order to achieve consistency. The placement of *Tarsius* and Macroscelidea are also difficult systematic problems in mammalian systematics as confirmed by the COII gene data as well as a host of previous morphological and molecular studies.

Our data do contradict, in several cases, more traditional hypotheses about mammalian relationships. First, the lack of a sister-group relationship between Dermoptera and Chiroptera and the failure of Archonta to form a monophyletic group are two clear cases where it would be hard to reconcile our results with recent morphological data. Second, the sister-group relationship of the old menotyphlan insectivores is not supported by most recent morphological data. In both of these cases, we suggest that more effort should be devoted to resolving these issues with additional molecular data as well as morphological analyses.

Finally, as suggested by Moritz and Hillis (1990), the conflicts between molecules and morphology have been overemphasized. This overemphasis, in part, comes from an unwillingness on both sides to consider the proposition that molecules and morphology can sometimes evolve in rather complex ways. A good example is the recent suggestion by molecular systematists that the guinea pig may not be a rodent, an assertion based on the observation of rather unusual modes of molecular evolution in certain caviomorph rodents relative to other rodent groups and mammals (Graur *et al.*, 1991). In order for this hypothesis to be correct, the large array of morphological features supporting rodent monophyly would have to be wrong (Allard *et al.*, 1991b). It is our contention that cases of such clear incongruence provide useful starting points for both sides to learn something fundamental with respect to character evolution, both molecular and morphological. In the case of the rodents, it requires the molecular systematist to accept the possibility that some lineages may not be evolving in a parsimonious fashion. Therefore, what a concerted attempt to unite data derived from both molecules and morphology offers to systematics is an objective way to examine character-state change along independently derived phylogenies. We advocate a stronger focus by molecular systematists on the many long-standing issues of mammalian evolution because mammals offer a great opportunity to evaluate the mode and tempo of molecular evolution.

ACKNOWLEDGMENTS

This research was supported by a grant from the National Science Foundation to R.L.H. (BSR-8918445) and a National Science Foundation Predoctoral Fellowship to R.M.A. Dr. Robert J. Baker of The Museum, Texas Tech University, and Dr. Duane Schlitter of the Section of Mammals, Carnegie Museum of Natural History, provided loans of frozen tissues, and Dr. Wesley M. Brown provided isolated mtDNA of *Tarsius* and *Lagothrix*. Invaluable laboratory assistance was provided by Beatriz Perez.

References

Adkins, R. M., and Honeycutt, R. L. 1991. Molecular phylogeny of the superorder Archonta. *Proc. Natl. Acad. Sci. USA* **88:**10317–10321.

Allard, M. 1990. Further comments on Goodman's maximum parsimony procedure. *Cladistics* **6:**283–289.

Allard, M. W., and Honeycutt, R. L. 1991. Nucleotide sequence variation in the mitochondrial 12S rRNA gene and the phylogeny of African mole-rats (Rodentia: Bathyergidae). *Mol. Biol. Evol.* **9:**27–40.

Allard, M. W., Ellsworth, D. L., and Honeycutt, R. L. 1991a. The production of single-stranded DNA suitable for sequencing using the polymerase chain reaction. *Biotechniques* **10:**24–26.

Allard, M. W., Miyamoto, M. M., and Honeycutt, R. L. 1991b. Are rodents polyphyletic? *Nature* **353:**610–611.

Ammerman, L. K., and Hillis, D. M. 1990. Relationships within archontan mammals based on 12S rRNA gene sequence. *Am. Zool.* **30:**50A.

Anderson, S., Bankier, A. T., Barrell, B. G., deBruijn, M. H. L., Coulson, A. R., Drouin, J., Eperon, I. C., Nierlich, D. P., Roe, B. A., Sanger, F., Schreier, P. H., Smith, A. J. H., Staden, R., and Young, I. G. 1981. Sequence and organization of the human mitochondrial genome. *Nature* **290:**457–465.

Anderson, S., deBruijn, M. H. L., Coulson, A. R., Eperon, I. C., Sanger, F., and Young, I. G. 1982. Complete sequence of bovine mitochondrial DNA: Conserved features of the mammalian mitochondrial genome. *J. Mol. Biol.* **156:**683–717.

Bailey, W. J., Slightom, J. L., and Goodman, M. 1992. Rejection of the "flying primate" hypothesis by phylogenetic evidence from the epsilon-globin gene. *Science* **256:**86–89.

Baker, R. J., Novacek, M. J., and Simmons, N. B. 1991. On the monophyly of bats. *Syst. Zool.* **40:**216–231.

Beard, J. M., and Goodman, M. 1976. The hemoglobins of *Tarsius bancanus*, in: M. Goodman and R. E. Tashian (eds.), *Molecular Anthropology*, pp. 239–256. Plenum Press, New York.

Beard, K. C. 1990. Gliding behavior and paleoecology of the alleged primate family Paromomyidae (Mammalia, Dermoptera). *Nature* **345**:340–341.

Beard, K. C., Krishtalka, L., and Stucky, R. K. 1991. First skulls of the early Eocene primate *Shoshonius cooperi* and the anthropoid–tarsier dichotomy. *Nature* **349**:64–67.

Beintema, J. J. 1985. Amino acid sequence data and evolutionary relationships among hystricognaths and other rodents, in: W. P. Luckett and J.-L. Hartenberger (eds.), *Evolutionary Relationships among Rodents: A Multidisciplinary Analysis*. Plenum Press, New York.

Beintema, J. J., and Lenstra, J. A. 1982. Evolution of mammalian pancreatic ribonuclease, in: M. Goodman (ed.), *Macromolecular sequences in Systematics and Evolutionary Biology*, pp. 43–73. Plenum Press, New York.

Bibb, M. J., VanEtten, R. A., Wright, C. T., Walberg, M. W., and Clayton, D. A. 1981. Sequence and gene organization of mouse mitochondrial DNA. *Cell* **26**:167–180.

Bonner, T. I., Heinemann, R., and Todaro, G. J. 1980. Evolution of DNA sequences has been retarded in Malagasy primates. *Nature* **286**:420–423.

Brown, W. M., Prager, E. M., Wang, A., and Wilson, A. C. 1982. Mitochondrial DNA sequences of primates: Tempo and mode of evolution. *J. Mol. Evol.* **18**:225–239.

Cartmill, M., and MacPhee, R. D. E. 1980. Tupaiid affinities: The evidence of the carotid arteries and cranial skeleton, in: W. P. Luckett (ed.), *Comparative Biology and Evolutionary Relationships of Tree Shrews*, pp. 95–132. Plenum Press, New York.

Cronin, J. E., and Sarich, V. M. 1980. Tupaiid and Archonta phylogeny: The macromolecular evidence, in: W. P. Luckett (ed.), *Comparative Biology and Evolutionary Relationships of Tree Shrews*, pp. 293–312. Plenum Press, New York.

deJong, W. W. 1985. Superordinal affinities of Rodentia studied by sequence analysis of eye lens proteins, in: W. P. Luckett and J.-L. Hartenberger (eds.), *Evolutionary Relationships among Rodents: A Multidisciplinary Analysis*, pp. 211–226. Plenum Press, New York.

deJong, W. W., and Goodman, M. 1988. Anthropoid affinities of *Tarsius* supported by lens alpha-A-crystallin sequences. *J. Hum. Evol.* **17**:575–582.

Dene, H. T., Goodman, M., and Prychodko, W. 1976. Immunodiffusion evidence on the phylogeny of primates, in: M. Goodman and R. E. Tashian (eds.), *Molecular Anthropology*, pp. 171–196. Plenum Press, New York.

Disotell, T. R., Honeycutt, R. L., and Ruvolo, M. 1992. Mitochondrial DNA phylogeny of the Old World monkey tribe Papionini. *Mol. Biol. Evol.* **9**:1–13.

Dutrillaux, B., and Rumpler, Y. 1988. Absence of chromosomal similarities between tarsiers (*Tarsius syrichta*) and other primates. *Folia Primatol.* **50**:130–133.

Easteal, S. 1988. Rate constancy of globin gene evolution in placental mammals. *Proc. Natl. Acad. Sci. USA* **85**:7622–7626.

Easteal, S. 1990. The pattern of mammalian evolution and the relative rate of molecular evolution. *Genetics* **124**:165–173.

Farris, J. S. 1983. The logical basis of phylogenetic analysis, in: N. I. Platnick and V. A. Funk (eds.), *Advances in Cladistics*, Vol. 2, pp. 7–36. Columbia University Press, New York.

Farris, J. S. 1988. *Hennig86: Program and Documentation*. State University of New York, New York.

Felsenstein, J. 1978. Cases in which parsimony and compatibility methods will be positively misleading. *Syst. Zool.* **27**:401–410.

Felsenstein, J. 1990. PHYLIP: Phylogenetic Inference Package, Version 3.3. Program and documentation. Department of Genetics, University of Washington, Seattle.

Gadaleta, G., Pepe, G., DeCandia, G., Quagliariello, C., Sbisa, E., and Saccone, C. 1989. The complete nucleotide sequence of the *Rattus norvegicus* mitochondrial genome: Cryptic signals revealed by comparative analysis between vertebrates. *J. Mol. Evol.* **28**:497–516.

Goodman, M., Romero-Herrera, A. E., Dene, H., Czelusniak, J., and Tashian, R. E. 1982. Amino acid sequence evidence on the phylogeny of primates and other eutherians, in: M. Goodman (ed.), *Macromolecular Sequences in Systematic and Evolutionary Biology*, pp. 115–191. Plenum Press, New York.

Goodman, M., Czelusniak, J., and Beeber, J. E. 1985. Phylogeny of primates and other eutherian orders: A cladistic analysis using amino acid and nucleotide sequence data. *Cladistics* **1:**171–185.

Gouy, M., and Li, W.-H. 1989. Phylogenetic analysis based on rRNA sequences supports the archaebacterial rather than the eocyte tree. *Nature* **339:**145–147.

Graur, D., Hide, W. A., and Li, W.-H. 1991. Is the guinea-pig a rodent? *Nature* **351:**649–652.

Gregory, W. K. 1910. The orders of mammals. *Bull. Am. Mus. Nat. Hist.* **27:**1–524.

Haeckel, E. 1866. *Generelle morphologie der organismen.* G. Reimer, Berlin. (Secondary citation from Luckett, 1980.)

Hartenberger, J.-L. 1985. The order Rodentia: Major questions on their evolutionary origin, relationships and suprafamilial systematics, in: W. P. Luckett and J.-L. Hartenberger (eds.), *Evolutionary Relationships among Rodents: A Multidisciplinary Analysis,* pp. 1–33. Plenum Press, New York.

Hayasaka, K., Gojobori, T., and Horai, S. 1988. Molecular phylogeny and evolution of primate mitochondrial DNA. *Mol. Biol. Evol.* **5:**626–644.

Hennig, W. 1966. *Phylogenetic Systematics.* University of Illinois Press, Urbana.

Hill, J. E., and Smith, J. D. 1984. *Bats: A Natural History.* University of Texas Press, Austin.

Hillis, D. M. 1991. Discriminating between phylogenetic signal and random noise in DNA sequences, in: M. M. Miyamoto and J. Cracraft (eds.), *Phylogenetic Analysis of DNA Sequences.* Oxford University Press, pp. 278–294. London.

Hixon, J. E., and Brown, W. M. 1986. A comparison of the small ribosomal RNA genes from the mitochondrial DNA of the great apes and humans: Sequence, structure, evolution, and phylogenetic implications. *Mol. Biol. Evol.* **3:**1–18.

Honeycutt, R. L., Edwards, S. V., Nelson, K., and Nevo, E. 1987. Mitochondrial DNA variation and the phylogeny of African mole rats (Rodentia: Bathyergidae). *Syst. Zool.* **36:**280–292.

Irwin, D. M., Kocher, T. D., and Wilson, A. C. 1991. Evolution of the cytochrome b gene of mammals. *J. Mol. Evol.* **32:**128–144.

Janecek, L., Honeycutt, R. L., and Davis, S. K. 1992. Mitochondrial gene sequences and the molecular systematics of the artiodactyl subfamily Bovinae. *Syst. Zool.* (in press).

Kay, R. F., Thorington, R. W., Jr., and Houde, P. 1990. Eocene plesiadapiform shows affinities with flying lemurs not primates. *Nature* **345:**342–344.

Keohavong, P., and Thilly, W. G. 1989. Fidelity of DNA polymerase in DNA amplification. *Proc. Natl. Acad. Sci. USA* **86:**9253–9257.

Kirsch, J. A. W. 1977. The comparative serology of Marsupialia. *Aust. J. Zool. Suppl. Ser.* **52:**1–152.

Koop, B. F., Tagle, D. A., Goodman, M., and Slightom, J. L. 1989. A molecular view of primate phylogeny and important systematic and evolutionary questions. *Mol. Biol. Evol.* **6:**580–612.

Kraft, R., Tardiff, J., Krauter, K. S., and Leinwand, L. A. 1988. Using mini-prep plasmid DNA for sequencing double stranded templates with sequenase. *Biotechniques* **6:**544–547.

Kraus, F., and Miyamoto, M. M. 1991. Rapid cladogenesis among the pecoran ruminants: Evidence from mitochondrial DNA sequences. *Syst. Zool.* **40:**117–130.

Le Gros Clark, W. E. 1971. *The Antecedents of Man,* 3rd ed. Edinburgh University Press, Edinburgh.

Li, W.-H., Luo, C.-C., and Wu, C.-I. 1985. Evolution of DNA sequences, in: R. J. McIntyre (ed.), *Molecular Evolutionary Genetics,* pp. 1–94. Plenum Press, New York.

Li, W.-H., Guoy, M., Sharp, P. M., O'Huigin, C., and Yang, Y.-W. 1990. Molecular phylogeny of Rodentia, Lagomorpha, Primates, Artiodactyla, and Carnivora and molecular clocks. *Proc. Natl. Acad. Sci. USA* **87:**6703–6707.

Luckett, W. P. 1980. The suggested evolutionary relationships and classification of tree shrews, in: W. P. Luckett (ed.), *Comparative Biology and Evolutionary Relationships of Tree Shrews,* pp. 3–31. Plenum Press, New York.

Luckett, W. P. 1985. Superordinal and intraordinal affinities of rodents: Developmental evidence from the dentition and placentation, in: W. P. Luckett and J.-L. Hartenberger (eds.), *Evolutionary Relationships among Rodents: A Multidisciplinary Analysis,* pp. 227–276. Plenum Press, New York.

McKenna, M. C. 1975. Toward a phylogenetic classification of the Mammalia, in: W. P. Luckett and F. S. Szalay (eds.), *Phylogeny of the Primates*, pp. 21–46. Plenum Press, New York.

McKenna, M. C. 1987. Molecular and morphological analysis of higher-level mammalian relationships, in C. Patterson (ed.), *Molecules and morphology in Evolution: Conflict or Compromise?* pp. 55–93. Cambridge University Press, London.

MacPhee, R. D. E., and Cartmill, M. 1986. Basicranial structures and primate systematics, in: D. R. Swindler, and J. Erwin (eds.), *Comparative Primate Biology*, Vol. 1, pp. 219–276. Liss, New York.

Martin, R. D. 1990. *Primate Origins and Evolution: A Phylogenetic Reconstruction.* Princeton University Press, Princeton, N.J.

Martinez, N. L. 1985. Reconstruction of ancestral cranioskeletal features in the order Lagomorpha, in: W. P. Luckett and J.-L. Hartenberger (eds.), *Evolutionary Relationships among Rodents: A Multidisciplinary Analysis*, pp. 151–190. Plenum Press, New York.

Mayr, E. 1986. Uncertainty in science, Is the giant panda a bear or a raccoon? *Nature* **323**:769–771.

Mindell, D. P., Dick, C. W., and Baker, R. J. 1991. Phylogenetic relationships among megabats, microbats, and primates. *Proc. Natl. Acad. Sci. USA* **88**:10322–10326.

Miyamoto, M. M., and Boyle, S. M. 1989. The potential importance of mitochondrial DNA sequence data to eutherian mammal phylogeny, in: B. Fernholm (ed.), *The Hierarchy of Life*, pp. 437–450. Elsevier, Amsterdam.

Miyamoto, M. M., and Goodman, M. 1986. Biomolecular systematics of eutherian mammals: Phylogenetic patterns and classification. *Syst. Zool.* **35**:230–240.

Moritz, C., and Hillis, D. M. 1990. Molecular systematics: Context and controversies, in: D. M. Hillis and C. Moritz (eds.), *Molecular Systematics*, pp. 1–10. Sinauer Assoc., Sunderland, Mass.

Novacek, M. J. 1987. Auditory features and affinities of the Eocene bats *Icaronycteris* and *Palaeochiropteryx* (Microchiroptera, insertae sedis). *Am. Mus. Novit.* **2877**:1–18.

Novacek, M. J. 1989. Higher mammal phylogeny: The morphological–molecular synthesis, in: B. Fernholm, K. Bremer, and H. Jornvall (eds.), *The Hierarchy of Life*, pp. 421–435. Elsevier, Amsterdam.

Novacek, M. J. 1992. Mammalian phylogeny: Shaking the tree. *Nature* **356**:121–125.

Novacek, M. J., and Wyss, A. R. 1986. Higher level relationships of the Recent eutherian orders: Morphological evidence. *Cladistics* **2**:257–287.

Novacek, M. J., Wyss, A. R., and McKenna, M. C. 1988. The major groups of eutherian mammals, in: M. J. Benton (ed.), *The Phylogeny and Classification of the Tetrapods*, Vol. 2, pp. 31–71. (Clarendon), Oxford University Press, London.

Pettigrew, J. D. 1986. Flying primates? Megabats have the advanced pathway from eye to midbrain. *Science* **231**:1304–1306.

Pettigrew, J. D., Jamieson, B. G. M., Robson, S. K., Hall, L. S., McAnally, K. I., and Cooper, H. M. 1989. Phylogenetic relations between microbats, megabats, and primates (Mammalia: Chiroptera and Primates). *Philos. Trans. R. Soc. London* **325**:489–559.

Pocock, R. I. 1918. On the external characters of the lemurs and of *Tarsius*. *Proc. Zool. Soc. London* **1918**:19–53.

Prager, E., and Wilson, A. C. 1993. Information content of immunological distances. *Methods Enzymol.* (in press).

Ruvolo, M., Disotell, T. R., Allard, M. W., Brown, W. M., and Honeycutt, R. L. 1991. Resolution of the African hominoid trichotomy by use of a mitochondrial gene sequence. *Proc. Natl. Acad. Sci. USA* **88**:1570–1574.

Saiki, R. K., Gelfand, D. H., Stoeffel, S., Scharf, S. J., Higuchi, R., Horn, G. T., Mullis, K. B., and Erlich, H. A. 1988. Primer-directed enzymatic amplification of DNA with a thermostable DNA polymerase. *Science* **239**:487–491.

Sanger, F., Nicklen, S., and Coulson, A. R. 1977. DNA sequencing with chain terminating inhibitors. *Proc. Natl. Acad. Sci. USA* **74**:5463–5467.

Sarich, V. M. 1985. Rodent macromolecular systematics, in: W. P. Luckett and J.-L., Hartenberger (eds.), *Evolutionary Relationships among Rodents: A Multidisciplinary Analysis,* pp. 423–452. Plenum Press, New York.

Sarich, V. M., and Cronin, J. E. 1976. Molecular systematics of the primates, in: M. Goodman and R. E. Tashian (eds.), *Molecular Anthropology,* pp. 141–170. Plenum Press, New York.

Schwartz, J. H., and Tattersall, I. 1987. Tarsiers, adapids and the integrity of strepsirhini. *J. Hum. Evol.* **16:**23–40.

Shoshani, J. 1986. Mammalian phylogeny: Comparison of morphological and molecular results. *Mol. Biol. Evol.* **3:**222–242.

Simmons, N. B., Novacek, M. J., and Baker, R. J. 1991. Approaches, methods, and the future of the chiropteran monophyly controversy: A reply to J. D. Pettigrew. *Syst. Zool.* **40:**239–244.

Simons, E. L., Holroyd, P. A., and Brown, T. M. 1991. Early Teritary elephant-shrews from Egypt and the origin of the Macroscelidea. *Proc. Natl. Acad. Sci. USA* **88:**9734–9737.

Swofford, D. L. 1990. PAUP: Phylogenetic Analysis Using Parsimony, Version 3.0. Program and documentation. Illinois Natural History Survey, Champaign.

Swofford, D. L., and Olsen, G. J. 1990. Phylogeny reconstruction, in: D. M. Hillis and C. Moritz (eds.), *Molecular Systematics,* pp. 411–501. Sinauer Assoc., Sunderland, Mass.

Szalay, F. S. 1985. Rodent and lagomorph morphotype adaptations, origins, and relationships: Some postcranial attributes analyzed, in: W. P. Luckett and J.-L. Hartenberger (eds.), *Evolutionary Relationships among Rodents: A Multidisciplinary Analysis,* pp. 83–132. Plenum Press, New York.

Tagle, D. A., Miyamoto, M. M., Goodman, M., Hofman, O., Braunitzer, G., Goltenboth, R., and Jalanka, H. 1986. Hemoglobin of pandas: Phylogenetic relationships of carnivores as ascertained with protein sequence data. *Naturwissenschaften* **73:**512–514.

Templeton, A. R. 1983. Phylogenetic inference from restriction endonuclease cleavage site maps with particular reference to the evolution of humans and the apes. *Evolution* **37:**221–244.

Thewissen, J. G. M., and Babcock, S. K. 1991. Distinctive cranial and cervical innervation of wing muscle: New evidence for bat monophyly. *Science* **251:**934–936.

Tindall, K. R., and Kunkel, T. A. 1988. Fidelity of DNA synthesis by the *Thermus aquaticus* DNA polymerase. *Biochemistry* **27:**6008–6013.

Wible, J. R., and Covert, H. H. 1987. Primates: Cladistic diagnosis and relationships. *J. Hum. Evol.* **16:**1–22.

Wible, J. R., and Novacek, M. J. 1988. Cranial evidence for the monophyletic origin of bats. *Am. Mus. Novit.* **2911:**1–19.

Wyss, A. R., Novacek, M. J., and McKenna, M. C. 1987. Amino acid sequence versus morphological data and the interordinal relationships of mammals. *Mol. Biol. Evol.* **4:**99–116.

A Molecular View of Primate Supraordinal Relationships from the Analysis of Both Nucleotide and Amino Acid Sequences

8

MICHAEL J. STANHOPE,
WENDY J. BAILEY, JOHN CZELUSNIAK,
MORRIS GOODMAN, JING-SHENG SI,
JOHN NICKERSON, JOHN G. SGOUROS,
GAMAL A. M. SINGER, and
TRAUTE K. KLEINSCHMIDT

Introduction

The fossil record suggests that the orders of eutherian mammals arose in a burst of adaptive radiation at the dawn of the Cenozoic (Savage and Russell,

MICHAEL J. STANHOPE, WENDY J. BAILEY, JOHN CZELUSNIAK, and MORRIS GOOD-MAN • Department of Anatomy and Cell Biology, Wayne State University School of Medicine, Detroit, Michigan 48201.   JING-SHENG SI and JOHN NICKERSON • Department of Ophthalmology, Emory University, Atlanta, Georgia, 30322.   JOHN G. SGOUROS, GAMAL A. M. SINGER, and TRAUTE K. KLEINSCHMIDT • Max Planck Institute for Biochemistry, Munich Germany.

Primates and Their Relatives in Phylogenetic Perspective, edited by Ross D.E. MacPhee. Plenum Press, New York, 1993.

1983). Possibly because of the apparent bushlike pattern of this radiation, establishing the course of phylogenetic branching that led to the orders and suborders of eutherian mammals has proven difficult. Although some regard this radiation as an almost simultaneous emergence of major clades (Simpson, 1978), most feel that such phylogenies are not actually bushlike (Gingerich, 1985; Novacek, 1990), although several splitting events may be close enough in time and in character to represent a considerable challenge for molecular and morphological analyses.

To date, morphological and molecular studies have produced little agreement on how to arrange the orders of eutherian mammals into clearly recognizable supraordinal clades. One suggested supraordinal clade that has received considerable attention is the proposed superorder Archonta, a group consisting of Primates, Scandentia (tree shrews), Dermoptera (flying lemurs), and Chiroptera (bats). The concept of Archonta was first promoted by Gregory (1910) and has received support from morphological studies by Szalay (1977) and Novacek and Wyss (1986), while others have rejected the monophyly of this group (Simpson, 1945; Cartmill and MacPhee, 1980; Luckett, 1980; Novacek, 1980). Disputes exist about relationships of orders and suborders within Archonta; certainly the most controversial proposal is the hypothesis that Megachiroptera (megabats) are more closely aligned with Primates than with Microchiroptera (microbats)—the "flying primate" hypothesis (Pettigrew, 1986, 1991a,b; Pettigrew *et al.*, 1989). Anatomical features shared between Dermoptera and Chiroptera have been suggested as synapomorphies strongly supporting these two orders as a subdivision within Archonta (Novacek, 1990). Still other studies place Primates closer to Lagomorpha and Rodentia (Miyamoto and Goodman, 1986; Czelusniak *et al.*, 1990; Penny *et al.*, 1991). An impressive array of anatomical as well as fossil evidence exists to support the concept of Glires, a superordinal grouping of rodents and lagomorphs (Gregory, 1910; Simpson, 1945; Novacek, 1985a; Luckett, 1985; Li and Ting, 1985) but little molecular evidence exists to corroborate this claim. In fact, there is no hypothesis of eutherian ordinal relationships that is widely accepted.

Congruence among data sets is generally regarded as the best means of resolving difficult phylogenetic questions (Penny *et al.*, 1982; Kluge, 1989; Miyamoto and Cracraft, 1991). Perhaps because molecular systematics is still in its infancy, this important principle of science has not yet been consistently applied to molecular data sets (see discussion by Miyamoto and Cracraft, 1991). Tests of congruence between different data sets will undoubtedly become more common now that PCR is a well-established technique in the field of molecular systematics. Here, we focus our attention on congruence between different molecular data sets we have collected that are pertinent to the issue of primate supraordinal relationships.

From the perspective of molecular evolution one of the major problems in addressing the systematics of eutherian mammals lies in the choice of the DNA locus. Although there are clearly a great many to choose from, it is not yet clear which molecular system will provide the best resolution. It is thought that many mammalian orders shared a common ancestry for a relatively short period. In slowly evolving DNA sequences, this short period would have permitted only a few molecular changes, to allow documentation of common ancestry. Alternatively, rapidly evolving DNA sequences would have accumulated a great many changes subsequent to the separation of the various taxa, obscuring possible relationships. A recent study involving restriction site analysis of rDNA concluded that this particular approach was unlikely to provide much insight into mammalian interordinal relationships (Baker *et al.*, 1991a). A study involving the mitochondrial cytochrome oxidase II (COII) gene, at the nucleotide sequence level, provided more encouraging results; however, because of the relatively rapid rate of evolution in mitochondrial genes, the authors found it necessary to employ various character weighting schemes in their analysis, such as consideration of only transversions (Adkins and Honeycutt, 1991, this volume).

Here we elaborate on the results of two nuclear genes we have been exploring as possible molecular systems for examining mammalian interordinal relationships (Bailey *et al.*, 1992; Stanhope *et al.*, 1992) and present an updated picture of mammalian relationships as revealed through a third system, the analysis of amino acid sequences. Although the emerging picture of mammalian phylogeny is by no means clear-cut, there are certain important congruences among results from these three systems, as well as from other reported studies. In examining the cladistic validity of a superordinal grouping of Primates, Scandentia, Dermoptera, and Chiroptera in Archonta, we include representatives from orders other than those traditionally considered to comprise this superorder, since there is no clear agreement on which eutherian groups represent a primate supraordinal clade. After first presenting the procedures involved in data collection, we next present a brief description of the overall picture resulting from these various data sets, followed by a more detailed treatment of each data set as it pertains to several current and controversial phylogenetic hypotheses.

Data Collection and Analysis

DNA Sequences

IRBP

The gene encoding interphotoreceptor retinoid binding protein (IRBP) is a large, single-copy gene located on chromosome 10 (Borst and Nickerson, 1988). The entire gene has been sequenced in bovine and human. Its first exon is between 3173 and 3180 base pairs in length and shows between bovine (Borst *et al.*, 1989) and human (Fong *et al.*, 1990) genes a sequence divergence of approximately 16%. In view of this large size and relative conservation of exon 1, sequencing of DNA fragments amplified by the polymerase chain reaction, PCR (Saiki *et al.*, 1988), using only a few sets of "universal primers," can generate extensive comparative data on different orders of mammals. Moreover, since IRBP is a single-copy gene, homologous IRBP sequences are in all likelihood also orthologous, i.e., derived from a common ancestral gene by the process of speciation and thus suitable for resolving cladistic relationships among species lineages.

PCR primers were constructed by choosing conserved regions of exon 1, in a comparison of the aligned human and bovine sequences. Four primers were designed (Fig. 1a):

> +IRBP217 5'ATGGCCAAGGTCCTCTTGGATAACTACTGCTT 3'
> −IRBP1531 5'CGCAGGTCCATGATGAGGTGCTCCGTGTCCTG 3'
> +IRBP379 5'CCTCGCCTGGTCATCTCCTATGAGCCCAGCAC 3'
> −IRBP1426 5'CAGGTAGCCCACATTGCCTGGCAGCAC 3'

The prefixes + or − in the primer names refer to the reading or complementary strands, respectively, and the numbers designate the 3' end of the primer in the published human sequence. Thirteen species representing nine different mammalian orders were included: the order Primates was represented by human (*Homo sapiens*), tarsier (*Tarsius syrichta*), and galago (*Galago crassicaudatus*); the order Dermoptera by flying lemur (*Cynocephalus variegatus*); Scandentia, tree shrew (*Tupaia glis*); Chiroptera, megabat (*Pteropus hypomelanus*) and microbats (*Tonatia bidens, T. silvicola*); Carnivora, cat (*Felis catus*); Artiodactyla, bovine (*Bos taurus*); Lagomorpha, rabbit (*Oryctolagus cuniculus*); Rodentia, mouse (*Mus domesticus*); Marsupialia, opossum (*Didelphis virginiana*). The human and bovine sequences were taken from the published accounts of the entire gene of these two species. The mouse sequence was obtained by screening a lambda EMBL3A genomic library, generated from BALB/c mice, with a bovine 900-bp IRBP cDNA fragment. All other sequences were obtained from PCR-amplified products. With the exception of the rabbit, the 1.3-kb section of exon 1 could be amplified with the pair of primers +IRBP217 and −IRBP1531. Amplification of this fragment from the primates, flying lemur, tree shrew, megabat, both species of microbat, and cat, using these primers, did not necessitate a separate annealing step, i.e., thermal cycling profile was 30 cycles of 95°C 1 min, 70°C 2 min. The opossum required a separate annealing step of 55°C. The rabbit template was obtained by amplification with +IRBP379 and −IRBP1426; amplification in this instance necessitated a 60°C annealing step. Sequencing template was prepared using asymmetric PCR (Gyllensten and Erlich, 1988), employing various combinations of the aforementioned primers. Asymmetric PCR products were purified using an anion exchange column (Qiagen). The sequence was obtained using the dideoxy chain-termination method; additional sequencing primers were designed as necessary. Sequence was obtained on both strands at least twice from separate asymmetric PCR preparations.

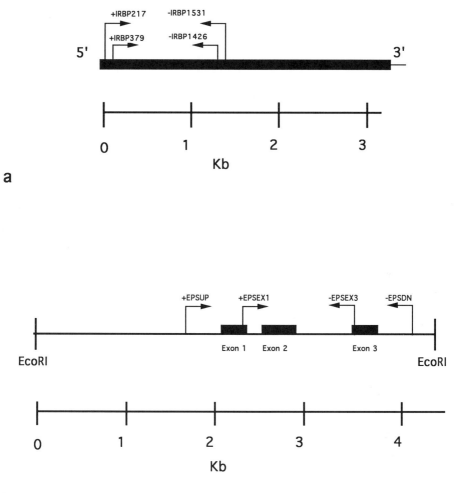

Fig. 1. Schematic diagram of (a) exon 1 of IRBP and (b) ε-globin locus, indicating position of PCR primers used in amplification of sequence upon which the analyses were based.

ε-Globin Gene

The ε-globin gene of mammals is the 5′-most member of the five gene loci that comprise the β-globin gene cluster. During the course of evolution of the β-globin cluster in placental mammals, the ε gene has been less prone to tandem duplication than have several of the other β-globin genes. Thus, phylogenetic studies focused on the ε locus, avoid the problem of comparing paralogous genes. In addition, most of the sequence data are noncoding and therefore not under the selective pressures that could result in convergent substitutions.

Analysis of the ε-globin gene included 17 species representing six orders. Seven of these sequences came from work already published: human (*Homo sapiens*) (Collins and Weissman, 1984); orangutan (*Pongo pygmaeus*) (Koop *et al.*, 1986); galago (*Galago crassicaudatus*) (Tagle *et al.*, 1988); rabbit (*Oryctolagus cuniculus*) (Hardison, 1983); tarsier (*Tarsius syrichta*) (Koop *et al.*, 1989); brown lemur (*Lemur fulvus*) (Harris *et al.*, 1986); and goat (*Capra hircus*) (Shapiro *et al.*, 1983). Four of these (human, orangutan, galago, and rabbit) are represented by a 4.1-kb *Eco*RI subclone (Fig. 1a) from lambda genomic libraries, while the remaining three (tarsier, lemur, and goat) were library subclones of 1.8 kb in length, spanning the region just upstream of the promoter, to about 100 bp downstream of the polyadenylation signal. Sequences from gibbon (*Hylobates lar*) and capuchin monkey (*Cebus albifrons*) were derived from the screening of lambda genomic libraries and are represented by the 4.1-kb *Eco*RI subclone. The remaining sequences were obtained by plasmid cloning and dideoxy chain-termination sequencing of PCR-amplified products. Clones of 1.8 kb in length were obtained from the common chimpanzee (*Pan troglodytes*), pygmy chimpanzee (*Pan paniscus*), gorilla (*Gorilla gorilla*), and rhesus monkey (*Macaca mulatta*), using the following two primers: +EPSUP 5′CACTGCTGACCCTCTCCTGACCTG 3′; −EPSDN 5′TCTC-CATCCCTCAGCCCATGAAC 3′ (Fig. 1b). A pair of primers designed within exon 1 and exon 3 (+EPSEX1 5′GGAAGAGGCTGGAGGTGAAGCCTTG-GG 3′ and −EPSEX3 5′AATAATCACCATCACGTTACCCAGG 3′, respectively, Fig. 1b) were used to obtain 1.2-kb clones from flying lemur (*Cynocephalus variegatus*), tree shrew (*Tupaia glis*), megabat (*Cynopterus sphinx*), and microbat (*Megaderma lyra*). Clones from three independent PCR reactions were sequenced on both DNA strands from each species. When sequence discrepancies existed between clones of the same species, the majority consensus was used for comparative analysis.

Data Analysis

Initial sequence alignments of the ε data were obtained using the pairwise alignment algorithm of Smith and Waterman (1981), as modified by Goodman *et al.* (1984), with the final alignment being completed by hand. Gaps were inserted when they increased similarity that could be attributed to common ancestry. The IRBP data were all coding sequences and thus alignments were straightforward. From the aligned sequence data, pairwise divergence values were calculated for each data set. These data were then used as input for analysis by the UPGMA (Sneath and Sokal, 1973) and neighbor-joining (NJ) (Saitou and Nei, 1987) methods. These trees then served as starting points for analysis by the parsimony method. The analysis proceeded by searching for the most parsimonious tree or trees by using a "global swap" program (PTRFC). PTRFC swaps branches by all possible bisections and pairwise reconnections in a manner similar to the procedure described by

Swofford and Olsen (1990). The program was run with numerous different starting trees. A tree of the lowest score found in the search was then submitted to an "all trees" program (PTRALL), which examines all of the trees for any particular designated number of terminal taxa. A terminal taxon can be one of the 13 or 17 species (IRBP and ϵ data sets, respectively) or a specified subtree in which no branch swapping is allowed. With the ϵ data set, and for the first series involving IRBP, PTRALL was run with eight terminal taxa. PTRALL examined all 10,395 trees formed by these eight taxa and the program CONSEN ordered the trees from lowest to highest score. From the ordered trees, program SURF22 determined the minimum number of substitutions necessary to break up any particular clade. Many such eight-branch runs were carried out, each with a different set of eight terminal taxa. Finally, for the IRBP data set, several ten-branch runs were performed, with the inclusion of instructions to reject trees in excess of a specified score. A total of 2,027,000 trees were examined on each of such runs.

Amino Acid Sequences

Data Sources and Analysis

The amino acid analysis was performed using two different data sets, both of which are updated versions of earlier alignments. The first is tandemly combined α- and β-hemoglobin chains from 145 vertebrate species. An earlier paper involving these data focused on the phylogenetic position of the gundi and at that time included 134 species (Beintema *et al.*, 1991). The 145 species included: 27 Primates, 10 Rodentia, 1 Lagomorpha, 1 Scandentia, 22 Carnivora, 15 Chiroptera, 3 Insectivora, 1 Tubulidentata, 13 Artiodactyla, 3 Cetacea, 2 Proboscidea, 1 Hyracoidea, 1 Sirenia, 5 Perissodactyla, 2 Endentata, 2 Monotremata, 2 Marsupialia, 23 Aves, 6 Reptilia, 1 Amphibia, 3 Osteichthyes, and 1 Chondrichthyes. The other data set is an updated version of the tandem alignment of β-hemoglobin, α-hemoglobin, myoglobin, αA lens crystallin, fibrinopeptide, cytochrome c, ribonuclease, and embryonic α-hemoglobin chains (Miyamoto and Goodman, 1986; Czelusniak *et al.*, 1990; see the latter for a detailed list of species and the sequences by which they are represented). The current version included 6 additional β-hemoglobin sequences, +6 α-hemoglobin sequences, +2 fibrinopeptides, +1 cytochrome c, +1 ribonuclease, and +1 embryonic α-hemoglobin.

A significant difference between the present chapter and previous analyses of both of these data sets is that the earlier works did not employ a global branch swapping approach. Our program PTRAA, which accommodates amino acid sequences and performs global branch swapping, was used for both of these data sets. With this improvement in analytical procedure we have greater confidence in our ability to find the most parsimonious tree, particularly with these large data sets.

Sequence Results

DNA Sequences

IRBP

Alignment of the IRBP sequences appears in Appendix I. In pairwise comparisons of these aligned sequences the eutherian species (with the exception of the *T. bidens–T. silvicola* comparison) diverged by 14–25% from one another (Table I). The opossum diverged by 33–38% from the eutherians. The NJ tree constructed from these data (Fig. 2a) did not support the view of an Archonta consisting of Primates, Scandentia, Dermoptera, and Chiroptera. It did, however, support a monophyletic grouping of Mega- and Microchiroptera, but with Carnivora as the bat sister group. The branch adjacent to the primates consisted of a tree shrew/flying lemur clade joined by rabbit. The tree had relatively long terminal branches with short internodes, probably reflecting the ancient, but comparatively short time course over which these branching events occurred. The UPGMA tree had a flying lemur/cat clade on the branch adjacent to the primates, followed on successive branches by rabbit, bovine, the chiropteran clade, tree shrew, mouse, and opossum.

Parsimony analysis was performed with and without the opossum, because it became clear that there was considerable homoplasy in the opossum sequence and that much stronger groupings were obtained in the absence of opossum than in its presence. The homoplasy caused unexpected groupings in the most and near-most parsimonious trees. In the most parsimonious trees (consensus shown in Fig. 2b), opossum joined tree shrew, but for only two additional substitutions (+2) it joined tarsier; for +3 it joined the rodent, for +4 it joined the carnivore, and for +6 it joined the megachiropteran. We

Table I. Sequence Divergence Matrix for the IRBP Data Set[a]

	Tsy	Gcr	Cva	Tgl	Phy	Tbi	Tsi	Fca	Bta	Ocu	Mdo	Dvi
Hsa	.156	.146	.171	.193	.206	.218	.206	.169	.185	.179	.231	.355
Tsy		.158	.167	.200	.206	.219	.208	.173	.192	.193	.242	.346
Gcr			.153	.187	.186	.200	.196	.157	.188	.182	.221	.344
Cva				.173	.177	.178	.174	.156	.183	.167	.223	.348
Tgl					.210	.234	.223	.182	.210	.202	.245	.354
Phy						.147	.137	.148	.184	.197	.226	.332
Tbi							.009	.167	.198	.211	.247	.367
Tsi								.158	.191	.204	.244	.361
Fca									.165	.168	.214	.326
Bta										.196	.244	.368
Ocu											.231	.368
Mdo												.383

[a]Divergence calculations were based on sequence represented in Appendix I. Values represent divergence corrected for superimposed mutations (Jukes and Cantor, 1969). Abbreviations are defined in Appendix I.

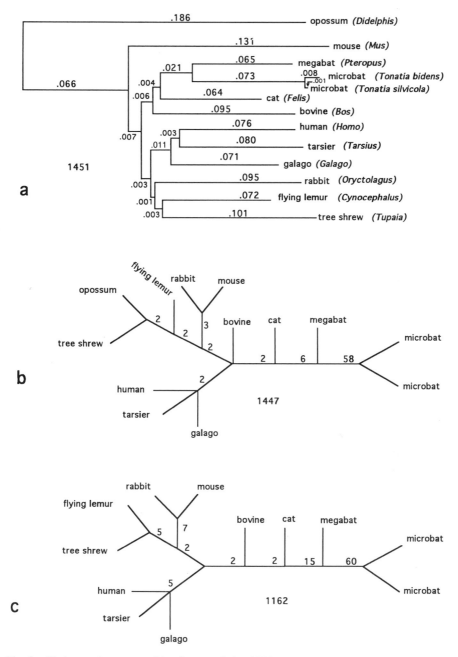

Fig. 2.   Phylogenetic trees resulting from analysis of IRBP sequences. (a) Neighbor-joining tree based on pairwise divergence values presented in Table I. Numbers on branches represent branch lengths calculated according to Saitou and Nei (1987). (b) Consensus maximum parsimony tree with strength of grouping numbers indicated along branches. These numbers indicate for each internode the minimum number of substitutions that must be added to the most parsimonious tree to eliminate the barrier, and thus allow movement of one or more sequences between the groups separated by the internode. (c) Consensus maximum parsimony tree and strength of grouping numbers when opossum is removed from data set.

surmise that this is symptomatic of the longest branches in the tree (the most divergent taxa) having accumulated a much larger proportion of homoplastic changes, probably caused by the fact that multiple sequence changes cannot be detected on long unbranched lineages.

The same overall branching order was obtained when the opossum was removed, but as mentioned, the groupings were more strongly supported (Fig. 2c). The most parsimonious tree supported a monophyletic Chiroptera, but, like the NJ tree, not as part of the proposed superorder Archonta. The overall topology was very similar to the NJ tree except that the relatively divergent mouse sequence (Table I) clearly showed some cladistic similarity to the rabbit, thus forming a Glires clade on the branch adjacent to the primates. The position of galago and tarsier relative to human was not resolved. The strongest grouping was the two congeneric species of microbat, requiring a minimum of 60 substitutions be added to the most parsimonious score to break up that clade. The next most strongly supported clade was the megabat with the microbats, requiring the addition of a minimum of 15 nucleotide substitutions to break up the group. The lagomorph/rodent clade was supported by a strength of grouping of 7. The tree shrew/flying lemur clade was supported by the same strength of grouping as the primates (5).

ϵ-Globin Locus

Alignment of the ϵ-globin sequences appears in Appendix II. Divergence values calculated from these data (Table II) suggested that the flying lemur was closer to the primates (22.5–29.4%) than it was to members of the other eutherian orders (33.1–39.2%) and that the mega- and microbat were more similar to each other than either was to any other species.

The UPGMA tree constructed from these divergence data placed flying lemur between anthropoid and strepsirhine primates, followed on successive branches by tarsier, tree shrew, a bat clade, goat, and rabbit. The NJ tree had a flying lemur/tree shrew/rabbit clade on the adjacent branch to Primates (Fig. 3a). When only the 1.2-kb section common to all species (see Appendix II) was used for divergence calculations, the resulting UPGMA tree was similar to that described for the entire data set, except that rabbit was on a branch between tree shrew and the bat clade. With this 1.2-kb data set the NJ tree placed flying lemur by itself adjacent to the primates, joined next by a tree shrew/rabbit clade, followed by the bat clade, and finally goat.

The most parsimonious tree was similar to the NJ trees except that there was a tree shrew/rabbit clade adjacent to Primates, with flying lemur on the next branch (Fig. 3b). The anthropoids were the clade most strongly supported, requiring the addition of 65 substitutions to the most parsimonious score to break up the group. The number of nucleotide substitutions (NSs) necessary to break up the bats (31) was similar to that supporting the strepsirhines (37). Other strongly supported groups were the catarrhine primates

Table II. Sequence Divergence Matrix for the ε Data Set[a]

	Ptr	Ppa	Ggo	Ppy	Hla	Mmu	Cal	Tsy	Gcr	Lfu	Cva	Tgl	Csp	Mly	Ocu	Chi
Hsa	.011	.010	.012	.032	.031	.043	.114	.252	.304	.212	.227	.329	.312	.346	.411	.348
Ptr		.002	.007	.033	.025	.041	.102	.243	.267	.198	.225	.333	.309	.345	.331	.341
Ppa			.007	.031	.024	.040	.103	.245	.268	.199	.226	.335	.310	.346	.332	.342
Ggo				.033	.024	.041	.104	.244	.265	.200	.230	.334	.315	.347	.333	.347
Ppy					.034	.047	.118	.258	.301	.216	.239	.339	.316	.348	.410	.353
Hla						.041	.110	.253	.301	.211	.236	.329	.311	.336	.406	.355
Mmu							.115	.262	.276	.213	.242	.344	.326	.361	.341	.367
Cal								.284	.344	.246	.267	.342	.342	.385	.423	.369
Tsy									.329	.257	.289	.373	.419	.420	.381	.406
Gcr										.192	.294	.387	.341	.392	.452	.398
Lfu											.234	.325	.331	.369	.325	.343
Cva												.331	.337	.355	.335	.392
Tgl													.404	.432	.406	.483
Csp														.269	.442	.363
Mly															.480	.423
Ocu																.424

[a]Divergence calculations were based on the sequence represented in Appendix II. Values represent divergence corrected for superimposed mutations (Jukes and Cantor, 1969). Abbreviations are defined in Appendix II.

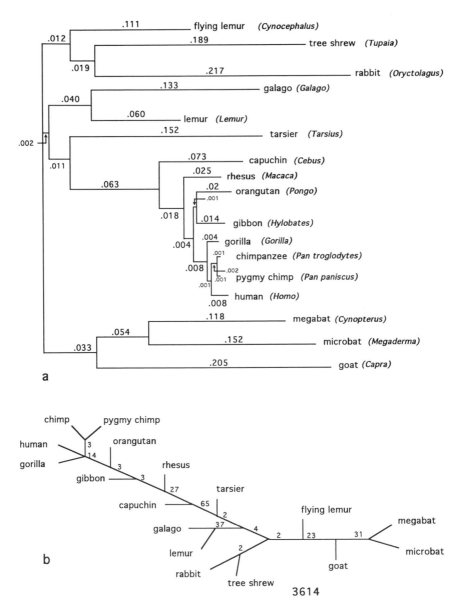

Fig. 3. Phylogenetic trees derived from analysis of ε-globin sequences. (a) Neighbor-joining tree based on pairwise divergence values presented in Table II. Numbers on branches represent branch lengths calculated according to Saitou and Nei (1987). (b) Maximum parsimony tree with strength of grouping numbers.

(27) and the wider group encompassing primates, flying lemur, tree shrew, and rabbit (23).

Amino Acid Sequence

α- and β-Hemoglobin

Cladistic relationships among mammalian orders deduced from parsimony analysis of the α and β sequences appear in Fig. 4a. Primates, Artiodactyla, and Insectivora were not monophyletic. Lemuridae and Lorisidae emanated from the same point as Scandentia, Rodentia, and Lagomorpha in the strict consensus tree (Fig. 4a), thus disrupting the monophyly of Primates. The two representatives of the Camelidae and the single representative from the Suidae did not group with the other artiodactyls, and the hedgehog did not group with the other insectivores. The key features of this tree pertinent to our discussion here are that Primates, Rodentia, Lagomorpha, and Scandentia form a supraordinal clade, joined next by Carnivora and then the Chiroptera/Insectivora/Tubulidentata grouping. Thus, these data indicate not only that Chiroptera is monophyletic, but also that this order is genealogically distant from Primates.

Tandem Alignment of Various Proteins

The picture emerging from analysis of the tandem alignment (Fig. 4b) is overall quite similar to that from just the α and β sequences. Primates, Rodentia, Lagomorpha, and Scandentia form a supraordinal clade joined next by a Carnivora/Insectivora/Chiroptera clade. Thus with this larger data set, the carnivores show a closer affinity with the insectivores and the chiropterans. Chiroptera remain monophyletic and genealogically distant from Primates. The principal difference between this tree and previous results involving this data set (Czelusniak *et al.,* 1990) is that in our construction of the consensus of all most parsimonious trees, employment of the "global swap" and CONSEN programs, and the removal of the "wandering" species (pika, pangolin, and aardvark) provided some additional resolution, resulting in the separation of the Carnivora/Insectivora/Chiroptera clade from the multichotomous branch involving the primate supraordinal clade. The most parsimonious trees upon which the original consensus involving the tandem alignment was constructed had a score of 4977. This data set included three species—pika, aardvark, and pangolin—that were identified in the CONSEN program as "wandering" species. Their exclusion increased the number of partitions in a consensus of 139 most parsimonious trees from 77 to 89. The tree depicted in Fig. 4b reflects those changes.

Fig. 4. Consensus maximum parsimony trees resulting from analysis of (a) α-, β-hemoglobin and (b) tandem alignment data sets. Dashed lines represent atypically placed taxa (e.g., Monotremata) and/or those species that disrupt the monophyly of a particular order (e.g., hedgehog).

Comparison and Discussion of Phylogenetic Hypotheses

Chiropteran Origins

The origin of Chiroptera has been an issue of controversy in mammalian systematics since Linnaeus (1758) first classified bats with the order Primates. Recently, this issue has come to the forefront again with the claims of Pettigrew and associates (Pettigrew, 1986, 1991a; Pettigrew *et al.*, 1989) that an array of neural characters shared between megabats and primates link these two groups. Thus, Pettigrew suggests a diphyletic origin for Chiroptera. If this astonishing proposal is correct, then flight evolved twice in the mammalian radiation. This would clearly be one of the most remarkable examples of convergent evolution ever recorded. This claim, however, is up against an impressive array of characters for bat monophyly (summarized and discussed in Baker *et al.*, 1991b). The details of this controversy have been presented elsewhere (Pettigrew, 1991a,b; Baker *et al.*, 1991b; Simmons *et al.*, 1991); here we discuss the evidence from IRBP, ε-globin, and the amino acid sequences for bat monophyly and against bat diphyly, as well as compare these results with other molecular studies on the origins of Chiroptera.

Bat Monophyly

Our phylogenetic analyses of both the IRBP coding DNA and the ε noncoding DNA strongly support a monophyletic Chiroptera. Both the distance matrix methods and the parsimony analysis of the IRBP data set indicated a monophyletic Chiroptera. Furthermore, the strength of grouping analysis indicated that Chiroptera was the ordinal grouping most strongly supported. We identified 22 sequence positions as sites with putative synapomorphies (shared derived substitutions), grouping *Pteropus* with the microbats (Table IIIa); 15 of these synapomorphies were synonymous substitutions, arguing against selected convergence as an explanation for the similarity in bat sequences. From the ordered trees we identified two trees with the same lowest score (+15) that disrupted chiropteran monophyly. In one of these, *Felis* and *Pteropus* exchanged places; in the other, *Felis* and *Pteropus* formed their own clade on a branch immediately adjacent to the microbats. Each of these alternative trees was supported by seven putative synapomorphies (Table IIIa). Using a one-tailed binomial test (also known as the "winning sites test"; Prager and Wilson, 1988) we calculated the probability that the most parsimonious arrangement and the two alternatives were equivalent. The number of sequence positions supporting bat monophyly was compared with the number of positions supporting the two alternatives. Support for chiropteran monophyly was highly significant ($p = 0.004$). A test with a similar theoretical basis, but one designed for comparing the three alternative dichotomous trees that can be formed by the four branches that emanate

Table III. Sequence Positions with Synapomorphic Characters Supporting Chiropteran Monophyly as Well as the Closest Competing Phylogenetic Alternatives for Both the IRBP and ε-Globin Data Sets

a. IRBP

	Positions supporting closest alternatives	
Positions supporting chiropteran monophyly Phy/*Tonatia* clade	Fca/*Tonatia* clade	Fca/Phy clade
160 161 167 228 343 344 382 391 451 466 577 589 592 658 826 871 931 1006 1049 1135 1146 1198	667 739 805 864 907 1177 1201	370 472 766 964 1060 1078 1195

b. ε-Globin

Positions supporting chiropteran monophyly Csp/Mly clade	Positions supporting closest alternatives Csp/Chi clade
2388 2396 2400 2405 2416 2477 2731 2738 2779 2782 2811–14 2815 2816 2846 2893 3219 3231 3441 3476 3494 3495 3496 3502–04 3511 3523 3536 3539–43 3583–87 3625 3626 3627 3628 3629 3630 3637 3688–89 3743 3763 3830	2390 2415 2448 2615 2629 2872 3199 3396 3685 3721–22 3750

from two adjacent interior nodes (Williams and Goodman, 1989), also resulted in highly significant support of chiropteran monophyly ($p = 0.0017$).

Like the IRBP data set, the ε-globin sequences also supported bat monophyly, using both the distance matrix methods and parsimony. The divergence value between the microbat and megabat was similar to divergence values within the order Primates, and the strength of grouping analysis indicated that the number of NSs necessary to break up the bats (31) was similar to that supporting the strepsirhines (37). Furthermore, over the 1200 nucleotide positions common to all species we identified 39 single or contiguous sequence positions with putative synapomorphies supporting chiropteran monophyly (Table IIIb). Only one of these positions was within a coding region (position 2477, within exon 2). Of these 39 positions supporting bat monophyly, 26 were identified that had no convergent nucleotides between the bats and other species. From the ordered trees, we identified the first tree to break up bat monophyly as one in which megabat joined the goat. The number of characters supporting that group was found to be 11. Application of the one-tailed binomial test in this instance (comparison of 39 putative synapomorphies versus 11) resulted in highly significant support of chiropteran monophyly ($p = 0.000045$).

Both sets of amino acid sequence data supported megabats grouping with microbats; the α- and β-hemoglobin sequences included 15 species of Chiroptera (4 megabats and 11 microbats), and the tandem alignment included 6 species of Chiroptera (2 megabats and 4 microbats). The relationships of the species comprising the chiropteran branch of the α- and β-hemoglobin tree appear in Fig. 5. The Megachiroptera formed a monophyletic clade, but the Microchiroptera did not; several microchiropterans joined at the base of the megachiropteran branch. Several family groupings were evident in this tree, including Megadermatidae (*Macroderma* and *Megaderma*), Vespertilionidae (*Myotis, Antrozous, Chalinolobus*), Molossidae (*Mormopterus* and *Tadarida*), and the proposed superfamily Vespertilionoidea (a grouping of molossids and vespertilionids). The earliest microchiropterans date back to approximately 50 million years ago (Novacek, 1985b), while the earliest megachiropteran is dated at about 30 million years old (Baker *et al.,* 1991b). This information from the fossil record concomitant with the evolutionary relationships suggested by the hemoglobin sequences (Fig. 5) suggests the possibility that Microchiroptera represents a grade rather than a clade.

In the recent exchange of views regarding the bat controversy, Pettigrew (1991a) suggested that because α- and β-hemoglobin sequences do not support "the undoubted monophyly of microbats," these data provide one with very little confidence in any attempts to decide between phylogenetic hypotheses. We suggest that Pettigrew's argument is fallacious, since there is no

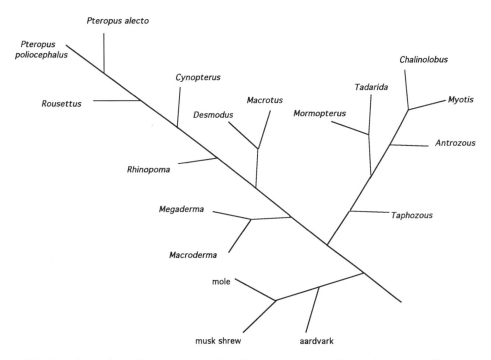

Fig. 5.   Relationships of species comprising chiropteran branch of α- and β-hemoglobin tree.

definitive evidence that microbats must represent a monophyletic group. Moreover, he claims that a subset of this α- and β-hemoglobin data set involving 52 species does not support the monophyly of Chiroptera. Our analysis of the same subset of species does support chiropteran monophyly (Fig. 6), i.e., both the full data set (Fig. 5) and the Pettigrew subset (Fig. 6) support chiropteran monophyly. Although we have not seen his actual alignment, we assume from what he indicates that he used exactly the same sequences that we did and thus we conclude that his analytical procedure did not allow him to reach the most parsimonious tree. We found the score of his tree to be 1376 and the score of our most parsimonious trees to be 1369 (i.e., 40 partitions in a consensus of 52 trees at 1369). We admit that this hemoglobin subset only weakly supports bat monophyly, since at one additional substitution to the most parsimonious consensus tree, the Vespertilionoidea split away from the rest of Chiroptera and the tree becomes even more "bushy." However, claims by Pettigrew that these data do not support bat monophyly are clearly misleading and are a misrepresentation of the data.

Although trees generated from amino acid sequences often consist of weakly supported clades (Czelusniak *et al.*, 1990), an additional measure of confidence in the data can be achieved by examining the degree of congruence with well-established phylogenetic relationships. For example, the tandem data set yields evidence on primate phylogeny that is highly congruent with morphological and other molecular data. These data support the taxonomic scheme suggested by Hill (1955) for the order Primates, with the division into Strepsirhini (Lemuroidea and Lorisoidea) and Haplorhini (Tarsioidea and Anthropoidea), and with Anthropoidea subdivided into Platyrrhini and Catarrhini and Catarrhini further subdivided into cercopithecoids and hominoids (Fig. 7a). Several other notable features of the amino acid data sets, illustrating congruence with morphological and/or molecular phylogenetic hypotheses, were evident. The α-, β-hemoglobin data set supported the monophyly of Rodentia (Fig. 7b; discussed in detail by Beintema *et al.*, 1991), including the guinea pig and the gundi, two species that have had either uncertain or controversial placements (Luckett and Hartenberger, 1985; Wood, 1985), but for which there is considerable morphological evidence for their inclusion within Rodentia (Luckett, 1985; Bugge, 1985; George, 1985). The monophyly of Carnivora was supported using both data sets, including the placement of the greater panda (and possibly the lesser panda) in the family Ursidae, in agreement with the bear/greater panda grouping found with immunological (Sarich, 1973), DNA hybridization (O'Brien *et al.*, 1985), and morphological (Mayr, 1986) data. Several other groupings within Carnivora followed recognized familial or subordinal relationships, including the Felidae, Canidae, and Pinnipedia (Fig. 7c). Both amino acid data sets support the inclusion of Sirenia, Hyracoidea, and Proboscidea in the superorder Paenungulata, in agreement with earlier classifications (e.g., Simpson, 1945; Novacek, 1986; Miyamoto and Goodman, 1986). The monophyly of the superorder Ungulata (orders: Artiodactyla, Perisso-

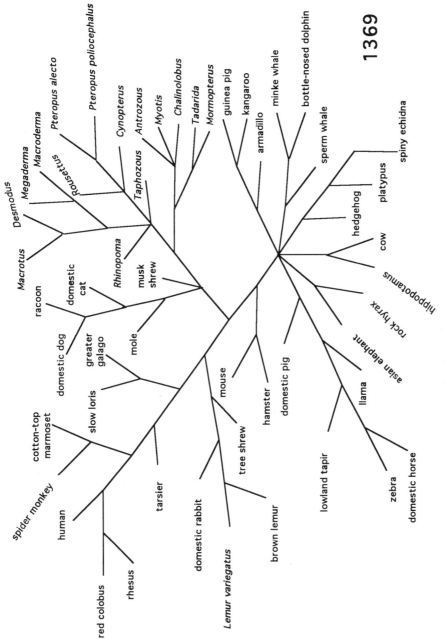

1369

Fig. 6. Our consensus maximum parsimony tree involving the same 52 species and α-, β-hemoglobin sequences discussed by Pettigrew (1991a).

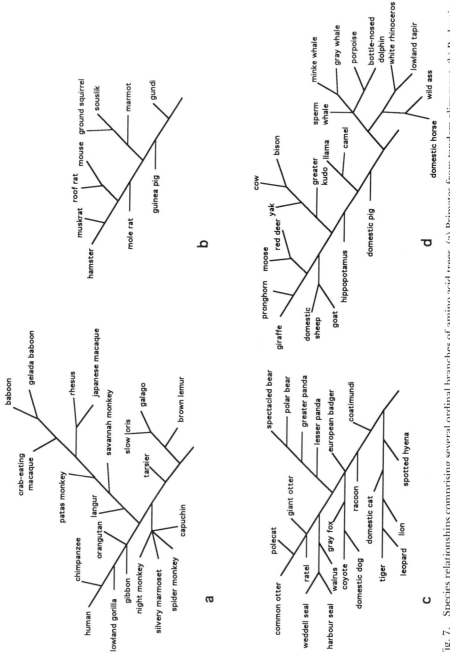

Fig. 7. Species relationships comprising several ordinal branches of amino acid trees. (a) Primates from tandem alignment. (b) Rodentia from α-, β-hemoglobin set. (c) Carnivora from α-, β-hemoglobin set. (d) Ungulata from tandem alignment.

dactyla, and Cetacea) is supported with the tandem alignment, with classical familial relationships for the most part being observed (Fig. 7d). Thus, although bat monophyly may only be supported by a single or very few amino acid substitutions, the data are overall congruent with many existing views of mammalian evolution and therefore should not be discounted, contrary to the views put forward by Pettigrew (1991a).

There is an ever-increasing array of congruent molecular support for bat monophyly. Both of our DNA data sets, one composed entirely of coding DNA sequence and the other principally of noncoding DNA, support chiropteran monophyly. These findings are in agreement with recent studies involving mitochondrial genes: 12 S rRNA (Ammerman and Hillis, 1992), cytochrome oxidase II (COII) (Adkins and Honeycutt, 1991, this volume), and 12 S rRNA combined with COI (Mindell *et al.*, 1991). By contrast, we are aware of no molecular study of DNA sequences in which both micro- and megabats were included that argues against bat monophyly (with the possible exception of the Adkins and Honeycutt analysis, when they weighed all nucleotide substitutions equally). Both of our amino acid data sets (as well as the α and β set composed of 52 species used by Pettigrew) support chiropteran monophyly. There is also immunological evidence to support the monophyly of Chiroptera (Cronin and Sarich, 1980). Clearly, the evidence is beginning to weigh heavily in one direction.

Bat Diphyly

In any analysis of chiropteran origins, claims of bat monophyly should be compared with the contrary hypothesis. Testing of our data with several of Pettigrew's hypotheses illustrates the degree to which the bat diphyly hypothesis is unparsimonious. In comparing the IRBP data set with alternative phylogenetic hypotheses, we fixed the arrangement of the taxa pertinent to a particular hypothesis and, using the PTRFC program, allowed the other species to reach their most parsimonious topology. The trees and their scores depicted in Fig. 8a–c for IRBP are thus the least costly arrangement that still support a Pettigrew hypothesis. The least parsimonious hypothesis is one in which Megachiroptera is the sister group to Primates, requiring 54 additional substitutions to the most parsimonious score. Having Dermoptera as the sister group to Primates, followed by Megachiroptera, or having Megachiroptera and Dermoptera as a clade on the branch adjacent to Primates, were each equally unparsimonious (+46). We were able to identify only three putative synapomorphies that would support either of these latter two hypotheses (positions 415, 518, and 873).

The same comparisons using the ε data set were similarly highly unparsimonious. The lowest scoring topologies that still support the Pettigrew hypotheses are presented in Fig. 8d–f. The least parsimonious arrangement had a Dermopteran/Megachiropteran clade as sister to Primates, requiring 122 additional substitutions to the most parsimonious score. When Megachirop-

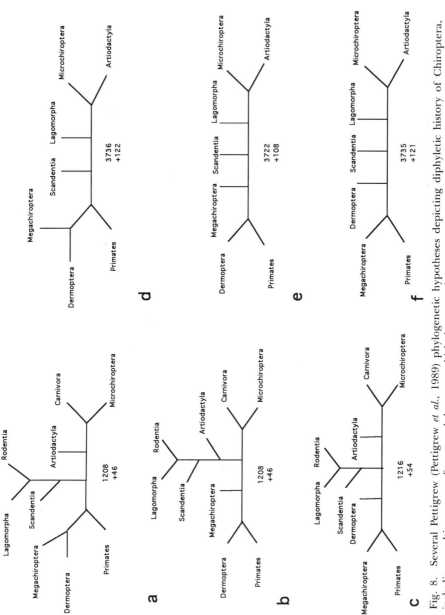

Fig. 8. Several Pettigrew (Pettigrew *et al.*, 1989) phylogenetic hypotheses depicting diphyletic history of Chiroptera, including resulting scores of trees and degree to which they exceed our most parsimonious arrangements for IRBP (a–c) and ε-globin (d–f) data sets.

tera was the sole primate sister group, the score was +121. Placing Dermop-tera closer to Primates was somewhat more parsimonious (+ 108). Only two positions were identified that would support either a megachiropteran/dermopteran clade (positions 2560 and 2821) or a megachiropteran/primate clade (positions 2861 and 2891) and only three that would support a mega-chiropteran/dermopteran/primate clade (positions 2830, 3213, and 3614).

Clearly, our results from the analysis of these coding and noncoding DNA sequences do not support the concept of bat diphyly. Instead, our data strongly suggest that morphological homoplasy is the likely explanation for the similarity in neural pathways common to megabats and primates.

Archonta

The concept of grouping primates, tree shrews, flying lemurs, and chi-ropterans in the superorder Archonta has fluctuated several times through-out this century from a position of favor to disfavor (e.g., Gregory, 1910; Simpson, 1945; McKenna, 1975; Szalay, 1977; Cartmill and MacPhee, 1980; Novacek and Wyss, 1986; Wible and Covert, 1987). Recently, Novacek and associates (Novacek and Wyss, 1986; Novacek, 1990) have suggested that there may be sufficient evidence to warrant some support for this superordinal grouping. They admit, however, that there are only a few characters specifi-cally supporting its monophyly. The Novacek hypothesis depicting archontan relationships has a Primates/Scandentia clade joined by a branch consisting of a Dermoptera/Chiroptera clade. Novacek suggests that the best supported subdivision within Archonta, based on a series of anatomical characters (some associated with flight and some not), is the grouping of Dermoptera and Chiroptera (Novacek, 1990). We compared our IRBP and ϵ data with the Novacek hypothesis and found the topology to be very unparsimonious (Fig. 9). With the IRBP data set, the least costly arrangement that still supported this hypothesis required the addition of 24 more substitutions over those of our most parsimonious tree. This increase in length was primarily the result of the inclusion of Chiroptera in the proposed superorder Archonta. The archontan hypothesis was also very unparsimonious with the ϵ data set, re-quiring the addition of 52 substitutions over the most parsimonious tree. The marked divergence of Chiroptera from Primates, Scandentia, and Dermop-tera, apparent in the analysis of both of these data sets, is in agreement with the results of the α- and β-hemoglobin and tandem alignment analyses (Fig. 4) as well as immunological data (Cronin and Sarich, 1980) and recent studies of mitochondrial gene sequences (Adkins and Honeycutt, 1991; Ammerman and Hillis, 1992). Thus, there is an increasing body of molecular data that sup-ports a much earlier split for Chiroptera than is suggested by the more clas-sical view. In addition, we are aware of no molecular data that strongly endorse a Dermoptera/Chiroptera clade. Our results from IRBP and the ϵ-globin locus, concomitant with those of Adkins and Honeycutt (this volume), strongly

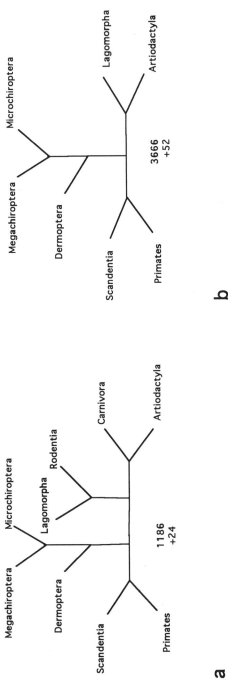

a

b

Fig. 9.   Hypothesis of a superorder Archonta as proposed by Novacek (1990) with resulting scores and degree to which they exceed our most parsimonious arrangement for (a) IRBP and (b) ε-globin data sets.

suggest that the features associated with flight suggested as synapomorphies for a Dermoptera/Chiroptera aggregate (Novacek, 1990) are in fact convergent homoplasies.

Since the beginning of this century, the possibility of a close phylogenetic relationship between tree shrews and primates has been a dominant issue in discussions of primate origins (see discussion in Martin, 1990). Recent paleontological information, however, suggests that early Eocene paromomyid plesiadapiforms share a close phylogenetic relationship with extant Dermoptera, and since plesiadapiforms are often identified as a sister group of Primates, Dermoptera may represent the closest living primate relative (Beard, 1990; Kay *et al.,* 1990). Our IRBP results do not resolve the issue of which order represents the primate sister group. The NJ tree depicted a rabbit/flying lemur/tree shrew clade as sister to Primates, and in the parsimony tree for the data set that excluded opossum the branch adjacent to Primates was composed of both a rabbit/mouse clade and a flying lemur/tree shrew clade. The flying lemur/tree shrew grouping broke at +5, to the most parsimonious score, with a rearrangement that had Dermoptera as the closest branch to Primates, with Scandentia as the next adjacent branch. A Dermoptera/Scandentia clade was also found by Adkins and Honeycutt (1991), in their study of the mitochondrial COII gene. In the ε data set, tree shrew grouped with rabbit. Placing flying lemur as the sister group of primates in this latter data set added only 2 to the most parsimonious score of 3614, in agreement with the weak strength of grouping for the rabbit/tree shrew/ primate clade (Fig. 3b). Both distance matrix methods suggested a close genetic relationship of Dermoptera with Primates; the UPGMA tree placed flying lemur between the anthropoids and the strepsirhines and the NJ tree had a flying lemur/tree shrew/rabbit clade on the branch adjacent to Primates.

The amino acid sequence evidence, although it does not include Dermoptera, suggests that there exists a primate supraordinal clade composed of Primates, Lagomorpha, Rodentia, and Scandentia. Both amino acid data sets support this clade, but the relationships between the orders are not clear. Although many of the clades within these amino acid trees are not strongly supported, there are two key features (in addition to their congruence with other sets of information and well-recognized relationships) that both of the data sets have: (1) they represent more densely the major clades of eutherian mammals than do most other molecular data sets and (2) they have proper outgroups. We have stressed elsewhere (Czelusniak *et al.,* 1990) the importance of phylogenetic density in any attempts to find correct phylogenetic relationships, and these alignments represent two of the largest bodies of comparative amino acid sequence data available for the analysis of eutherian phylogeny.

There is, thus, congruence between the various data sets (nucleotide and amino acid) for a close relationship of Lagomorpha, Rodentia, Scandentia, and Dermoptera with Primates, but the question of which mammalian order represents the primate sister group remains equivocal. There are, however, several significant clues emerging: (1) that the closest primate relative is probably not Chiroptera, (2) that tree shrews are no more closely related to primates than are flying lemurs, and (3) that Lagomorpha and Rodentia possibly occupy a close phylogenetic relationship with Primates and may in combination with Dermoptera and Scandentia comprise a primate supraordinal clade. This last point is perhaps the most controvertible, primarily because of the inclusion of Rodentia in this group. Several other molecular studies place Rodentia as a much earlier split in eutherian radiation (e.g., Li *et al.*, 1990; Bulmer *et al.*, 1991; Easteal, 1990). All of these studies, however, involve only a few taxa from a few orders, and we feel that they may suffer from a lack of phylogenetic density. In addition, we believe that the rapid rate of nucleotide substitution in the rodent lineage (Wu and Li, 1985; Li *et al.*, 1990) is undoubtedly confounding its accurate phylogenetic placement.

Glires

The concept of Glires, a superordinal grouping of lagomorphs and rodents, was first proposed by Gregory (1910), was subsequently endorsed by Simpson (1945), and has considerable morphological and paleontological support (summarized in Novacek, 1990). An impressive array of anatomical evidence has prompted Novacek and co-workers to suggest that Glires is perhaps the best candidate for a eutherian superordinal grouping (Novacek and Wyss, 1986; Novacek, 1990). There are, however, very little molecular data that would support this hypothesis. Our IRBP data provide a rare example of molecular support for the long-standing proposed relationship between lagomorphs and rodents. The strength of grouping in our most parsimonious tree for the Glires clade was 7, next to the bats the most strongly supported group. The Glires clade was supported by a total of 13 putative synapomorphies (positions 239, 343, 353, 535, 781, 814, 826, 886, 892, 976, 1049, 1052, and 1119). There were two trees with the same lowest score that broke the lagomorph/rodent clade. One of these joined *Cynocephalus* with *Oryctolagus;* the other joined *Mus* with *Tupaia*. Both of these alternatives were supported by 6 putative synapomorphies (*Cynocephalus/Oryctolagus* clade: 246, 289, 461, 481, 490, and 751; *Mus/Tupaia* clade: 228, 283, 338, 1027, 1069, and 1117). Although the "winning sites test" was not significant in this instance (p = 0.0835), the data are nonetheless highly suggestive of a Glires clade. Fewer positions were available for comparison in this case since the rabbit sequence was only represented by 935 bp (compared with the 1.1 or 1.2 kb of most other species). It is possible that when additional sequence data are obtained from along this exon, the Glires concept may be more strongly supported.

Our ϵ data set did not include a rodent; however, in other phylogenetic studies of ϵ DNA sequence, representing five orders of mammals (marsupials, rodents, lagomorphs, artiodactyls, and primates), a Glires clade was weakly supported when only the coding sequence was utilized (Koop and Goodman,

1988; Easteal, 1990), but was not supported with the noncoding region (Easteal, 1990). When the coding regions of α, β, and ε were combined with noncoding regions of β and ε, Glires was not supported (Easteal, 1990). The majority of the phylogenetic trees resulting from the analysis of globin genes in the Easteal study depicted a Primates/Lagomorpha clade. A close phylogenetic relationship between Lagomorpha and Primates has been suggested in other molecular studies (e.g., Penny *et al.*, 1991) and is consistent with our view of lagomorphs as a member of a primate supraordinal clade.

In earlier molecular analyses of eutherian phylogeny, Miyamoto and Goodman (1986) and Czelusniak *et al.* (1990) emended the classical concept of a superorder Glires to include Primates, Rodentia, and Lagomorpha (Miyamoto and Goodman, 1986) and later Scandentia (Czelusniak *et al.*, 1990). Although this application of the term does not correspond directly to the traditional concept, it emphasizes the proposed close relationship of these orders to Primates. Whatever label one wishes to attach to such a superorder, a good case can now be made for inclusion of the order Dermoptera. Although the precise relationships between these orders (such as the possibility of a traditional Glires) are highly uncertain, we interpret the present overall evidence from amino acid and nucleotide sequences as providing some support for a primate supraordinal clade comprised of Primates, Dermoptera, Scandentia, Lagomorpha, and Rodentia.

Conclusions

It is clear from the information presented and discussed above that there is a growing body of molecular evidence, from several distinct molecular systems, for chiropteran monophyly. Not only is there congruent evidence against the "flying primate hypothesis," but there is also a good deal of evidence to suggest that Chiroptera is not a member of the proposed superorder Archonta, and in fact occupies a position genealogically distant from Primates. We suggest instead that a more likely primate supraordinal clade consists of Primates, Dermoptera, Lagomorpha, Rodentia, and Scandentia, although the precise relationships among these orders are highly uncertain.

Congruence between molecular data sets is certainly one of the best means of evaluating different phylogenetic hypotheses, and with the advent of PCR this has proven to be logistically quite feasible. Mammalian interordinal relationships, however, represent a particularly challenging problem. Short internodes and relatively long terminal branches are likely to be typical of all trees depicting the phylogenetic relationships of mammalian orders. Increasing the phylogenetic density, by better representing each order, will reduce the length of the terminal branches and thus counteract the tendency for these long unbranched lineages, with their high proportion of homoplasy, to cluster together. Concomitant with the emergence of these larger data sets, there must also be more rigorous phylogenetic analyses of the molecular data. As Templeton (1992) has recently pointed out in his reanalysis of the data presented in Vigilant *et al.* (1991), a single heuristic run of some phylogenetic computer program is extremely inadequate when one is dealing with large data sets.

Appendix I: IRBP Sequence Alignment

Gaps are indicated with an asterisk; nucleotides the same as in the human sequence are indicated with a period.

Hsa: *Homo sapiens* (human)
Tsy: *Tarsius syrichta* (tarsier)
Gcr: *Galago crassicaudatus* (galago)
Cva: *Cynocephalus variegatus* (flying lemur)
Tgl: *Tupaia glis* (tree shrew)
Phy: *Pteropus hypomelanus* (megabat)
Tbi: *Tonatia bidens* (microbat)
Tsi: *Tonatia silvicola* (microbat)
Fca: *Felis catus* (cat)
Bta: *Bos taurus* (bovine)
Ocu: *Oryctolagus cuniculus* (rabbit)
Mdo: *Mus domesticus* (mouse)
Dvi: *Didelphis virginiana* (opossum)

```
                                                                                                                        120
Hsa    CCCGGAGAACCTGCTGGGCATGCAGGAAGCCATCCAGCAGGCCATCAAGAGCCATGAGATTCTGAGCATCTCAGACCCGCAGACGCTGGCCAGTGTGCTGACAGCCGGGGTGCAGAGCTC
Tsy                       ....................G.T...A......T....................C..CA.T........A.....
Gcr                       ..........G..........G......G...T....A......CAG.............T..A...........
Cva                       A..T.G..........G..............T..A.......CA............TT..............
Tgl                                                  ....CAC........G..................
Phy                                                               .....A.....T..
Tbi                                                  ....CA...C.......TT...........
Tsi                                                                 ..............
Fca                       .......T.G......C..CGC.....G.....T.....T......CA.......GA.T.............
Bta    ...A........A....G.......G.......G........A..T..G.........TCT...........T.....T....CA...T.......T...........
Ocu
Mdo    ...T.........A...A......A.C......TG.......G..G.....T..C.........G.................T..........CAA.........T..A..C.....T..
Dvi    ...A..A.....AA.........A....T...TG.A..A..A.........TGGG.....C...GAT.....T..T..C..A.T............C.........A..T.....G.TG.
```

```
                                                                                                                        240
Hsa    CCTGAACGATCCTCGCCTGGTCATCTCCTATGAGCCCAGCACCCCCGAGCCTCCCCCACAAGTCCC***AGCACTCACCAGCCTCTCAGAAGAGGAACTGCTTGCCTGGCTGCAAAGGGG
Tsy    .T.......C..........A...................GTT........AG....C...***..ATA.......T..TA...............G.............AT..
Gcr    ...A..T..C......G....................T....G.....GG..G.C...***.A........A....A..C.G...........G.....
Cva    .T....T..C...................A..........TT...G.....AG....C...***G...T.......A....A.GC.......G.....C...........G.AA..
Tgl    ........C..................................TG.T.GT...A.C...AG....C...CCT....T.TGA.......A.CC.G....G.....CA..CAT.....GGA...
Phy    TT....T..C.........................TG...T.AT.G......AG..G.CG..***.........CA...A..C....A..........TCAA.....G.AA..
Tbi    .T....T..C.....T..................TG...T.ACAG....T.AGA..A..A.***.......A....A..C...........CA.T....A...
Tsi    .T....T..C.....T..................TG...T.ACAG....T.AGA..A..A.***.......A....A..C...........CA.T....A...
Fca    .T.......C..............T..A..C.........T....G.....GG....C...***.......A....A..A.......A........C...........G.A...
Bta    .T....T..C..........................T..G.C.T..G.G..CT..***....G...G.A....A...CT.......A.CA.C..AG.......GGAT...
Ocu                          .GGC..C...***.....C..G..A...TA.CC.G.........G..T......G.AA..
Mdo    .T....GT..C..A..A...CT........T..C.........T....T...G.....AG....CA..***...TG.......A....A.CCG...A.........G...CA.A.A..G...AA
Dvi    ...C..T..C..A..G..A..G.......T...........TT..T...A.....AG..T.....***.AA.T.TG...A.G..A.TC.G............CA..CT....C..GCAAAT
```

```
                                                                                                    360
Hsa  CCTCCGCCATGAGGTTCTGGAGGGTAATGTGGGCTACCTGCGGGTGGACAGCGTCCCGGGCCAGGAGGTGCTGAGCATGATGGGGGAGTTCCTGGTGGCCCACGTGTGGGGGAATCTCAT
Tsy  .....A................C..C.................TGA.A....A.....AC....A..A.AAC....A.CA.......A......C...A...G...C.
Gcr  .......T.............C...........T......A...GA.A...T....A.........A.C....CT.........C........G.....
Cva  .A...A.........C....C..C.......T....A..TGATC....C......A....A....A.C..........A.....A.C...A.C.G.....
Tgl  .A............C...........AC....GA.C...AAA.........A.CA.C....C........A......C...AT...G.....
Phy  .A....C..T..........C..........A....GA.A.....AAT....A..AG....A.C....GC...T.A....AAA.TA.C...A..G..T..
Tbi  .A...A.T.C..CA.C.....A.C.........T........TGA.A...A.........A.......AAC.T...A.....A.TA.C...A..G...
Tsi  .A...A.T.C..CA.C.....A.C.................TGA.A...A.........A.......AAC....C.........A.TA.C...A..G...C.
Fca  .A........C..........C........T......GA.A....AA.......A.G....A.C....GC........AG...C..A...G...
Bta  ...T.......A....A.C..............GA.A.........A.......A.C.A..AGC.......A....C...A...G...G.
Ocu  TA...AA.......A.C..ACAC...........TGA.C...........A.C....C.G..........T..C....C.A....
Mdo  .A............T....C..C.........A..A...GA.C...T...A....A.....GA.C........T.A...AG...T......A.C.G....
Dvi  GA..AAGT.CC.....A..........C......T.....AA.A.T..TA.A....T.........G.AGAG.AAG.C.........T.....AATA..A.C...AAA..G.....
```

```
                                                                                                          960
Hsa  CACTCTGCGCAGCGCCCTTCCAGGGGTAGTCCACTGCCTCCAGGAGGTCCTGAAGGACTACTACACGCTGGTGGACCGTGTGCCCACCCTGCTGCAGCACTTGGCCAGCATGGACTTCTC
Tsy  G.....T.........T.G..G.....C....G......G........CG...C..A.........A.A...C....A......G.........GC...C................
Gcr  ...........G...C...T..G.G....G......CG...C...............C....GG......C..C...............
Cva  ...C....C.T....G...A..G..GG...T.G......C....G................C....TG...A...C...C...............
Tgl  ...A......CA...G..G...G..G...C.T..GA....C...C..A.........C......C...G.G....GC...G.....G...........
Phy  ...C.......A...G..G..A..CA.G.GGC....A....C....C...........A..........C....GG....A..A..T...C...............TG.
Tbi  ...A.....G....G...AA..CA.G..GCA....G......C...C..........C......AC...TG....C...C...C............A.
Tsi  ...A.....G....G...AA..CA.G..GCA....G......C...C..........C......AC...TG....A...C...C............A.
Fca  ..........C.G...G......A..T...GC...G......C.G..C..AC......C...............G......C...C..A...A.............
Bta  ..TG.....G......G......A..CA.T..GC....T....CG...CGC..G..................G......AGC...C...GC.....C.G.
Ocu  ...C...C.G......G..G...C.G...G.C...A....A.AC...CC...A..................G.......T...C....C............
Mdo  ...C....TC.T....G.......T...T.C.A.T...A.C..CC.......T....AT.A......A....GGG.........A....
Dvi  ...C.....G..AAGA..G..A.CCA.T..AC.T...AT...A...T.AC..A.T.........C.......T..G..A...G.G....C..T...C.CA.TGT...C...AT..
```

```
                                                                                                          1080
Hsa  CACGGTGGTCTCCGAGGAAGATCTGGTCACCAAGCTCAATGCCGGCCTGCAGGCTGCGTCTGAGGATCCCAGGCTCCTGGTGCGAGCCATCGGGCCCACAGAAACTCCTTCTTGGCCCGC
Tsy  ................G..............C.............T.....C.........C..TG..AA....A...GGCT..G.GA.T...A
Gcr  ...A......G.....G........G....A..T...T..C.A.C....A.T..A...G.TG.TA.A....G.............C...G.G.A
Cva  .T.T....C......G..C.....G......C........A..T..C....C..........A..G..AG..AA...G....CT....CG....T.A
Tgl  .GA.........T.....G.GC.....G..G.........T......A.T..C......C...G.....G..AG..AA...G...G.CT.CGTGGA...T.A
Phy  .G.A.............C.....T.......C.T..T......T..C..C...........TA...GG...CT...G..A
Tbi  .T.........T...G.....A...G......C.T..TT........T........C.T.......A..GC...A......A...CT...T.G...T.A
Tsi  .T.........T...G.....A...G......C.T..T.........T........C.T.......A..GC...A......A...CT...T.G...T.A
Fca  .G.A..A......G....G..C..........A....C.T...A..A.T........C.........G..G.GA....AG..CG.CT....GC....A
Bta  .T................C..........T........T.............T........C.......A...AG.TGG..A.A...A..G.CT.....T.A
Ocu  ...........T..A....C....T..............G......T..T........C...........T...G.....G........G...G..T.A
Mdo  AG.T........T..A..G...G..A..........T..T....T..G..................A.G....C..A...G...CT.CT.C..AA.A..T.A
Dvi  .T.A...C..A.A........T...CG..........A..A.....A...T..A.....C..........C..T..G.TTC.GAA...TGAG...GTCA.AA..GACT.T.A
```

```
                                                                                                          1200
Hsa  GCCCGACGCTG***CAGCCGAAGACTCACCA***GGGGTGGCCCCAGAGTTGCCTGAGGACGAGGCTATCCGGCAAGCACTGGTGGACTCTGTGTTCCAGGTGTCGGTGCTGCCAGGCAA
Tsy  .G.T.GG.TC.*********.....C.G.GGGG....C...T.C.G.G.............A.GC.....G...C.A....T.........AC.........G.....
Gcr  .G.T.GAA...*********...GG.CT.G***...A......G.................G.............C.....A...A.....G.....
Cva  .G.T..AA...*********.T...C.G...***.A...T...CACAA....C...T....TC.A..G..C...TG.CAC.........T.....G.....
Tgl  .G...GA....*********.CA.TG.C..G***..A.A......G.T..A..CACA......CTC..T..G..C..T............C.....T...T...C.
Phy  .G.T.GA....*********.CT...C.C...***..A...T....CAG....A...T..AT.GC......C..T........A.....T....A...
Tbi  ...T.GA..C.*********.T...CT....***.A.A.......CAG....A...T..T.GC.....GA.C....T.....T.......A.....G..T.
Tsi  ...T.GA..C.*********.T...CTC...***.A.A.......CAG....A...T..T.GC.....GA.C....T.....T.......A.....G..A.
Fca  .GTT.G..AG.*********.T..GC.T..G***.A..A.T...CAC.G...C.GA.CT......GC....G..C....T..A..T.......A.....
Bta  .GAA..A...*********...AC.T...***.A..C..T..G..AG...C............G.T...GG..T..........T...T.....G.....
Ocu  .G.G.CA.T..*********.G...C.T...***.A..CA.....C
Mdo  .A.T.GGC..A*********.T..G..C...***.CA.CCA....C..G....CAC...A..A.A.GC....AGG..C.............T.......C.....G.....
Dvi  AGTTAG.AAG.AGG******C...TA.TA..***.CT.......ACTCAC.......T...A.TCCCAGA.......C.....A.............A..A...T.......T..
```

```
                                            1255
Hsa  TGTGGGCTACCTGCGCTTCGATAGTTTTGCTGACGCCTCCGTCCTGGGTGTGTTG
Tsy  C.....G.............C..C....A..T....G..G....
Gcr  C...........T..C..C....C..T...G..G....
Cva  .............T..C.G...A..T.......G...AG.A
Tgl  ...C...........C.TG...C........G.
Phy  C...........
Tbi  ......T.......T..C..A...T...T....T..G...AG.CAC..
Tsi  ......T........T..C..A...T...T....T..G.
Fca  .........
Bta  C................C.....C....T.....T.......AG...C..
Ocu
Mdo  ...............T....A.....A......G.T..G...AGAC.C...
Dvi  C........TT..A.G..T...GAG.........T...G..A..T.CACCC.G.
```

Appendix II: ε-Globin Sequence Alignment

Gaps are indicated with an asterisk; nucleotides the same as in the human sequence are indicated with a period.

Hsa: *Homo sapiens* (human)
Ptr: *Pan troglodytes* (common chimpanzee)
Ppa: *Pan paniscus* (pygmy chimpanzee)
Ggo: *Gorilla gorilla* (gorilla)
Ppy: *Pongo pygmaeus* (orangutan)
Hla: *Hylobates lar* (gibbon)
Mmu: *Macaca mulatta* (rhesus monkey)
Cal: *Cebus albifrons* (capuchin monkey)
Tsy: *Tarsius syrichta* (tarsier)
Gcr: *Galago crassicaudatus* (galago)
Lfu: *Lemur fulvus* (brown lemur)
Cva: *Cynocephalus variegatus* (flying lemur)
Tgl: *Tupaia glis* (tree shrew)
Csp: *Cynopterus sphinx* (megabat)
Mly: *Megaderma lyra* (microbat)
Ocu: *Oryctolagus cuniculus* (rabbit)
Chi: *Capra hircus* (goat)

```
                                                                                                                     120
Hsa   GAATT*CCTGGTTTTGTCTGTGTTAGCCAATGGTTAGAATATATGCTCAGAAAGATACCATTGGTTAATAGCTGAAAGAAAATGGAGTAGA****AATTCAGTGGCCTGGAATAATAACA
Ppy   ......*..................G.A.............C.........C.........A.................****................................
Hla   .....C..........................C.........C.......C...A.................****...............................
Cal   .....C...C..A........G.G.....A......T.....A.C.......TG..........A.................****......AA..............
Gcr                           CT.........TTC..AA...GA.C..A......G...TAA..****......T.........C..CT.
Ocu   .G.A.TT...AC...A...A.C..CTT...****************************************..A....C..GAAAG.G..GTAGAG.....A.AAT...........CT.

                                                                                                                     240
Hsa   ATTTGG**GCAGTCATTAAGTCAGGTGAAGACTTCTGGAATCATGGGAGAAAAGCAAGGGAGACATTCTTACTTGCCACAAGTGTTTTTTTTTTTTTTTTTTTTTATCACAAACATAAGAAA
Ppy   ......**.......................................................................................******.................
Hla   ......**.................................................................................****A.................
Cal   G.....**......G..........................C.G......G.T....A.........*..........A.........C........AA..TG..G.G.......
Gcr   .....AGAG....C.........A.....CA.....AA....GA...A****.T...T.C.TG...T..G............**********......CT.T.......
Ocu   G...AGG.G........T.A..T.A.....*.....ACT..T.GA.T.T...G.A.GATA.AGT.G...C.TG...T.G..A.....GCA..*************...CCTT..G...G

                                                                                                                     360
Hsa   ATATAATAAATAACAAAGTCAGGTTATAGAAGAGAGAAACGCTCTTAGTAAACTTGGAATAT**GGAATCCCCAAAGGCACTTGACTTGGGAGACAGGAGCCATACTGCTA*********
Ppy   ...............G....................T..................**.........A.............GAGGTTCTT
Hla   .C.............G.................T...........T...**.........A.............*********
Cal   .C...T.....G..G.....C......TTAA....GG.......**.....T.....T.............A.............*********
Gcr   .G.........GC..G..AG.....A..TA.....GA.T...C......TA.......AT...GAT.T..G.....A...G..A..G..TA.T..A.TG.T.G*********
Ocu   TA..C.....G..C..GC..AG.......A.CA.....GA.T..GG.......A...C..**T..GATG*...G.*....G....A......AA.......G...A.*********

                                                                                                                     480
Hsa   **AGTGAAAAGACGAAGAA*CCTCTAGGGCCTGAACAT**AGGAAATTGTAGGAACAGAAATTCCTAGATCTGGT*GGGGCAAGGGGAGCCATAG*GAGAAAGAAATGGTAGAAATGGA
Ppy   CT********************...........A..GC................C.........*.............*.........A.........
Hla   **..............*.........A..GC.........T.C.......*.........A.........
Cal   **...A.......C....*.....AA..A..GC.........G.......C.....TC*A.........A..A...T.*.G.........A........****
Gcr   **...A..*G.G.C..A..*.....G....A.C.A..GC.A.......A.....C...C..G.....AA....C.ATA....ACC..*.....A.A.......****
Ocu   **.ACA....G...CC....G.AC.C.T.T.ACC.GA..GC.A....C.A.......A......T.C.TG.A.A.CAA.A...A..A..AAC...*AG........G.A......****
```

```
                                                                                              600
Hsa  TGGAGACGGAGGCAGAGGTGGGCAGATCATGAGGTCAAGAGATCGAGACCATCCTGGCAAACATGGTGAAATCCCGTCTCTACTAAAAATAAAAAAATTAGCTGGGCATGGTGGCATGCG
Ppy  ......G...................C........................................................................................G.........
Hla  ......G...............C................T...C...........A.......C.......C..............
Cal  ************************************************************************************************************************
Gcr  ************************************************************************************************************************
Ocu  ************************************************************************************************************************

                                                                                              720
Hsa  CCTGTAGTCCCAGCTGCTCGGGAGGCTGAGGCAGGAGAATCGTTTGAACCCAGGAGGCGAAGGTTGCAGTGAGCTGAGATAGTGCCATTGCACTCCAGTCTGGCAACAGAGTGAGACTCC
Ppy  ..............A...............................A...........C...........
Hla  ..........A....C................................C...........G.......T
Cal  ************************************************************************************************************************
Gcr  ************************************************************************************************************************
Ocu  ************************************************************************************************************************

                                                                                              840
Hsa  GTCTCAAAAAAAAAAAAAAG*********AAAGAAAGAAAAGAAAAAGAAAAAAGAAAAAATAAATGGATGTAGAACAAGCCAGAAGGAGGAACTGGGCTGGGGCAATGAGATTATGGTG
Ppy  ...................AAAAAAAAAG..A............*G.........T......................A..............
Hla  ...............***************************.........G....G...............A..............
Cal  ************************************************.....G...TG.....G.A...........T......C....
Gcr  **********************************************.A.....G...TT...T...*C.A.......A.A.CTC.CT......A...
Ocu  ***************************************************CAT.C.GG...TT.......TAT.A....A.TTAA..AAT.CT....A..A.T

                                                                                              960
Hsa  ATGTAA*GGGACTTTT*ATAGAATTAACAATGCTGGAATTTGTGGAACTCTGCTTCTATTATTCCCCCAATCATTACTTCTGTCACATTGATAGTTAAATAATTTCTGTGAATTTATTC*
Ppy  ......*.........*.............................................................C.........C.......*
Hla  ......*.........*.......................***.......................C.........C.......*
Cal  ......*..C......*C.......G.................C...........T..T....GC...........C............C.......*
Gcr  .C...GCAA......TT............A.G...C....C.TTC...G..GC...*......T......C.A..T..C..TC.C..CG....C.C..C....*.TC..G..**
Ocu  .....GT..A.TG...*C.G.......A.G........C..C.GT...*....C.......T.T.TG....C.AC.T.T..CTT....C.A...T...C....CC.CTC.TT

                                                                                              1080
Hsa  CTTGATTCTAAAA**TATGAGGATAATGACAATGGTATTATAAGGGC***AGATTAA**GTGATATAGCATAAGCAATATTCTTCAGGCACATGGATCGAATTGAATACACTGTAAATCC
Ppy  .............**...........***.....**.........G.............A...C..........
Hla  .....C......***..............***.....**.......G...G..........
Cal  ...........AA**......................***.C.....**.CAG...C...G..G...T..............A..A.....
Gcr  *..C..AG.C...AAA..A..G.......C..TA.....CT...CAC**...G.A*AA...C....C.GAG..C.......CT...T....G..C......T.*.A.C...G.
Ocu  ...T....C.****T.CC.T......C...T....G.CTCA.TACAG..CAGTGAAA.A.C.A..GGA.T**.TCT....A.G.A*..TCAAT.ATTC.C.*......*.A....GT

                                                                                              1200
Hsa  CAACTTCCAGTTTCAGCTCTACCAAGTAAAGAGCTAGCAAGTCATCAAAATGGGGACATACAGAAAAAAAAAA*****************GGACAC*******TAGAG*GAA******
Ppy  .....................C.........C...........****.....******************......*******.....*...******
Hla  .................G................C...........****.....******************......*******.....*...******
Cal  .......T..C...T.A....G.......A.........C..CA.....*..A..........AAAAAAAAAAAAAAAAAA......AC*****.....*..TTTGAA
Gcr  ..G..*....T....TC.A..TA.........A.......T.C..******...G.C.TT**********************......G..CTATGGC....CT..******
Ocu  GG.T..A....TGT...C.A..T.C....AT....C.........C.TCA...GGC.T..A...C..****************......G..******T...*A...******

                                                                                              1320
Hsa  ****************************TAA**TATACCCTGACTCCTAGC****CTGATTAATATA*TCGATTC*AC*******TTTTTTCTCTGTTTGATGACAAATTCTGGCTTT*AA
Ppy  ************************...**................****..A.......*.T....*.*.*******...................
Hla  ************************...**................****..A.......*............*..
Cal  GTCTTTGCTATCTGTGGCTACAAACAC...AA..C.T....T.....****..A...G.....*.T....*.*.*******...........T..*..
Gcr  ************************CAC..A*.T..A......T.CAT****.......*..A.AAAT..*******..A......A.......C....*...T....T..
Ocu  ********************T.TG.A...**..A.CCTGGTGA..ACAGC.C.C.C.A..CAAAT.AATTAAC..A..C..A.......T.T....C....G..*..

                                                                                              1440
Hsa  ATAATTTTAGGATTTTAGGCTTCTCAGCTCCCTTCCCAGTGAGAAGTATAAGCAGGACAGACAGGCAAGCAAGAAGAGAGCCCCAGGCAATACTCACAAAGTAGCCAGTGTCCCCTG**T
Ppy  ....C.................T.........................................................A...........**.
Hla  .........................T.........................................................A...........**.
Cal  ..............G....T..GA..........T........................*..........C.C...CA............CA**.
Gcr  ..........CA*********.........T...CA.TGA.........G....C...........**...A**..T...T.AG..A...*.C........G....**..T....CTG.
Ocu  ..........*********.G....T..G..CATGGA.T.*.......T...AG.....A..C.AA...AG....****T....A....TC.CTGG....GCT.TC...*********

                                                                                              1560
Hsa  GGTCATAGAGAAATGAAAAGAGAGAGGATTCCCTGGAAGCACTGGATGTAATCTT*TTCTGTCTGTCCTCTCTAGGGAATCACCCCAAGGTACTGTACTTTGGGATTAAGGCTTTAGTCC
Ppy  ...........................G...*.........G..............T.................
Hla  ........****........T.......*..................T..................T.
Cal  .....C......*****.....G....A...C.C.GC........*.........T.....................G..............
Gcr  ...................C.G....G...T.......CC*...C..GGT.TT...T..C..................C.A.AT
Ocu  *****************************.....A...GA...C...G**.......G.....A.....T..T..**...........A...G..T...*.....
```

```
                                                                                                      1680
Hsa  CACTGTGGACTACTTGCTATTCTGTTCAGTTTCTAGAAGGAACTATGTACGGTTTTTGTCTCCCTAGAGAAACT*AAGGT*ACAGAAGTTTTGTTTACAATGCACTCCTTAAGAGAGCTA
Ppy  ....................................................................T...........*.....*.............................
Hla  ...................T....................T................G..*.....*..........A.............
Cal  ..........C.........AG....GT.....T.........A....T...*..........T..........T................T.......T.
Gcr  .GGG...A.T.TT......C...........G..C....T.....A.........A....*..A.*C...G....G.G*.....GT..G..ACC....CA..
Ocu  T....CA..TCC....TC.C...C......C...C..G..T...C...C.T..CCCC......*A.G..A.....*...A.CT..A...***......G..G.T....***.G.GA....*

                                                                                                      1800
Hsa  GAACTGGGTGAGATTCTGTTTTAACAGCTTTA**TTTTCTTTTCCTTGGCC*****CTGTTTTTGTCAA**********CTGTCACCACCTTTAAGGCAAATGTT*******AAATGCGC
Ppy  ...................................**.................*****...................**********..............A..******....
Hla  ....A............................**.........T........*****............**********.C...............CA..******......T
Cal  .....T...**********************************************...,,,**********...........T.....TA...A..*******...CAT..
Gcr  .......A.A.A.C..CA.G.C.........ACCC...TG....*.......AG*GG....C........TTGTCACTATC.ATAAT...AA...T.AA.TC..***********T.TG..
Ocu  C..G...A...TG.....CC.C...CA....CAT....TG.G.GTG..TGTGCTGTG.TGC.....TTT**********.CT.G...TT....TT.T..CCAACCTTGACCC....AT.*

                                                                                                      1920
Hsa  TTTGGCTGAAACTTTTTTTTCC*******TATTTTGAGATTT*GCTC*****CTTTATATGAGGCTTTCTTGGAAAAGG*AGAATGGGAGAGATGGATATCATTTTGGAAGATGATGAAGA
Ppy  ...................*******...............*....*****...G...................*.............................
Hla  ...................*******.............*....*****..C................*...............................
Cal  ...............TT*****C.........C...*.....*****...TG...T...............*..........G.........C............
Gcr  .G*******....G*......*******...CC.GAG.C..A*.T..A********T......A..**........GG....A...A...TT.C.T....T...G....***.T
Ocu  ********.....AC...AG.CCTAACACT..CCAGTC..AT...AGCTG..C.C.....AT...............*G.......AT.......GC...C.....A**...T.CAT

                                                                                                      2040
Hsa  GGGTAAAAAAGGG***GACAAATGG*************AAATTTGTGTTGC*AGATAGATGAGGAGCCAACAAAAAAGAG***CCTCAGGATCCAGCACACATTATCACAAACTTAGTGT
Ppy  ...........***............**************..........A..*..............T.G.***...T...................
Hla  .............***............**************..............*.................G.***..................
Cal  .T..T....G**..............*************..........A..............G.***T.C....TGA...G....T.......
Gcr  ...G...G...A.AGT..GT.GA..**************G...G...CA.G.A*G.GAG...C.CA..GA..G...****G.***T....C...CC......T..T****..
Lfu                                                                          .........T...T****..T.
Ocu  ..AA...G...A.AGT..A..CATCGTAATGTATGACA..T.G...TC...TT.GAG..***..A......G.GT..G.ATGT....C....GAA.GC.............C....A
Chi                                                                          ..GG......T.C..C.A.A.

                                                                                                      2160
Hsa  CCATCCATCACTGCTGACCCTCTCCGGACCTGACTCCACCCCTGAGG*ACACAGGTCAGCCTTGACCAATGACTTTTAAGTACCATGGAGAACAGGGGGCCAGAACTTCGGCAGTAAAGA
Ptr  ....................................G....G..................................................
Ppa  ....................................G................................................
Ggo  ....................................*................................................
Ppy  ...................A.............A.G.....................A..........................G....
Hla  ...................G....................................C............................
Mmu  ...................*....................................................T............
Cal  .......C.........T.T.CA........G..G....A............C....T....G...*...A....
Gcr  G.C.A.T.............G.T...A.G.......*..*..G..C.T...T.........C..C....C.......A....T.......A....GG..G
Lfu  ..G..A.T.............T.....G..........G..C.T...T.........C..C..A..CA..G.A...........A....GG...
Ocu  A.CA....******......A...GGT.G...G.......A..A.G....TTC...T.......CA....TG.AA.......C..A..A..GGG..
Chi  A....**..CT.......A..AG.T.............G....CCT.A..........CA..G..A.G..G..G.A......G....G..A......

                                                                                                      2280
Hsa  ATAAAAGGCCAGACAGAGAGGCAGCAGCACATATCTGCTTCCGACACAGCTGCAATCACT****AGCAAGCTC****TCAGGCCTGGCATCATGGTGCATTTTACTGCTGAGGAGAAGGC
Ptr  .....................................................****.........****...A...................
Ppa  .....................................................****.........****..........................
Ggo  .............................G....A..................****.........****....C..................
Ppy  ...........C.........................................*...C****.....A...****..................
Hla  ...........C.........................................C****.........****...T.................
Mmu  .........T.A........T...............T...........G...C****.........****....................
Cal  .........CG....A....C...........................A....C****.........****...A...A.........C.....T
Tsy                                          ......****.....T.G****C..A...A...........C.........A..A..T.
Gcr  ...........C.T...A.AT........T.C.C..AT...T.GT....TG....C****.........****C..A.T..A..C.............A...
Lfu  ...........C......A......A..C....T.......TG..G.C****.....CCGCC..A.T..A..C..................T.
Ocu  .........AG.CTT..A.........A.G......T.....TT..TG...GA.CACC.....****C..A.G..A..C........C.....C......A...TG
Chi  ..........C.GCATCCA...........G.CT......T..TG.TT...TG.....C****T.T.......****CAC.A.T..A.................C...........
```

```
                                                                                                            2400
Hsa  TGCCGTCACTAGCCTGTGGAGCAAGATGAATGTGGAAGAGGCTGGAGGTGAAGCCTTGGGCAGGTAAGCATTGGTTCTCAATGCATGGGAAT**GAAGGGTGAATATTACCCTAGCAAGT
Ptr  .................................................................................................**......................
Ppa  .................................................................................................**......................
Ggo  .................................................................................................**......................
Ppy  .................................................................................................**.............G........
Hla  .........................A.......................................................................**......................
Mmu  ........................................A.....................................................T...GC**....................
Cal  ....A............................................................................................**..........G....TT.T.....
Tsy  .T...T.........G.T...............T..A.............T.................C...........A...A...**...A.....G.TG..G......
Gcr  .ATTA...TG.........G.A...G....A.A...........A..............A..C.........G..........**A....G......A..T..G....AC
Lfu  CA..A..CTG.........G.....G.................C...............A....T...G...........**A...AG.......A....G...A.
Cva  ..................................................C.....C.....TG....GA**......GT...G..TT.G.T..A.
Tgl  ..................................................AT...A.......GGAGA...........GG.T.*.G...AA
Csp  .................................................G.A.C...G.T.........GA**.G.T.A...CC...A.G...GAA..AC
Mly  .................................................G.AGC...G.T.........A...A**...CA...G...A...T.G.A..AC
Ocu  CATTA...G..AG.A....G..C..G....CA.C..T...A..........T...........T.............A.T...T.GA**...A..CA.....G.....G.T.GAC
Chi  ...TA.....G.....G....AG.C................C..G..TC..........GAA.G...ACT...TG.GGGA..*******.T.........G.G...G....A.
```

```
                                                                                                            2520
Hsa  TGATTGGGAAA*GTCCT*CAAGATTTTTT**GCATCTCTAATTTT*GTATCTGATATGGTGTCATTTCATAGACTCCTCGTTGTTTACCCCTGGACCCAGAGATTTTTTGACAGCTTTGG
Ptr  ...........*.....*...........**..G........*............................................................
Ppa  ...........*.....*...........**..G........*............................................................
Ggo  ...........*.....*...........**...........G............................................................
Ppy  ...........*.....*...........**.A.........*...............G............................................
Hla  ..........G*.....*...........**...........*...............G............................................
Mmu  ...........*.....*...........**...........*...T...........G............................................
Cal  ...........*T...*..G...G..*A..........*...........T.....................................A......
Tsy  ...CCCA..T.*A...*T..A......*GA.T....T......*T.G...C.......C.T.C..G............T..................A......
Gcr  ...CCA.....*.....*A.......**GA.....A.......*CC.C..T......A.G.A......G...T.....C..................C.....G.C.....
Lfu  ...CCA.....*.....*T.......**GA.........*CC....C....C....C..C..C..G..........C................A.......C....T.AT.....
Cva  ...CCA.....*.....*G.........GA.TG.T.T......*CACC...C.C....C......G.....G......C................G..C......A.....
Tgl  ...CCA....A.....*..........*A...G........*TA.....AC...C..C..G.T..G...........T..................G..C......A.....
Csp  ...C.A....ATCTT.*T.......A*T.AG.....G.....*CC..T.TCC.....TA...C....G..T.G....C..................G..C......A.....
Mly  ...C.A....*...T**T............C*T.AG.....G....TC..C..T.....TC.C.C......G..T.G..C..C................G..C......T.....
Ocu  ...CCA.....*T.G.*..AG.....***.T.......G.....*C...T..C.G.T..CC...CAT.....G...T.....C..................A.....C......A.....
Chi  C.GCCA.A...*A.T..T...A.A.C.****.AG.TG..G.....*CC.....C...T.TC...C.......G.....G.......C..................G..C....T.......
```

```
                                                                                                            2640
Hsa  AAACCTGTCGTCTCCCTCTGCCATCCTGGGCAACCCCAAGGTCAAGGCCCATGGCAAGAAGGTGCTGACTTCCTTTGGAGATGCTATTAAAAACATGGACAACCTCAAGCCCGCCTTTGC
Ptr  ....................................................................................................
Ppa  ....................................................................................................
Ggo  ....................................................................................................
Ppy  .......A.......................................................................................A..A.......
Hla  ...........................................................................................A..A.......
Mmu  .......A..............................T............................................AT.A.......
Cal  ...T..A..C................................C..........................................A..A.T.......
Tsy  ........C...T.......TA...A..................C........G.............T.........G...GGT..G.....
Gcr  ........C...G.........A.....................A........C........A..G.C..........GGT.......
Lfu  ........C...G.........A.....................C........G..G.C..........GGT.......
Cva  C........T..G.........A...........A..................................G..T..........GG.......
Tgl  C.....T..C....A.........A.................C.................C.......G...T....AAGT..T...A
Csp  C........C..G.........AA......T...A.........C.................G...T..........AAGT.......
Mly  C.T...C..G..........TA...T..................G...C....C.......G.T.......T..GG.A.......
Ocu  C..T....C...T...........A...A..................A...............C..C.G..........T...GGT.......
Chi  C........C..G.........AA....A...............C.................C.......A...G..TT..........AGGT.....C..
```

```
                                                                                                              2760
Hsa  TAAGCTGAGTGAGCTGCACTGTGACAAGCTGCATGTGGATCCTGAGAACTTCAAGGTGAGTTCAGGTGCTGGTGATGTGATTTTTT****GGCTTTATATTTTGA**CATTAAT**TGAA
Ptr  ..............................................................G.................****...........**......**....
Ppa  ..............................................................G.................****...........**......**....
Ggo  ..............................................................G.................****...........**......**....
Ppy  ..............................................................G....A..C....A*****...........**......**....
Hla  ..............................................................C..G...............****...........**......**....
Mmu  ..............................................................T..TG.....A........****.......G.....**......**.A..
Cal  ......A.........A....................................G........CATG....C...........T***........A...**......**...G
Tsy  ...A.....................................C........A........G......A...A.AG.C.C..A..C.C...T***..A..T..C...**......**....
Gcr  ...............C.........................A........AAA..C.AC.AG.C.C..******......*.C.C....GA..A...**G...
Lfu  .................A.....C.................G..A.A..C.C....CC.C...T***......T..C.....GA..A...C**G...
Cva  .................G.....C.................C........A..A..C.C.A..T.....T***.TT...T..C....GT..A...**G...
Tgl  A......AGC.A..........T.........C..........A...T..AAA..C.C..C*.T.C...TTTT......T..C..AGAAT.G..G**G...
Csp  ...T..........................................G........AAA..T.C...TC.....TTTTTCA...C.A******GA*..A...**AA...
Mly  ...A..........................................G........AAA..T.C...G...C..TTTTCTT...TAT..CA.GA*TTA...C**G...
Ocu  ...........................A.............A.....T..A..C.C..A.C.C...CT***..T......AC..AGCA..A...**G.G.
Chi  ...............T....C...................G..........AAG..T.C..C.T.CCC..T***......T..CC...***..A...AA..G.
                                                                                                              2880
Hsa  *GCTCATAATCTTATTGGAAAGACCAACAAAGATCTCAGAAATCATGGGTCGAGCTTGATGTTAGAACAGCAGACTTCTAGTGAGCATAACCAAAACTTAC*ATGATTCAGAACTAGTGA
Ptr  *.....G.................................................................................................*T...............
Ppa  *.....G.................................................................................................*T...............
Ggo  *.....G.................................................................................................*T...............
Ppy  *...T.G......................G.......................................................................A..*TG...........
Hla  *...T.G.....................................A.......................................................A..*T...............
Mmu  *...T.GG...............T......................T.................................................G..A..*C...............
Cal  *..G.G..........C.T..T.........G....AT...A....T......................*....G....GG..A.*T...........G.........
Tsy  *..G.C.C.A.G....A...T..GTG..........AT.......G.......A.......C...G.....GG..CA..TT.A...................A.
Gcr  *.G.T.G.C.A.G...A..G..T..........CG.......A...A....C.T..G.C...A..AGA...T..C...AG.......G..G..CA..*T.........TG..G..
Lfu  *...T...C.A.A......AC....T......T.........C..*****.C........C.......GGA........C.G.G.......G..CA.T*T.........
Cva  *...**G.C.A.A......GT............CTC.A...T.A.....GT....C..GA..GAG......G..CA.T*C.............
Tgl  *.TAG.G.C.T.G....A.....T........G.T..C..A...**T.A..CAG......GGA..A.AC..G..G.G.ACT....T..CG.A*T.A........AC.....
Csp  *.TAG.GA.GTT.T.GT.G..G..A..T...G...........G..T......****TA..G.A......GG.....TC..C...AG.T...C...GG.CAC.*T.......AG.A.....
Mly  *.T.G.GTT.T.G....AG..A..T..........T..G...G...A..****TA..G...A..GGA......C..C....TC.....G.G.CACT*T.A.......A.......
Ocu  *TA.A.G.C.A.G......G..T.........T......G...A.A.ATGTT...T.......GATCTCC.A.GG.CA.C.A..AG......*******TT.A.......A....A..
Chi  A.T.G.GTG.T..............T.G*.....C.............A.A..A.A..A.G......G.GGA.........C....G......*..G.GC.CAC**T.........G..........
                                                                                                              3000
Hsa  CAGTAAAGGACTACTAACAGCCTGA********************************************************************************
Ptr  .......................****************************************************************************
Ppa  .......................****************************************************************************
Ggo  .......................****************************************************************************
Ppy  .......................****************************************************************************
Hla  ......G................****************************************************************************
Mmu  .......................****************************************************************************
Cal  .......AG.....GG........*************************************************************************
Tsy  ......GA.....G.....A..*************************************************************************
Gcr  ..A.....AGA..G...T..G..A.********************************************************************
Lfu  .........GA.....G...G..********************************************************************
Cva  ..A.......G.....G......G*********************************************************************
Tgl  T..***********.......********************************************************************
Csp  ........AGAAG.A.GT.......****************************************************************
Mly  ......G.AGTGG...G.....A..*************************************************************
Ocu  G*.....A.G...T.....T..*.****************************************************************
Chi  ..*.....AG...TGGG......T.CTGTGCATGCATGGCTAAGTCGCTTCAGGTGTCAGACTCTTTGTGACCCCATGGCTGTAGCCACCAGGTCCCTCTGTCCATGGGATTCTCCAGGC
```

```
                                                                                                        3120
Hsa  ********************************************************************************************************************************
Ptr  ********************************************************************************************************************************
Ppa  ********************************************************************************************************************************
Ggo  ********************************************************************************************************************************
Ppy  ********************************************************************************************************************************
Hla  ********************************************************************************************************************************
Mmu  ********************************************************************************************************************************
Cal  ********************************************************************************************************************************
Tsy  ********************************************************************************************************************************
Gcr  ********************************************************************************************************************************
Lfu  ********************************************************************************************************************************
Cva  ********************************************************************************************************************************
Tgl  ********************************************************************************************************************************
Csp  ********************************************************************************************************************************
Mly  ********************************************************************************************************************************
Ocu  ********************************************************************************************************************************
Chi  TAGGATACAGGTATGTGTTGCCATTTCTTTTTCCAGGGGATCTACCCAGCCCAAGGATCATATCTGTATCTCTTACATCTCCTTCAATAGCAGGCATGTTCTTTATCACTAGCACCATGA
```

```
                                                                                                        3240
Hsa  ************ATTGGCTTAA**CTTTTCAGGAAATCTTGCCA*GAACTTGATGTGTTTAT*CCCAGAGAATTGTATTATAGAATTGTAGACTTGTGAAAGAAGAA*TGAAATTTGGCT
Ptr  ************.........**................*.............*.....................................................*...........
Ppa  ************.........**................*.............*.....................................................*...........
Ggo  ************.........**................*.............*.....................................................*...........
Ppy  ************.........**................*.............*..TG.................................................*...........
Hla  ************......C..**................*.............*..TG.....G...........................................*...........
Mmu  ************......C..**................*.............*..G...................................................*...........
Cal  ************..GA.....**........A.*...............*..TG.....CA........A..............C...G...*.....*****..
Tsy  ************...TT....AA......*..C..G.TA..T.***....A.........T.*..TG..T...AA...........C.A......T..C........A.ATG.....A.
Gcr  ************...TT....CCA........CAT...T..A.....G....GT..T.T..TG.......AA...........T..A.C...C.C..C..AG....A..C.......
Lfu  ************...TA....AAT......CT..T...A...GA.C..T.T.........*..TG.....AAC.......T...AAC....C..C...G..T..A...C.......A.
Cva  ************...TT....AA.....A..A...T...T..TT.*...T.G..CA....*..TG.......AA***........A.C.....*..C...G....A.T.......
Tgl  ************...T......AA......*...A...A.G.T..*....A..AA....G..*..TG....T.AA......C....*G...AG****************..G....CA..
Csp  ************...G..T..C...AA......GCA..C.T..T..*....GA........*.T.T....AAC..CA...CA.CA.G....TT..AG...A..
Mly  ************..*GC.T......AA......*A.GGC.T..T.*...A.A.........G*...****************...G....A.C..T..ACTT....G..*..G......A..
Ocu  ************************...AT......A...*..**************....*..T*....AA.G...T.........CT....AC...GA..A.AG.....T..
Chi  TGAGCATCCATAG..T......AAG.....T...C.TC..T..*.....G.....A....C*...........***..CA....A.C.T.T...TCTG.G....A......C.....
```

```
                                                                                                        3360
Hsa  TTTGGTAGATGAAAGTCCA*TTTCAAGGAAATAGAAATGCCTTATTTTATGTGGGTCATGATAATTGAGGTTTAG***AAGAGATTTTTGCAAAAAAAA*****************TAA
Ptr  ...................*.....................................................A**....................*****************...
Ppa  ............G......*.....................................................A**....................*****************...
Ggo  ...................*.................................C...................A**...................******************...
Ppy  ...................*.....................................................A**..............AAAAAAAAAAA**********
Hla  ...................*.....................................................A**..C............AAAAAAAAAAAAAAAAAA...
Mmu  ...................T.....................................................A**....................*****************...
Cal  .............*..TG.................*...A..............C.............G**......G....A.......*****************T...
Tsy  ....A.TAGA.G..AG..TG.C...A...GG......T...T*.A..........T.A...G....T.A...TCAA..T.A.A....GGGGG...TA*****************...
Gcr  .............G.*..TG.C...A..GG.C...T...*......A.......T.......CC.A.C...G**..A.....GTT.....TTTTAAA***************GG
Lfu  .............G.*..TA.C...A...GA.C...T.T.............C.....C..A...A**G.A.AGA...C.GTG..G..TTTAAA*************G.
Cva  .........A...GT*..G.C...GC...GA.............C.....C...A....***G.A.ATA....GCCG..CTTT***************TA..
Tgl  ....CA...T..GA*T.TG.C...A.G..AT*****.T...*..........A..C.A..G.C...AA..******************************AAA..
Csp  ...........A..G..*..A.GAGA............T..C....*.C.A.A......G**..........G.AG...*****************
Mly  ..C.TA.....A..G**TATGAC......GGCT.C...T............****.C...G..G..A....G**......................AA****************T
Ocu  G..A.........G.*..TA.C...T...GG.........*.AATGC...CA.T**AT...TT.TTC...T.A.G.AC**T..GAC..GA.TAGGG.CG.TT*****************TATG
Chi  ....AA...AT*........G.C..T....GGG....*****....CC.....An..CC...G.C..A.......G**....****.A.TTGGG.G..TAAT***********TATTAGC
```

```
                                                                                                      3480
Hsa  AAGATTTGCTCAAAGAAA*AATAAGACACATTTTCTAAAATATGTTAAATTTC*************************CCATCAGTATTGTGACCAA*GTGAAGGCTTGTTTCC*GAA
Ptr  ...............*.....................*************************....................*...............*...
Ppa  ...............*.....................*************************....................*...............*...
Ggo  ...............*.....................*************************....................*...............T*...
Ppy  *......T.....*....G...................*************************....................*...............T*...
Hla  .......T........*....................*************************....................*...............T*...
Mmu  .......T........*.............G..A....*************************..............A.....*...............T*...
Cal  .......T........*......A..............A................................A...........*........T.....T*.T.
Tsy  ..T....GGGT..A.***.T...A.G.T........C...G.AA.AT..GACAAA*TTCTTTAAAATATGCTAAATTTC.TG........C...ATTG..T..A..T.AA.T****
Gcr  T......T........*...CAG..A..C.........*****...CT**************************.TG......C...TG...*A.TG.......AGT*AC.
Lfu  C......T..G....*..CCA..A..C.........CA.....T**************************.TG........*A..G..A..T..C.T*.C.
Cva  ..CC..G...G....*..G.T........C......A........C.*************************.TGG.............*...G...........T.C.
Tgl  ..ACA..**....A..T**T.........T....G...G.****...A.A.AAAATTCTCTAAAATATATTAAATTCT.TG...A.T....T...C*A..T..A...A.CTT*.C.
Csp  C...A.GT....A..T*******...AC....TA.........C...A.**************************TTGATG...A........*...G.......C..CAG*.T.
Mly  T...CAGAA.GTCTC...A....***...AC..C.........C...A.**************************TTG...C..........*...G.......AC.CTG*.C.
Ocu  CGAGA..CT..TG..A...*...CA...A.......G..A..A......CT**************************C........*A.AT.AA.......C.T*.CT
Chi  C....CAT..........*..TGAT..ATA.C..A.GG.********...A.**************************..A.C....T.G*...G......A..GTT*.C.

                                                                                                      3600
Hsa  TTTGTTGGGGATTTTAAACTCCCGCT*GAGAACTCTTGCAGCACTCACATTCTACATT*********TACAAAAATTAGACAA*********TTGCTTAAAGAAAAACAGGGAGAGAG*GGA
Ptr  ......A..................*............*********........*********..................*...
Ppa  ......A....................*********........*********..................*...
Ggo  ......A....................*********........*********..................*...
Ppy  ......A.........T...*......*********.T..G....*********C................*...
Hla  ......A..........*.........*********.G......*********C..A.............*...
Mmu  ......A..........*.........T....C.........*********.G......*********C......T.........*...
Cal  ..ACA.......T...A..T........A....GG...A........G...T.********C.G...***********************
Tsy  *****************..A.C***A......C...T...G.A...TG.********..TGG.T..A..*********..T......G...**.A...A.A*...
Gcr  ..G...A......C......A..*..A........AT.A..G.A..CTG.T.********...G....A.T****************.T.G..*........A.*..
Lfu  ..G...A......C........A.........C......G.A..C..TG.********...G.......G*...T..G..*...A....A*.A.
Cva  ...G.....C..G.......*A..GA....A..CT.........G..TT.********....T........TCAGTAAA....**......TG....G*******
Tgl  ..G.C..AA....A...A..TA..*..A...C........G.T.A..G.AT.T.********....G.T...TGG.********...G*CG..TG.G..AGAAAG.GA.AT**.
Csp  ..G.....AA****GACTT.AA***..AATTC....GAG...CA.G.A.C...GA.A.CCTCTGAG...GG.......A..A********..AGAG.G..*****.......G***
Mly  ..G.....AAGGAGACTTAAA***.AAACTC....GAG...C.A.GA..C..GA.A.CCTATGATC..G...C.T...G********.AAA.*G.GAG.*****.....C..G***
Ocu  ..G...........*****A..*.C.....A...GG.....TG....C..A..********...TG....C..A.********.....AT.A..T..G...G.AAA.A...AA.A.
Chi  ..GAA.T.A.GG....CTAAG.TCA.TCT*...AACCCAT...GC.CTGA*A.C.TA.*****GAA..*T.........*************************.GG.....A.AG.****

                                                                                                      3720
Hsa  ACCCAATAATACTGGTAAAATGGGGAAGG**GGGTGAGGGT*GTAGGTAGGT*AGAATGTTGAATGTAGGGCT**CATAGAA*TAAAA*TTGAA******CCTAAGCTCATCTGAA****
Ptr  G....................G..**....*........*......C.......**..A..*...*..A.******....****
Ppa  G....................G..**....*........*......C.......**..A..*...*..A.******....****
Ggo  G....................G..**....*........*......C.......**..A..*...*..A.******....****
Ppy  G..................A..GA.G*....A.*....G...*......**..A..*...*......******....T....****
Hla  G..................G..*...A.*.........*......**..A..*...*..A.******....****
Mmu  G****..............A.GA.A*.........A.*.C....*........**..A.*...A.******....T.....****
Cal  ****..A...........GA.A*..............**..A.*.AC..A..******..C........T****
Tsy  GG.A.T.GGC.A...G......A...GAAA**...TGA..ATT.TT.......*...A..C........**..A..G.*......******....T..T..****
Gcr  GAGA...G.C.A...GG...AA..GATA*.A...A.*A...T..A.C*.AG..........A...T.**..A.A..AA...AC.A..*TTAGG...T.T.T..T.A.****
Lfu  GG.A...C.A...T......GA.AG***********....C*..G.............**..AC.A.G..*......******....T.T..A.****
Cva  **.A.....A.A...G......*GA.A*........A.*A..C......CG...A....AC....TGGT..A.GGGG.***..A.******..A...T.T....TG****
Tgl  .AAA....G.G******....A..A.GA.A*A....A.*A..C.......C*T...A....**.C..A.GATT...A......***************TT......G....G****
Csp  **.A......AGA...AG.....AGT.AG**...G..A.*A......C*............G..A..*.....**..A...AAT..**..A..******TG..G.G*AC.AG...**CT
Mly  **.A......A.A.*..G..G...AGT.AG**...G.AA.*A......C*......T....AA.....**..A.A.G..**.....TTTAAGGAC...G.A...T****
Ocu  GATA....TG****...G......T......A***.AT.A.A.*A.......TC*.............C.G.A.T..**...A..A.G..*...T******..*.TT..TG...TGG****
Chi  **.A.C...A.A.A..G....*A...GA.A**..CA...ATA.....C..AC*.A...A...T...G......**....G.T.T...*..A..******TTG....GA...AGCTC.TCTG
```

```
                                                                                                    3840
Hsa  **TTTTTTGGGTGGGCACAAACCTTGGAACAGTTTGAGGTCAGGG*TTGTCTAGGAA*TGTAGGTATAAAGCCGTTTTTGTTTGTTTGTTTG****TTTTTTCATCAAGTTG**TTTTCG
Ptr  **.........A...................................*.........*.............................................TTTG............**...T.
Ppa  **.........A...................................*.........*............................................****...........**....T.
Ggo  **.........A...................................*.........*.............................................TTTG............**......
Ppy  **...A..A..A...................................*.........*...........A................................****...........**......
Hla  **C........A.................................A.*...A.....*...........A................................****...........**......
Mmu  **....C....A...................................*...A.....*...........................................****........G....**......
Cal  **.........A..................................T*******************************...................GTC.A...**...A.
Tsy  **.G.A..TACA.CA....A....A.G.........A.*.C.C.A...T*........C...A.C...*****************************************************
Gcr  **CA.A....A****************************...*....*.....**...**..G...C..C..T..*************....CCCTTA...CA.T*C..C.T.
Lfu  **C..A....A.A..T.G.C......G..........A.*.A..AG...**..**.C........A...A****************....C.TG...A..C**A..T.
Cva  **...A..A.A.A....C....C.G..T.........A.*..A..AG...G*........T.T...****************A...CCTG.G..T.GT**....T.
Tgl  **CA.AC...A....G..CA...TA.G.......TA..GG.TT*....AGA...C*........G.C.T.A.*****************....TG.T.AA.C**...T.
Csp  AG...A...TA.A....C.T......A...G.A...GCA.AG...G.AG*...G......T.T.T....*****************....A.....T..T.CACTGC...T.
Mly  **...A..TA.A.A...TTT.......AA...CACA.AG..CA.AG...G*...A...T...A.*********************....C.TT..T.TAATTA...G.
Ocu  ***...AA...A....T....GT.A..........A.*..T..AG...*.......C.T..T.A...C.T.C.T.G.AAA.T******************C...T.
Chi  AG...A...TA.A..T...C..A...GA....A..A.GT..AC...G*G..TGG*.T....CT.....A.************************T.TCT..AACTC...TA

                                                                                                    3960
Hsa  GAAACTTCTACTCAACATGCC*****TGTGTGTTATTTTGTCTTTTGCCTAACAGCTCCTGGGTAACGTGATGGTGATTATTCTGGCTACTCACTTTGGCAAGGAGTTCACCCCTGAAGT
Ptr  .....................*****...................................................................................................
Ppa  .....................*****...................................................................................................
Ggo  .....................*****...................................................................................................
Ppy  .....................*****.A..A..............................................................................................
Hla  .....................*****.A.........................C.......................................................T..............
Mmu  ..............T*****.A...........T...........................................................................................
Cal  .......A...T..T...T*****.A...........................A...T...................................A...T..C.
Tsy  *****....G.**.T.......*****.A.A.T.C..C..T....CA.................T...C.........T...T.........C.............
Gcr  .............A*TTC...C.T*C.........CA*...........A..T.............C...........T......A...T...T.
Lfu  A.....T.G...T...ATTTT.T.C.T.C........CA.............A.....................T.......A........T.
Cva  .........A.......T.*****.A.T...C.....CA..................................
Tgl  ...C..C..GG.A...C..T.*****.C...A......CCA................................
Csp  .....................*****.ACT..C......CA............................
Mly  ....T................*****.C.T...C......G...A............................
Ocu  .G...AC......T.......*****.A...A.C.........A...........A...C..C....CG......C...T........A..A...T..G..G..
Chi  .C..AC.T****....T..G.CTACC.AAT..C.....C......C.CA..C.............C..T.....T.............T.......GA..A............C..

                                                                                                    4080
Hsa  GCAGGCTGCCTGGCAGAAGCTGGTGTCTGCTGTCGCCATTGCCCTGGCCCATAAGTACCACTGAGTTCTCTT*CCAGTTTGCAGGTGTTCC*****TGTGACCC**TGACACCCTCCTTC
Ptr  ................................................................*..........CA......*****......**...........................
Ppa  ................................................................*..........CA......*****......**...........................
Ggo  ................................................................*..........CA......*****......**...........................
Ppy  .......................................A........................*..........CAA.....*****......T**..........................
Hla  ...............................................................*...A.......******.....*****......**.........................
Mmu  ................................................................*..........CA......*****.A....**...........................
Cal  .....................................G..C..........A....*....C..TG*.....*****.......**.....
Tsy  A....T...............GA..T...C..A....T..C.........ACT...T...ACGA.TA.....*****....TT.T**CA......T...
Gcr  ..................A......GG..G..T.C..T..A....C..........C..T..T.......CA.CA..T.TT***...TG..TT....T..TT....
Lfu  A.................A......GC..T...C..T.......C..........CC..T..T.T.....CA....CT.TG***...TT..**.....T....
Ocu  .........T.........C.....G..T.......T.......C....T....AC......C.......CTTT*A..G..CCTAC.CATC...***AG.GT......
Chi  ...................G...T..C..T.......C.........A......TA..A..CA********..ATTTTGTGTC...***AGTG..T......

                                                                                                    4200
Hsa  TGCACAT*GGGGACTGGGCTTGGCCTTGAGAGAAAGCCTTCTGTTTAATAAAGTACATTTTCTTC****AGTAAT***CAAAAATTGCAATTTTATCTTCTCCATCTTTTACTCTTGTGT
Ptr  .......*..................................................****.....***.......................................................
Ppa  .......*..................................................****.....***.......................................................
Ggo  .......*..................................................****.....***.......................................................
Ppy  .......*..................................................****.....***.....CC.............................................G..
Hla  .......*..................................................****.....***...C...............................................
Mmu  ...G.*.................T......C.........C.G.........****.....***.....G.............A.A..G.....
Cal  ....T.*.AAT.....C...........G..G................****.....***........C.T..C.............G....G......
Tsy  .....*..C......G..A....CT.AG.................T...CCAA.****.....***..*GG.C.............CT...T.......
Gcr  ..A...*........*..A..TTT.C..A..C.TAG....TC.......G..C..T...A..AGTA.A...AAAA....CCA...C....T..T.TTGCT...C..TG.A...
Lfu  ......A*.A...........TT.......C.TAG.........C.A..****.A...***A.............G..........G..CC...A...
Ocu  ..T..C.T........A....CA.......TG.C.CAGT..........A......C.A..****...***......A..ATG.C.C.....T.G.A.......A.....
Chi  ...C.C.T...ACTG...T..........T..ACCCAGA..............A...C.A..****...G..***.......AA...G..C.....T...A
```

```
                                                                                                                                    4320
Hsa   TAAAAGGAAAAAGT***GTTCATGGGCTGAGGGATGGAGAGAAACATAGGAAGAA***CCAAGAGCTTCCTTAAGAAATGTATGGGGGCTT**GTAAAATTAATGTGGATGTTATGGGAG
Ptr   .*............***.......................
Ppa   .*............***.......................
Ggo   .*............***.......................
Ppy   .............***........A..................***...............................**.....................C.A...
Hla   .*............***........................***TT...............................**....................
Mmu   .*............***.......................
Cal   .*.G...*.....***..C........A....G.................***T.....A.....A.T...........A......**.................
Tsy   C*....A......G**T.....A.........A....A.G.....AA.G..GAAAT.........T.....GG..G.......A..C**T.C...GG..AA..............
Gcr   .*...AA...GG.GGGG.............A.A.**....A...AG.G.******************.T....A..G...GC....AACCAGACC...AG..G...........A.A..
Lfu   .*..........G***....C........A....GA...G.GATA.G.A..AC***..A.....T.......G.A.G....AT.C**A.CG..GG..AC...........A..
Ocu   .T...AA.....AA***CCA..CAT.T.C.T*...T....C.G.G.G..AC.T.GG****....AA.*****.....*.C.T..A.TAG.ATAT..GG...T..C***.AA.A.A.....
```

References

Adkins, R. M., and Honeycutt, R. L. 1991. Molecular phylogeny of the superorder Archonta. *Proc. Natl. Acad. Sci. USA* **88**:10317–10321.

Ammerman, L. K., and Hillis, D. M. 1992. A molecular test of bat relationships: Monophyly or diphyly? *Syst. Biol.* **41**:222–232.

Bailey, W. J., Slightom, J. L., and Goodman, M. 1992. Rejection of the "flying primate" hypothesis by phylogenetic evidence from the ε-globin gene. *Science* **256**:86–89.

Baker, R. J., Honeycutt, R. L., and Van Den Bussche, R. A. 1991a. Examination of monophyly of bats: Restriction map of the ribosomal DNA cistron. *Bull. Am. Mus. Nat. Hist.* **206**:42–53.

Baker, R. J., Novacek, M. J., and Simmons, N. B. 1991b. On the monophyly of bats. *Syst. Zool.* **40**:216–231.

Beard, K. C. 1990. Gliding behaviour and palaeoecology of the alleged primate family Paromomyidae (Mammalia, Dermoptera). *Nature* **345**:340–341.

Beintema, J. J., Rodewald, K., Braunitzer, G., Czelusniak, J., and Goodman, M. 1991. Studies on the phylogenetic position of the Ctenodactylidae (Rodentia). *Mol. Biol. Evol.* **8**:151–154.

Borst, D. E., and Nickerson, J. M. 1988. The isolation of a gene encoding interphotoreceptor retinoid-binding protein. *Exp. Eye Res.* **47**:825–838.

Borst, D. E., Redmond, T. M., Elser, J. E., Gonda, M. A., Wiggert, B., Chader, G. J., and Nickerson, J. M. 1989. Interphotoreceptor retinoid-binding protein: Gene characterization, protein repeat structure, and its evolution. *J. Biol. Chem.* **264**:1115–1123.

Bugge, J. 1985. Systematic value of the carotid arterial pattern in rodents, in: W. P. Luckett and J.-L. Hartenberger (eds.), *Evolutionary Relationships among Rodents: A Multidisciplinary Analysis*, pp. 355–379. Plenum Press, New York.

Bulmer, M., Wolfe, K. H., and Sharp, P. M. 1991. Synonomous nucleotide substitution rates in mammalian genes: Implications for the molecular clock and the relationships of mammalian orders. *Proc. Natl. Acad. Sci. USA* **88**:5974–5978.

Cartmill, M., and MacPhee, R. D. E. 1980. Tupaiid affinities: The evidence of the carotid arteries and cranial skeleton, in: W. P. Luckett (ed.), *Comparative Biology and Evolutionary Relationships of Tree Shrews*, pp. 95–132. Plenum Press, New York.

Collins, F., and Weissman, S. 1984. The molecular genetics of human hemoglobin. *Prog. Nucleic Acid Res. Mol. Biol.* **31**:315–439.

Cronin, J. E., and Sarich, V. M. 1980. Tupaiid and Archonta phylogeny: The macromolecular evidence, in: W. P. Luckett (ed.), *Comparative Biology and Evolutionary Relationships of Tree Shrews*, pp. 293–312. Plenum Press, New York.

Czelusniak, J., Goodman, M., Koop, B. F., Tagle, D. A., Shoshani, J., Braunitzer, G., Kleinschmidt, T. K., de Jong, W. W., and Matsuda, G. 1990. Perspectives from amino acid and nucleotide sequences on cladistic relationships among higher taxa of Eutheria, in: H. H. Genoways (ed.), *Current Mammalogy*, Vol. 2, pp. 545–572. Plenum Press, New York.

Easteal, S. 1990. The pattern of mammalian evolution and the relative rate of molecular evolution. *Genetics* **124**:165–173.

Fong, S.-L., Fong, W.-B., Morris, T. A., Kedzie, K. M., and Bridges, C. D. B. 1990. Characterization and comparative structural features of the gene for human interstitial retinol-binding protein. *J. Biol. Chem.* **265**:3648–3653.

George, W. 1985. Reproductive and chromosomal characters of ctenodactylids as a key to their evolutionary relationships, in: W. P. Luckett and J.-L. Hartenberger (eds.), *Evolutionary Relationships among Rodents: A Multidisciplinary Analysis*, pp. 453–474. Plenum Press, New York.

Gingerich, P. D. 1985. South American mammals in the Paleocene of North America, in: F. G. Stehli and S. D. Webb (eds.), *The Great American Biotic Interchange*, pp. 123–135. Plenum Press, New York.

Goodman, M., Koop, B. F., Czelusniak, J., Weiss, M. L., and Slightom, J. L. 1984. The η-globin gene: Its long evolutionary history in the β-globin gene family of mammals. *J. Mol. Biol.* **180**:803–823.

Gregory, W. K. 1910. The orders of mammals. *Bull. Am. Mus. Nat. Hist.* **27**:1–524.

Gyllensten, U. B., and Erlich, H. A. 1988. Generation of single-stranded DNA by the polymerase chain reaction and its application to direct sequencing of the HLA-DQA locus. *Proc. Natl. Acad. Sci. USA* **85**:7652–7656.

Hardison, R. C. 1983. The nucleotide sequence of the rabbit embryonic globin gene β4. *J. Biol. Chem.* **258**:8739–8744.

Harris, S., Thackeray, J. R., Jeffreys, A. J., and Weiss, M. L. 1986. Nucleotide sequence analysis of the lemur β-globin gene family: Evidence for major rate fluctuations in globin polypeptide evolution. *Mol. Biol. Evol.* **3**:465–484.

Hill, W. C. O. 1955. *Primates*, Vol. 2. Wiley–Interscience, New York.

Jukes, T. H., and Cantor, C. R. 1969. Evolution of protein molecules, in: H. N. Munro (ed.), *Mammalian Protein Metabolism*, Vol. 2, pp. 21–123. Academic Press, New York.

Kay, R. F., Thorington, R. W., Jr., and Houde, P. 1990. Eocene plesiadapiform shows affinities with flying lemurs not primates. *Nature* **345**:342–344.

Kluge, A. G. 1989. A concern for evidence and a phylogenetic hypothesis of relationships among *Epicrates* (Boidae, Serpentes). *Syst. Zool.* **38**:7–25.

Koop, B. F., and Goodman, M. 1988. Evolutionary and developmental aspects of two hemoglobin β-chain genes (ϵ^M and β^M) of opossum. *Proc. Natl. Acad. Sci. USA* **85**:3893–3897.

Koop, B. F., Miyamoto, M. M., Embury, J. E., Goodman, M., Czelusniak, J., and Slightom, J. L. 1986. Nucleotide sequence and evolution of the orangutan ε-globin gene region and surrounding Alu repeats. *J. Mol. Evol.* **24**:94–102.

Koop, B. F., Sieminiak, D., Slightom, J. L., Goodman, M., Dunbar, J., Wright, P. L., and Simons, E. L. 1989. Tarsius δ- and β-globin genes: Conversion, evolution, and systematic implications. *J. Biol. Chem.* **264**:68–79.

Li, C.-K., and Ting, S.-Y. 1985. Possible phylogenetic relationships: Eurymylid–rodent and mimotonid–lagomorph, in: W. P. Luckett and J.-L. Hartenberger (eds.), *Evolutionary Relationships among Rodents: A Multidisciplinary Analysis*, pp. 35–58. Plenum Press, New York.

Li, W.-H., Gouy, M., Sharp, P. M., O'Huigan, C., and Yeng, Y.-W. 1990. Molecular phylogeny of Rodentia, Lagomorpha, Primates, Artiodactyla, and Carnivora and molecular clocks. *Proc. Natl. Acad. Sci. USA* **87**:6703–6707.

Linnaeus, C. 1758. Systema naturae per regna tria naturae, secundum classes, ordines, genera, species, cum characteribus, differentiis, synonymis, locis, Tomus I, editio decima, reformata, pp. 18–47. Holmiae, Impensis Direct. Laurentii Salvii, Stockholm.

Luckett, W. P. 1980. The use of reproductive and developmental features in assessing tupaiid affinities, in: W. P. Luckett (ed.), *Comparative Biology and Evolutionary Relationships of Tree Shrews*, pp. 245–266. Plenum Press, New York.

Luckett, W. P. 1985. Superordinal and intraordinal affinities of rodents: Developmental evidence from dentition and placentation, in: W. P. Luckett and J.-L. Hartenberger (eds.), *Evolutionary Relationships among Rodents: A Multidisciplinary Analysis*, pp. 227–276. Plenum Press, New York.

Luckett, W. P., and Hartenberger, J.-L. (eds.) 1985. *Evolutionary Relationships among Rodents: A Multidisciplinary Analysis.* Plenum Press, New York.

McKenna, M. C. 1975. Toward a phylogenetic classification of the Mammalia, in: W. P. Luckett and F. S. Szalay (eds.), *Phylogeny of the Primates: A Multidisciplinary Approach,* pp. 21–46. Plenum Press, New York.

Martin, R. D. 1990. *Primate Origins and Evolution: A Phylogenetic Reconstruction.* Chapman & Hall, London.

Mayr, E. 1986. Uncertainty in science: Is the giant panda a bear or a raccoon? *Nature* **323:**769–771.

Mindell, D. P., Dick, C. W., and Baker, R. J. 1991. Phylogenetic relationships among megabats, microbats, and primates. *Proc. Natl. Acad. Sci. USA* **88:**10322–10326.

Miyamoto, M. M., and Cracraft, J. 1991. Phylogenetic inference, DNA sequence analysis, and the future of molecular systematics, in: M. M. Miyamoto and J. Cracraft (eds.), *Phylogenetic Analysis of DNA Sequences,* pp. 3–17. Oxford University Press, London.

Miyamoto, M. M., and Goodman, M. 1986. Biomolecular systematics of eutherian mammals: Phylogenetic patterns and classification. *Syst. Zool.* **35:**230–240.

Novacek, M. J. 1980. Cranioskeletal features in tupaiids and selected eutherians as phylogenetic evidence, in: W. P. Luckett (ed.), *Comparative Biology and Evolutionary Relationships of Tree Shrews,* pp. 35–93. Plenum Press, New York.

Novacek, M. J. 1985a. Cranial evidence for rodent affinities, in: W. P. Luckett and J.-L. Hartenberger (eds.), *Evolutionary Relationships among Rodents,* pp. 59–81. Plenum Press, New York.

Novacek, M. J. 1985b. Evidence for echolocation in the oldest known bat. *Nature* **315:**140–141.

Novacek, M. J. 1986. The skull of leptictid insectivorans and the higher-level classification of eutherian mammals. *Bull. Am. Mus. Nat. Hist.* **183:**1–111.

Novacek, M. J. 1990. Morphology, paleontology, and the higher clades of mammals, in: H. H. Genoways (ed.), *Current Mammalogy,* Vol. 2, pp. 507–543. Plenum Press, New York.

Novacek, M. J., and Wyss, A. R. 1986. Higher-level relationships of the recent eutherian orders: Morphological evidence. *Cladistics* **2:**257–287.

O'Brien, S. J., Nash, W. G., Wildt, D. E., Bush, M. E., and Benveniste, R. E. 1985. A molecular solution to the riddle of the giant panda's phylogeny. *Nature* **317:**140–144.

Penny, D., Foulds, L. R., and Hendy, M. D. 1982. Testing the theory of evolution by comparing phylogenetic trees constructed from five different protein sequences. *Nature* **297:**197–200.

Penny, D., Hendy, M. D., and Steel, M. A. 1991. Testing the theory of descent, in: M. M. Miyamoto and J. Cracraft (eds.), *Phylogenetic Analysis of DNA Sequences,* pp. 155–183. Oxford University Press, London.

Pettigrew, J. D. 1986. Flying primates? Megabats have the advanced pathway from eye to midbrain. *Science* **231:**1304–1306.

Pettigrew, J. D. 1991a. Wings or brain? Convergent evolution in the origins of bats. *Syst. Zool.* **40:**199–216.

Pettigrew, J. D. 1991b. A fruitful, wrong hypothesis? Response to Baker, Novacek, and Simmons. *Syst. Zool.* **40:**231–239.

Pettigrew, J. D., Jamieson, B. G. M., Robson, S. K., Hall, L. S., McAnally, K. I., and Cooper, H. M. 1989. Phylogenetic relations between microbats, megabats and primates (Mammalia: Chiroptera and Primates). *Philos. Trans. R. Soc. London Ser. B* **325:**489–559.

Prager, E. M., and Wilson, A. C. 1988. Ancient origin of lactalbumin from lysozyme: Analysis of DNA and amino acid sequences. *J. Mol. Evol.* **27:**326–335.

Saiki, R. K., Gelfand, D. H., Stoeffel, S., Scharf, S. J., Jiguchi, R., Horn, G. T., Mullis, K. B., and Erlich, H. A. 1988. Primer-directed enzymatic amplification of DNA with a thermostable DNA polymerase. *Science* **239:**487–491.

Saitou, N., and Nei, M. 1987. The neighbour-joining method: A new method for reconstructing phylogenetic trees. *Mol. Biol. Evol.* **4:**406–425.

Sarich, V. M. 1973. The giant panda is a bear. *Nature* **245:**218–220.

Savage, D. E., and Russell, D. E. 1983. *Mammalian Paleofaunas of the World*. Addison–Wesley, Reading.

Shapiro, S. G., Schon, E. A., Townes, T. M., and Lingrell, J. B. 1983. Sequence and linkage of the goat ε[I] and ε[II] β-globin genes. *J. Mol. Biol.* **169**:31–52.

Simmons, N. B., Novacek, M. J., and Baker, R. J. 1991. Approaches, methods, and the future of the chiropteran monophyly controversy: A reply to J. D. Pettigrew. *Syst. Zool.* **40**:239–243.

Simpson, G. G. 1945. The principles of classification and a classification of mammals. *Bull. Am. Mus. Nat. Hist.* **85**:1–350.

Simpson, G. G. 1978. Early mammals in South America: Fact, controversy and mystery. *Proc. Am. Philos. Soc.* **122**:318–328.

Smith, F., and Waterman, M. S. 1981. Identification of common molecular subsequences. *J. Mol. Biol.* **147**:195–197.

Sneath, P. H. A. and Sokal, R. R. 1973. *Numerical Taxonomy*. Freeman, San Francisco.

Stanhope, M. J. Czelusniak, J., Si, J.-S., Nickerson, J., and Goodman, M. 1992. A molecular perspective on mammalian evolution from the gene encoding interphotoreceptor retinoid binding protein, with convincing evidence for bat monophyly. *Mol. Phyl. Evol.* **1**:148–160.

Swofford, D. L., and Olsen, G. J. 1990. Phylogeny reconstruction, in: D. M. Hillis and C. Moritz (eds.), *Molecular Systematics*, pp. 411–515. Sinauer Assoc., Sunderland.

Szalay, F. S. 1977. Phylogenetic relationships and a classification of the eutherian Mammalia, in: M. K. Hecht, P. C. Goody, and B. M. Hecht (eds.), *Major Patterns in Vertebrate Evolution*, pp. 315–374. Plenum Press, New York.

Tagle, D. A., Koop, B. F., Goodman, M., Slightom, J. L., Hess, D. L., and Jones, R. T. 1988. Embryonic ε- and γ-globin genes of a prosimian primate (*Galago crassicaudatus*): Nucleotide and amino acid sequences, developmental regulation, and phylogenetic footprints. *J. Mol. Biol.* **203**:439–455.

Templeton, A. R. 1992. Human origins and analysis of mitochondrial DNA sequences. *Science* **255**:737.

Vigilant, L., Stoneking, M., Harpending, H., Hawkes, K., and Wilson, A. C. 1991. African populations and the evolution of human mitochondrial DNA. *Science* **253**:1503–1507.

Wible, J. R., and Covert, H. H. 1987. Primates: Cladistic diagnosis and relationships. *J. Hum. Evol.* **16**:1–22.

Williams, S. A., and Goodman, M. 1989. A statistical test that supports a human/chimpanzee clade based on noncoding DNA sequence data. *Mol. Biol. Evol.* **6**:325–330.

Wood, A. E. 1985. The relationships, origin and dispersal of the hystricognathous rodents, in: W. P. Luckett and J.-L. Hartenberger (eds.), *Evolutionary Relationships among Rodents: A Multidisciplinary Analysis*, pp. 475–513. Plenum Press, New York.

Wu, C.-I., and Li, W.-H. 1985. Evidence for higher rates of nucleotide substitution in rodents than in man. *Proc. Natl. Acad. Sci. USA* **82**:1741–1745.

Phylogeny through Brain Traits

9

Interordinal Relationships among Mammals Including Primates and Chiroptera

J. I. JOHNSON and J. A. W. KIRSCH

Introduction: Value of Neural Characters

Our use of brain characters for the reconstruction of mammalian phylogenies began with an analytical survey of the variable degree of ipsilateral versus contralateral projections to the somatic sensory thalamus and cerebral cortex, in a phylogenetic sample of species (Bombardieri *et al.*, 1975). We found that the distribution of ipsilateral versus contralateral dominance reflected a bifurcation within placentals that matched a then-current scheme based on paleontological data (Lillegraven, 1969; McKenna, 1969). In subsequent years the scheme lost favor with paleontologists, who concluded that most major placental groups arose from a polytomy that defied analysis (e.g., McKenna, 1975), particularly cladistic analysis into steps of successive bifurcations.

A subsequent study (Switzer *et al.*, 1980) in which we expected to find

J. I. JOHNSON • Anatomy Department, and Neuroscience Program, Michigan State University, East Lansing, Michigan 48824.   J. A. W. KIRSCH • Zoological Museum and Department of Zoology, University of Wisconsin-Madison, Madison, Wisconsin 53706.
Primates and Their Relatives in Phylogenetic Perspective, edited by Ross D.E. MacPhee. Plenum Press, New York, 1993.

differences in the olfactory bulbs of marsupials and placentals found instead a difference that distinguished all monotremes, marsupials, and some placentals from a large group of other placentals. This bifurcation of the placentals segregated the same placental groups as did the data from Bombardieri *et al.* (1975), but with a reversal of primitive and derived character states. The olfactory bulb data showed Insectivora, Primates, Dermoptera, Chiroptera, and Rodentia with a derived state of bulb architecture lacking in Monotremata, Marsupialia, Carnivora, Artiodactyla, Perissodactyla and Hyracoidea. In contrast, the carnivore–artiodactyl group possesses a derived state of thalamocortical sensory projections that is absent in the primate–rodent group and also in monotremes and marsupials.

This combination of findings alerted us to the potential value of neural data in analyzing difficult problems in evolutionary analysis.

The 15-Character Trees, 1982

We therefore combined several neural characteristics that we knew were related to mammalian phylogeny, with several others that we suspected might be, to see what sort of phylogenetic schemes could be produced from purely neural data. Quantitative algorithms had been devised for producing phylogenetic trees from data ordered numerically in ordinal sequences. We used one of these algorithms, developed successively by Wagner, Farris, and Kluge (Kirsch, 1983), to produce informative trees from the observed states of only 15 neural characters in as few as 38 representative species (Fig. 1, redrawn from Kirsch *et al.*, 1983). We were intrigued to find out that even from this limited data set, the algorithm fairly neatly grouped the species into familiar ordinal groupings. Formalizing these groupings yielded a tree suggesting some interesting interordinal relationships (Fig. 2). For example, Primates are firmly situated on that branch of the placental bifurcation that includes rodents, tree shrews, and insectivores. (No data were available from Chiroptera.)

The 24-Character Trees, 1991

More flexible algorithms then became available, and we have been using the PAUP (Phylogenetic Analysis Using Parsimony) program developed by Swofford (1984), version 2.4.1. This program, unlike the Farris algorithm, allows the use of incomplete data sets. This property, along with additional observations in other species, several of these from the published literature, enabled us to expand our set of neural characters to 24 (A through X, Table I), and our species population to 49 (Table II), the maximum number of taxa

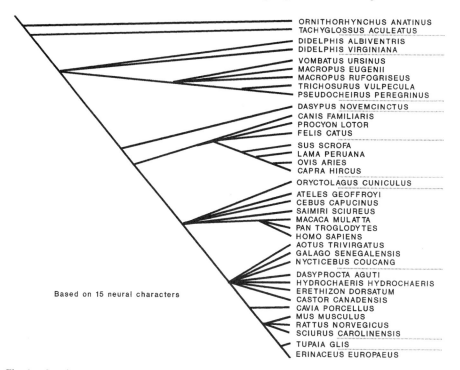

ORNITHORHYNCHUS ANATINUS
TACHYGLOSSUS ACULEATUS
DIDELPHIS ALBIVENTRIS
DIDELPHIS VIRGINIANA
VOMBATUS URSINUS
MACROPUS EUGENII
MACROPUS RUFOGRISEUS
TRICHOSURUS VULPECULA
PSEUDOCHEIRUS PEREGRINUS
DASYPUS NOVEMCINCTUS
CANIS FAMILIARIS
PROCYON LOTOR
FELIS CATUS
SUS SCROFA
LAMA PERUANA
OVIS ARIES
CAPRA HIRCUS
ORYCTOLAGUS CUNICULUS
ATELES GEOFFROYI
CEBUS CAPUCINUS
SAIMIRI SCIUREUS
MACACA MULATTA
PAN TROGLODYTES
HOMO SAPIENS
AOTUS TRIVIRGATUS
GALAGO SENEGALENSIS
NYCTICEBUS COUCANG
DASYPROCTA AGUTI
HYDROCHAERIS HYDROCHAERIS
ERETHIZON DORSATUM
CASTOR CANADENSIS
CAVIA PORCELLUS
MUS MUSCULUS
RATTUS NORVEGICUS
SCIURUS CAROLINENSIS
TUPAIA GLIS
ERINACEUS EUROPAEUS

Based on 15 neural characters

Fig. 1.   Species tree computed by the Farris algorithm based on states of 15 neural characters in 37 taxa, redrawn from Kirsch *et al.* (1983). Although not entirely justified by the branching pattern, this tree allows the arrangement of species into recognized orders, separated here by dotted lines. The orders are shown in Fig. 2.

accommodated in this version of PAUP. We have restricted our use of PAUP 2.4.1 to the default options; in such exploratory studies we thought it wise not to go too far into data manipulation before learning much more about our characters. Used in this way, PAUP 2.4.1 produced the species tree shown in the next figure (Fig. 3) from the data in Table II. Here again, in general, species are grouped into the traditional orders. However, our overly catholic accumulation of all sorts of characters, including ones that are almost certainly convergent and others that appear to have no consistent relation to phylogeny, along with peculiar things that PAUP 2.4.1 apparently does with blank entries in data cells, led to a few (6 of the 49) unbelievable placements (lowercase entries and dashed lines in Fig. 3).

The first group of problematic placements are those of three edentates (*Tamandua, Choloepus,* and *Myrmecophaga*). Here the PAUP 2.4.1 program evidently, based on derived states of characters G and L associated with large brains, substituted derived-state entries for empty data cells in characters such as N and O, ascribing to these species features of somatic sensory cortex otherwise peculiar to artiodactyls. Then the association of *Tamandua* with *Sus*

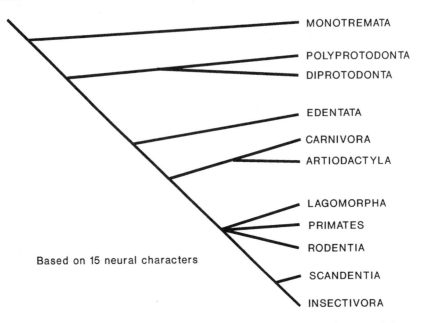

Fig. 2.   Orders tree drawn from Fig. 1, based on the distribution of states of 15 neural characters in 37 species, redrawn from Johnson (1986). In this and all following figures, we consider the subclass Metatheria to consist of three orders: Polyprotodonta, Diprotodonta, and Paucituberculata (Kirsch, 1977); no data from Paucituberculata were available for the tree in this figure. This tree shows an early divergence of edentates from the other placentals, then a major bifurcation of placentals into ferungulates (Carnivora + Artiodactyla) and archontan–glyroid groups (Lagomorpha, Primates, Rodentia, Scandentia, and Insectivora).

is likely the result of their common intermediate-derived state in character C, which occurs in a few disparate species with a well-developed olfactory apparatus.

The other group of odd placements, those of *Trichechus, Tursiops,* and *Pteropus,* are all species with many empty data cells along with derived states in both characters Q and U. As discussed in detail in Johnson *et al.* (in press), the parallels in progressive derivation in these two characters are striking. From nonneural data we conclude that the parallel losses of the accessory olfactory formation, and of lamination in the auditory dorsal cochlear nucleus, are probably convergences consequent upon departure from a terrestrial habitat, rather than evidence for common ancestry. Except for these problematic six, the ordinal groupings and their relative placements are again quite encouraging and interesting.

Our initial batch of characters was dominated by those we knew to discriminate monotremes, marsupials, and placentals from one another since they were characters we had been familiar with for a long time; not surprisingly, all of our trees show these groups as cleanly and consistently separated and grouped. To deal with the problematic placements within placentals

**Table I. Brief Descriptions of Neural Characters and States
Used for Phylogenetic Analysis**

A (arteriovenous anastomoses—primitive—, or loops—derived—). The derived state of paired arterioles and venules in loops distinguishes marsupials from other mammals.

B (bifurcating optic tract to widely separated thalamic terminals—derived—, or closely neighboring terminal regions—primitive—). The derived state distinguishes monotremes.

C (course of olfactory tract fibers under—primitive—, or through—derived—the accessory olfactory formation). The character from Switzer *et al.* (1980) suggesting a basic bifurcation in placentals; one major branch of placentals shows the derived state.

D (deep—derived—or superficial—primitive—position of the optic tract in the collicular tectum). The primitive state occurs in reptiles, birds, monotremes and in the armadillo *Dasypus novemcinctus.*

E (emergence of the facial nerve dorsal—primitive—, or ventral—derived—to the trigeminal sensory column). The primitive state occurs in reptiles, monotremes, and marsupials; the derived states 1,2,3 distinguish placentals.

F (failure—derived—, or persistence—primitive—of oil droplets in retinal cones). The primitive state is seen in reptiles, birds, marsupials, and the monotreme *Ornithorhynchus anatinus.* The derived state distinguishes placentals, and, according to Young and Pettigrew (1991), the monotreme *Tachyglossus aculeatus.*

G (gathering of olfactory bulb mitral cells into a monolayer—derived—, or scattered with no monolayer—primitive—). The primitive state occurs in reptiles, birds, and monotremes; the derived state in marsupials and placentals.

H (hemispheres connected by a corpus callosum—derived—, or no corpus callosum present—primitive—) The primitive state is seen in reptiles, birds, monotremes, and marsupials; the derived state in placentals.

I (internal capsule route of commissural connections in the fasciculus aberrans—derived—, or no such pathway present—primitive—). The derived state occurs only in diprotodont marsupials.

J (jugal or twin dropleted retinal cones present—derived—, or no such dropleted pairs present—primitive—). The derived state is reported only in Australian polyprotodont and diprotodont marsupials.

K (koniocortex which is sensory cerebral cortex containing barrel formations—derived—, or barrels absent—primitive—). The derived state is seen in certain families of diprotodont marsupials, lagomorphs, dermopterans, and rodents.

L (lissencephaly, a smooth cerebral surface—primitive—, or gyrencephaly, a convoluted cerebral surface—derived—). The derived states are usually correlated with body size, but not in rodents (all sciuromorphs and myomorphs have the primitive state) or carnivores or primates (all show the fully derived state 2 regardless of body size). An intermediate or polymorphic derived state 1 occurs in certain diprotodont marsupials where some specimens are lissencephalic and others gyrencephalic.

M (medial—derived—, or lateral—primitive—position of the ventral nucleus of the inferior olive). The primitive state occurs in monotremes and marsupials; the derived states are seen in placentals.

N (no postcranial—derived 1,2—, or a complete postcranial—primitive 0—, representation in primary SI somatosensory cortex) (Johnson *et al.*, 1982a). Complete absence of postcranial projections is the most derived state—2; partial postcranial representations are entered as derived—1; and complete body representations are scored primitive—0. This is a trait distinguishing a number of artiodactyls from other mammals.

O (oral projections to primary SI somatosensory cortex are contralateral—primitive 0—, or ipsilateral—derived 1,2,3—). Completely contralateral projections are scored primitive—0. Successive grades of ipsilateral projections are scored: derived—1 if there is a minimal presence of ipsilateral projections, more derived—2 if there is a substantial ipsilateral projection,

(continued)

Table 1. (*Continued*)

and most derived—3 if there are predominantly ipsilateral projections. This trait delineates the major branching of placental mammals, described more fully in our earlier articles (Bombardieri *et al.*, 1975; Switzer *et al.*, 1980; Johnson *et al.*, 1982a,b).

P (posterior SI somatosensory cortex with 2d body representation—derived 1—, or with only one representation—primitive—). This feature discriminates most anthropoid primates (all but the marmosets, Callitrichidae) from other mammals.

Q (quashed, i.e., eliminated, main and accessory olfactory bulbs). Accessory bulbs are missing (—derived 1—) in Old World anthropoid primates, sirenians, cetaceans, and most bats; both bulbs have been lost (—derived 2—) in some cetaceans. Several members of these diverse groups have all, to some degree, left a terrestrial habitat for one more aquatic, arboreal, or aerial (see Johnson *et al.*, 1993).

R (Rindenkerne, cell clusters in sixth layer of neocortex). These cell clusters (—derived—) occur only in sirenians.

S (somatosensory SI cortical representation of the body with forelimb digits pointing forward—primitive—, or backward—derived—). The derived state is seen only in bats, among the animals studied thus far. The two species for which data have been published include examples of both megabats and microbats.

T (tectopetal connections to the superior or anterior colliculus of one side predominantly from the contralateral retina—primitive—, or progressively more projections from the ipsilateral retina—derived—). The minimally derived state 1 occurs in carnivores. The more derived state 2 is seen in scandents and dermopterans, and the most derived state 3 is characteristic of primates and megachiropterans. All other mammals studied show the primitive state 0.

U [unlaminated dorsal cochlear nucleus (unlaminated—primitive—, laminated—derived—, progressive loss of lamination—more derived—)]. Monotremes exhibit the primitive state, most other mammals show derived states 1 or 2. More derived states 3, 4, and 5 are seen in an interesting group of animals adapted to nonterrestrial habitats (see Johnson *et al.*, 1993, for discussion of parallels with character Q).

V (visible separation of claustrum from cortex—derived—, or no separation—primitive—). The states of this character are to some degree correlated with gyrencephaly, but deviations from this correlation occur among primates, insectivores, and marsupials wherein the claustrum is more separated than would be predicted from the degree of cortical convolution.

W (with—derived—, or without—primitive—, a secondary somatosensory region in cerebral cortex), the primitive condition occurs in monotremes and the derived state in all other mammals studied.

X (external cuneate nucleus clearly separated from the cuneate–gracile nuclear complex—derived—, or not separated—primitive—). The primitive condition is seen in monotremes, sirenians, one sloth, and one shrew. The derived state occurs in all other mammals examined.

on the tree of Fig. 3, we resorted to two stratagems to obtain a cleaner picture. First, we filled in some blank data cells by the following rule: when in one large taxon all members that have been studied for a character show that character in a particular state, and all members of a sister taxon show the character (when studied) in a contrasting state, we applied those states to all members of the taxa whether or not it had been observed in a particular species. Second, since the program for the most part groups species into the traditional orders, to avoid the problems induced by multiple convergent states, we used orders rather than species as the taxa for analysis and entered for each order the most primitive state observed for each character within that order.

Table II. Distribution of States of 29 Characters in 50 Selected Taxa Arranged in Traditional Orders[a]

	A	B	C	D	E	F	G	H	I	J	K	L	M	N	O	P	Q	R	S	T	U	V	W	X	P4	P6	P18	P19	P23	(P6b)
Hypothetical ancestor	0	0	0	0	0	0	0	0	0	0	0	0	0	0	0	0	0	0	0	0	0	0	0	0	0	0	0	0	0	0
Monotremata																														
Ornithorhynchus anatinus	0	1	0	0	0	0	0	0	0	0	0	0	0	0	?	0	0	?	?	0	0	0	0	0	0	0	1	0	?	0
Tachyglossus aculeatus	0	1	0	0	1	0	0	0	0	0	2	0	0	0	0	0	0	0	0	0	0	0	0	0	0	0	1	0	?	0
Polyprotodonta																														
Didelphis virginiana	1	0	0	0	0	1	0	0	0	0	2	0	0	0	0	0	0	0	0	0	1	2	2	1	0	0	0	0	0	0
Sminthopsis marina	?	0	0	0	0	?	1	0	0	?	0	0	0	?	?	?	0	?	?	?	1	2	?	1	1	0	0	0	0	1
Isoodon obesulus	?	0	0	0	0	?	0	0	0	?	2	0	0	?	?	?	0	?	?	?	1	2	?	1	0	0	0	0	0	0
Dasyurus viverrinus	?	0	0	0	0	?	0	0	0	1	2	0	0	?	?	?	0	?	?	0	1	2	?	1	0	0	0	0	0	0
Diprotodonta																														
Pseudocheirus peregrinus	1	0	0	0	0	0	0	1	1	1	0	0	0	0	0	0	0	0	0	0	1	3	1	1	1	0	0	0	0	1
Trichosurus vulpecula	1	0	0	0	0	0	0	1	1	1	1	0	0	0	0	0	0	0	0	0	2	3	1	1	1	0	0	0	0	1
Vombatus ursinus*	?	0	0	0	0	0	0	1	1	?	2	0	0	0	0	0	0	0	0	0	2	2	1	1	1	0	0	0	0	1
Macropus eugenii*	?	0	0	0	0	?	0	1	1	?	2	0	0	0	0	0	0	0	0	0	2	3	1	1	1	0	0	0	0	1
Paucituberculata																														
Lestoros inca	?	0	0	0	0	?	1	0	0	?	0	0	0	?	?	?	0	?	?	?	1	2	?	1	0	0	0	0	1	0
Edentata																														
Tamandua tetradactyla	?	0	3	1	2	?	1	0	0	?	2	2	3	?	?	0	0	?	?	?	2	2	?	1	1	0	0	0	0	0
Myrmecophaga tridactyla	?	0	3	1	3	?	1	0	0	?	2	2	3	?	?	0	0	?	?	?	2	2	?	1	1	0	?	0	0	0
Choloepus hoffmanni	?	0	3	1	3	?	1	0	0	?	2	2	3	0	?	0	0	?	?	?	2	2	1	0	0	1	1	0	?	0
Dasypus novemcinctus	0	0	0	0	3	1	1	0	0	0	2	2	3	0	0	0	0	?	?	?	1	1	1	1	0	0	?	0	?	0
Lagomorpha																														
Oryctolagus cuniculus	0	0	3	1	3	1	1	1	1	0	1	2	3	0	?	0	0	0	0	0	1	0	1	1	1	0	0	0	0	0
Macroscelidea																														
Elephantulus rufescens	?	0	3	1	3	?	1	0	0	?	2	0	2	?	?	0	0	0	?	?	1	0	?	1	0	1	0	?	?	0

(continued)

Table II. (*Continued*)

	A	B	C	D	E	F	G	H	I	J	K	L	M	N	O	P	Q	R	S	T	U	V	W	X	P4	P6	P18	P19	P23	(P6b)
Insectivora																														
Blarina brevicauda	?	?	0	3	2	3	1	1	0	0	0	0	2	?	?	?	0	?	?	2	1	1	?	0	0	0	1	0	1	0
Scalopus aquaticus	?	?	0	3	2	3	1	1	?	0	0	0	2	?	?	?	0	?	?	1	1	1	?	1	0	0	0	0	0	0
Erinaceus europaeus	0	0	3	1	3	1	1	1	0	0	0	0	2	0	0	0	0	0	0	1	1	1	1	1	0	0	0	0	0	0
Hemicentetes hemispinosus	?	0	3	1	3	?	1	1	0	?	0	0	2	?	?	?	0	?	?	1	1	1	?	1	0	1	1	0	0	0
Scandentia																														
Tupaia glis	0	0	3	1	?	3	1	1	0	0	0	0	3	0	0	0	0	0	2	2	2	1	1	1	1	0	0	0	0	0
Primates																														
Galago senegalensis	?	?	3	1	3	1	1	1	0	0	0	2	3	0	0	0	0	3	3	3	2	3	1	1	0	0	0	0	0	1
*Saimiri sciureus**	?	?	3	1	3	1	1	1	0	0	0	2	3	0	1	0	0	3	3	3	3	3	1	1	0	0	0	0	0	1
Aotus trivirgatus	?	?	3	1	3	1	1	1	0	0	0	2	3	0	0	1	0	3	3	3	3	3	?	1	0	0	0	?	?	1
Macaca mulatta	0	?	?	1	3	1	1	1	0	0	0	2	3	0	1	1	1	3	3	3	3	3	1	1	0	1	1	0	0	1
Pan troglodytes	?	?	?	1	3	1	1	1	0	0	0	2	3	0	1	?	1	3	3	4	4	3	1	1	1	1	0	0	1	1
Dermoptera																														
*Cynocephalus volans**	?	0	3	1	3	?	1	1	0	?	1	2	3	?	?	0	0	?	2	2	2	2	?	1	1	0	0	0	1	1
Microchiroptera																														
*Myotis lucifugus**	?	0	?	1	3	1	1	1	0	0	0	0	2	0	?	0	1	0	1	1	1	1	1	1	0	1	1	1	0	0
Megachiroptera																														
*Pteropus giganteus**	?	0	?	1	3	1	1	1	0	0	0	2	2	0	?	0	1	3	3	4	4	1	1	1	1	1	0	0	0	1
Rodentia																														
Cavia porcellus	0	0	3	1	3	1	1	1	0	0	1	2	3	0	0	0	0	0	0	0	0	0	1	1	0	0	0	0	?	0
Dasyprocta leporina	?	0	3	1	3	?	1	1	0	?	0	2	3	0	0	0	0	0	0	1	1	2	1	1	0	0	0	1	?	0
Erethizon dorsatum	?	0	3	1	3	?	1	1	0	?	0	2	3	0	0	0	0	0	?	2	2	2	1	1	0	1	1	1	0	0
Castor canadensis	?	0	3	1	3	1	1	1	0	?	0	2	3	0	0	0	0	?	?	0	1	0	?	1	0	0	0	0	0	0

Sciurus carolinensis	?	0	3	1	3	1	1	1	0	0	1	0	3	0	0	0	0	0	1	1	1	1	0	0	0	?	1
Rattus norvegicus	0	0	3	1	3	1	1	1	0	0	1	0	3	0	0	0	0	0	1	1	0	1	0	0	0	0	0
Cetacea																											
Tursiops truncatus	?	0	?	1	3	?	1	1	0	?	0	2	3	?	?	?	5	3	?	?	0	1	0	1	?	0	0
Sirenia																											
Trichechus manatus	?	0	?	1	3	1	1	1	0	0	2	3	?	?	4	0	?	?	0	0	1	0	0	0	0		
Perissodactyla																											
Equus burchelli*	0	0	0	1	3	1	1	1	0	0	2	3	?	0	2	3	?	1	1	0	0	0	1	0	0	0	1
Artiodactyla																											
Sus scrofa	0	2	1	3	1	1	1	0	0	2	3	1	2	2	3	1	1	1	0	0	0	0	0	1			
Tayassu tajacu	?	0	?	3	?	1	1	0	?	0	2	3	?	?	2	3	?	1	1	0	0	0	0	0	1		
Lama glama	?	0	0	1	3	1	1	1	0	0	2	3	2	0	2	3	?	1	1	0	1	0	0	0	1		
Ovis aries	0	0	0	1	3	1	1	1	0	0	2	3	3	0	2	3	1	1	1	0	0	0	0	0	1		
Carnivora																											
Canis familiaris	0	0	1	3	1	1	1	0	0	2	3	0	1	1	3	1	1	1	0	0	0	0	0	1			
Felis catus	0	0	1	3	1	1	1	0	0	2	3	0	2	1	1	3	1	1	1	0	0	0	0	0	1		
Procyon lotor	?	0	?	1	3	1	1	1	0	0	2	3	0	2	1	3	1	1	1	0	0	0	0	0	1		
Callorhinus ursinus	?	0	0	3	?	1	1	0	?	0	2	3	?	?	3	3	?	1	1	0	0	0	0	0	1		
Phoca vitulina	?	0	?	1	3	1	1	1	0	?	0	2	3	?	?	4	3	?	1	1	0	0	0	0	0	1	
Hyracoidea																											
Procavia capensis	?	0	0	1	3	?	1	1	0	?	0	2	3	0	0	0	2	3	1	1	0	1	0	0	0	?	1

aCharacters A–X are from Johnson et al. (1982a,b, 1993); characters 4, 6, 18, 19, 23 are derived from Pettigrew et al. (1989). The original scoring of character P6 by Pettigrew et al. (1989) is listed under (P6b). Some of the "species" listed are in fact "species-amalgams" constructed to maximize completeness of the data sets. Such a species-amalgam is formed by summing the filled data cells from two closely related species, such that two incomplete data sets for real species form a more complete data set for the artificial species-amalgam. They are indicated in the table by an asterisk, and the species included in the amalgams are as follows: Vombatus ursinus includes Lasiorhinus latifrons; Macropus eugenii includes Thylogale billardieri; Saimiri sciureus includes Cebus capucinus and C. albifrons; Cynocephalus volans includes C. variegatus; Myotis lucifugus includes Macroderma gigas; Pteropus giganteus includes P. poliocephalus; Equus burchelli includes E. caballus. Exactly which states are derived from which species can be found in Johnson et al. (1993).

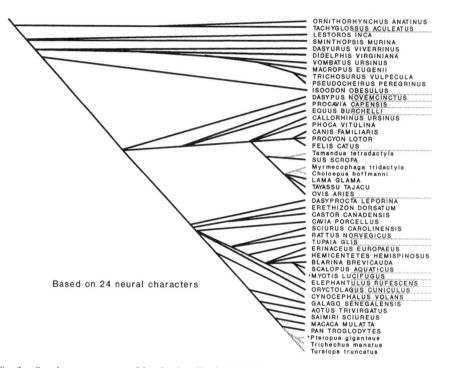

ORNITHORHYNCHUS ANATINUS
TACHYGLOSSUS ACULEATUS
LESTOROS INCA
SMINTHOPSIS MURINA
DASYURUS VIVERRINUS
DIDELPHIS VIRGINIANA
VOMBATUS URSINUS
MACROPUS EUGENII
TRICHOSURUS VULPECULA
PSEUDOCHEIRUS PEREGRINUS
ISOODON OBESULUS
DASYPUS NOVEMCINCTUS
PROCAVIA CAPENSIS
EQUUS BURCHELLI
CALLORHINUS URSINUS
PHOCA VITULINA
CANIS FAMILIARIS
PROCYON LOTOR
FELIS CATUS
Tamandua tetradactyla
SUS SCROFA
Myrmecophaga tridactyla
Choloepus hoffmanni
LAMA GLAMA
TAYASSU TAJACU
OVIS ARIES
DASYPROCTA LEPORINA
ERETHIZON DORSATUM
CASTOR CANADENSIS
CAVIA PORCELLUS
SCIURUS CAROLINENSIS
RATTUS NORVEGICUS
TUPAIA GLIS
ERINACEUS EUROPAEUS
HEMICENTETES HEMISPINOSUS
BLARINA BREVICAUDA
SCALOPUS AQUATICUS
•MYOTIS LUCIFUGUS
ELEPHANTULUS RUFESCENS
ORYCTOLAGUS CUNICULUS
CYNOCEPHALUS VOLANS
GALAGO SENEGALENSIS
AOTUS TRIVIRGATUS
SAIMIRI SCIUREUS
MACACA MULATTA
PAN TROGLODYTES
•Pteropus giganteus
Trichechus manatus
Tursiops truncatus

Based on 24 neural characters

Fig. 3.   Species tree computed by the Swofford PAUP 2.4.1 program from the distributions of states of 24 neural characters in 50 taxa (49 species plus a hypothetical ancestor). Tree length = 101; consistency index = 0.436. Data on the distributions of character states are included in Table II (characters A through X). Species were selected to include those for which most data were available, while including as many representatives as possible of each of the traditional orders. With six exceptions (the exceptions are listed in lowercase with dashed branches leading to them), species again can be grouped into the traditional orders. More details on the characters and the distributions of their states can be found along with alternative renditions of this tree in Johnson *et al.* (1993).

These procedures led PAUP 2.4.1 to give us the next tree [Fig. 4; along with the previous figure it is drawn from data in Johnson *et al.* (in press), included here within Tables II and III]. This is our current brain tree, and again shows primates and Chiroptera on the same major branch of a clean bifurcation of most of the placentals.

One of the topics that generated this volume is the hypothesis that Chiroptera is diphyletic. Separating the Mega- and Microchiroptera as subject taxa, but using the same data as to the distribution of character states, yields the next tree (Fig. 5). Micro- and Megachiroptera do indeed diverge, the former showing insectivore affinity according to these brain traits; but the fruit bats, rather than approaching primates, are instead classed with the Sirenia and Cetacea. Rather than propose a "flying sea cow" scenario, we prefer to think that PAUP 2.4.1, in the absence of other relevant information,

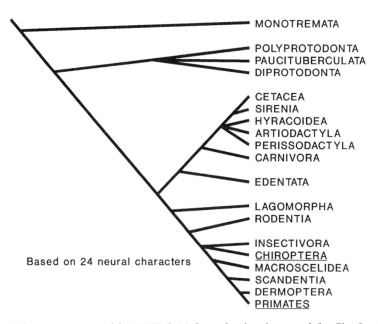

Fig. 4. Orders tree computed by PAUP 2.4.1 from the data base used for Fig. 3, with two modifications (see Table III): values entered were those for the most primitive state of a character found within a given order, and some values were assumed for unavailable data. Tree length = 64; consistency index = 0.609. Once again there appears a basic bifurcation of the placentals, this time with edentates joining the branch containing the ferungulates.

was overly impressed again by the previously noted convergent traits related to moving into aerial and aquatic life-styles.

A recent comprehensive account of mammalian interordinal relations based on more traditional characters (Fig. 6, a tree based on data from Novacek *et al.,* 1988) shows a seven-way multifurcation (the "bushy base" described by Shoshani, 1986). Our tree (Fig. 4) is free of such multifurcative bushy nodes (polytomies). In other respects it agrees in many details with the "traditionally based" tree from Novacek *et al.* (1988—both the tree they present and the tree we constructed from their data in Fig. 6): the subclasses are neatly separated; Lagomorpha and Rodentia are near neighbors; and Scandentia, Primates, Dermoptera, and Chiroptera are associated in a common distal branching as are Hyracoidea and Sirenia.

The 89-Character Tree, with 65 Nonneural Traits Added from Novacek et al. (1988)

As Simmons points out in Chapter 1, neural characters are best used in conjunction with other characters for serious analysis of relationships. For-

Table III. Distribution of States of 29 Neural Characters in 21 Orders[a]

	A	B	C	D	E	F	G	H	I	J	K	L	M	N	O	P	Q	R	S	T	U	V	W	X	P4	P6	P18	P19	P23	(P6b)
Ancestor	0	0	0	0	0	0	0	0	0	0	0	0	0	0	0	0	0	0	0	0	0	0	0	0	0	0	0	0	0	0
Monotremata	0	1	0	0	0	0	0	0	0	0	0	0	0	0	0	0	0	0	0	0	0	0	0	1	0	1	1	0	0	0
Polyprotodonta	1	0	0	0	0	0	0	0	0	0	0	0	0	0	0	0	0	0	0	0	1	2	1	1	0	0	0	0	0	0
Diprotodonta	1	0	1	1	0	0	1	0	1	1	0	0	0	0	0	0	0	0	0	0	1	2	1	1	1	0	0	0	1	1
Paucituberculata	1	0	0	1	0	0	0	0	?	0	0	0	0	?	0	0	0	0	?	?	1	2	?	0	0	0	0	0	1	0
Edentata	0	0	0	0	2	2	1	1	0	0	2	2	3	0	0	0	0	0	0	?	1	1	1	1	0	0	0	0	0	0
Lagomorpha	0	0	0	1	3	1	1	1	0	0	0	3	3	0	1	0	0	0	0	?	1	0	0	0	0	0	0	0	0	0
Insectivora	0	0	3	3	3	1	1	1	0	0	0	2	3	0	0	0	0	0	0	?	1	1	1	1	0	0	0	0	0	0
Macroscelidea	0	0	3	3	3	1	1	1	0	0	0	2	2	?	?	0	0	0	0	?	1	0	?	0	0	0	0	0	?	0
Scandentia	0	0	3	3	3	1	1	1	0	0	0	3	3	0	0	0	0	0	0	2	2	1	1	1	1	0	0	0	0	1
Primates	0	0	3	3	3	1	1	1	0	0	0	3	3	0	0	0	0	0	0	3	1	3	1	1	1	0	0	0	0	1
Dermoptera	0	0	3	3	3	1	1	1	0	1	2	3	3	?	?	0	0	?	?	2	1	2	?	1	1	0	0	0	1	0
Microchiroptera	0	0	3	3	3	1	1	1	0	0	0	1	1	0	?	0	0	1	1	0	4	0	1	1	1	0	0	1	1	1
Megachiroptera	0	0	?	1	3	1	1	1	0	0	2	2	2	0	?	0	1	1	1	3	4	1	1	1	0	0	0	0	0	0
Rodentia	0	0	3	3	3	1	1	1	0	0	0	3	3	0	0	0	0	0	0	0	1	0	1	1	0	0	0	0	0	0
Cetacea	0	?	?	?	3	1	1	1	0	0	2	2	3	?	?	0	0	0	?	?	5	3	?	?	0	1	1	1	?	0
Perissodactyla	0	0	0	0	3	1	1	1	0	0	2	2	3	?	?	0	0	0	?	2	2	3	?	1	0	0	0	0	0	1
Artiodactyla	0	0	0	1	3	1	1	1	0	0	2	2	3	1	2	0	0	0	0	2	2	3	1	1	1	1	0	0	0	0
Carnivora	0	0	0	2	3	1	1	1	0	0	2	2	3	0	1	0	0	0	0	1	1	3	1	1	1	0	0	0	0	1
Hyracoidea	0	0	0	3	3	1	1	1	0	0	2	2	3	0	0	1	1	1	0	?	2	3	1	1	0	1	0	0	0	1
Sirenia	0	0	?	3	3	1	1	1	0	0	2	2	3	?	?	0	0	?	0	?	4	0	?	0	0	0	1	0	0	0

[a]Characters A–X are from Johnson et al. (1982a,b, 1993); characters P4, P6, P18, P19, P23 are derived from Pettigrew et al. (1989). The original scoring of character P6 by Pettigrew et al. (1989) is listed under (P6b). For purposes of this analysis, Megachiroptera and Microchiroptera are considered as separate orders. Ancestor designates a hypothetical ancestral order. Scores listed are the most primitive states known to occur in that order. Underlined states are assumed rather than observed.

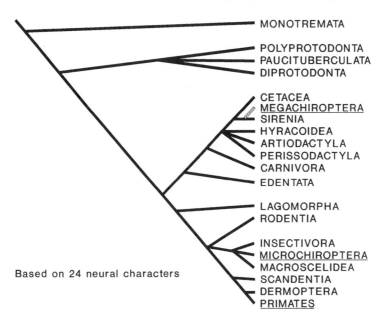

Fig. 5.   Tree derived from the same character set as Fig. 4, but with Micro- and Megachiroptera considered as separate orders. Tree length = 69; consistency index = 0.585. The microchiropterans assume the insectivore-related position held by the Chiroptera; the megachiropterans do not join the primates but appear near the Sirenia and Cetacea, because of their derived states of olfactory and auditory neural characters (Q and U) seen in many species adopting aquatic and aerial habitats. The same characters are responsible for the problematic placement of members of these taxa in the species tree of Fig. 3.

tunately, Novacek *et al.* (1988) have presented their data in a form that enables adding their data from 67 morphological characters (including two of our brain traits, characters F and H) to our 24 brain traits, yielding a total of 89 characters in all. From this assemblage of data, PAUP 2.4.1 produced the tree of Fig. 7. Separating or amalgamating the Mega- and Microchiroptera did not change this tree; we show them here separated. Our 24 brain traits were not overwhelmed by the 65 nonneural characters, and the polytomous bushy base has been replaced by orderly bifurcations. This may be caused in some part by the extra weight of our ordered multistate characters in contrast to the exclusively two-state traits reported by Novacek *et al.* This initial exercise in combining brain traits with other morphological features suggests that these neural features can contribute some resolutions to long-standing problems. With regard to the Chiroptera, the Mega- and Microchiroptera, far apart on the brain tree, are reunited in the 89-character tree, and are related to the Archonta rather than, respectively, to the Insectivora and Sirenia as the brain traits alone suggested.

This tree accommodates the extensive bank of data accumulated and evaluated by others (Novacek *et al.*, 1988), while altering their trees by elim-

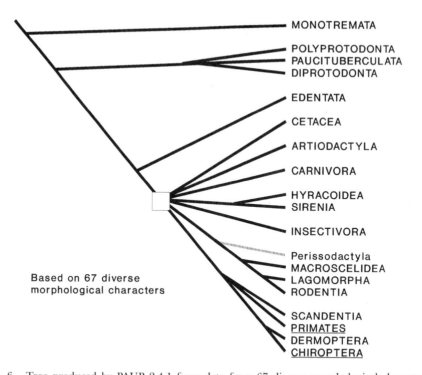

MONOTREMATA
POLYPROTODONTA
PAUCITUBERCULATA
DIPROTODONTA
EDENTATA
CETACEA
ARTIODACTYLA
CARNIVORA
HYRACOIDEA
SIRENIA
INSECTIVORA
Perissodactyla
MACROSCELIDEA
LAGOMORPHA
RODENTIA
SCANDENTIA
PRIMATES
DERMOPTERA
CHIROPTERA

Based on 67 diverse
morphological characters

Fig. 6. Tree produced by PAUP 2.4.1 from data from 67 diverse morphological characters presented by Novacek *et al.* (1988). Tree length = 71; consistency index = 0.930. This tree differs from the tree illustrated in their article (their Fig. 3.3) in some respects. We did not include their data concerning orders for which we have not observed any relevant brain characteristics (Pholidota, Tubulidentata, Proboscidea). In their analysis, Marsupialia was considered a single order: for consistency with our other figures, we show it here as three orders diverging from a fictitious common point. We used all default options in our single run of one program (PAUP 2.4.1): they used a variety of programs and options, relying to some degree on a consensus option of PAUP (CONTREE) to arrive at their published tree. The peculiar placement of Perissodactyla does not occur in their tree, nor is it mentioned in their text. In our run it results from the derived state of their character 43 (canals in the external alisphenoid for maxillary nerve branches) shared by Perissodactyla with Macroscelidea, Lagomorpha, and Rodentia according to their Table 3.2. This tree does share with the tree they present (and with many published trees, see Shoshani, 1986) the multifurcation (box) at the base of the major placental radiation, where seven lines diverge from a common point.

inating all of the polytomies but one. This was accomplished with no deliberate or selective weighting of characters, other than that consequent upon multiple states which were determined independently of weight considerations. There was no selection of the neural characters; the whole grab bag that we have assembled was included with no sorting or evaluation. No fiddling with relative parsimonies or consensuses was required. It is a very raw tree in this respect. Our only data management was to treat orders as taxa and assume a few reasonable entries to substitute for missing data.

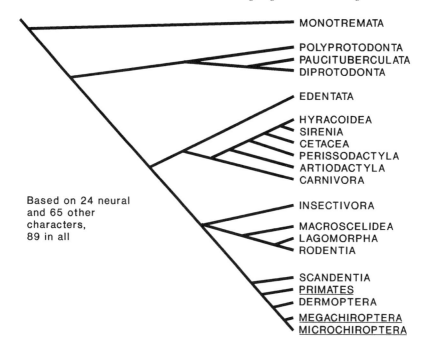

Fig. 7.   Tree computed by PAUP 2.4.1 from states of 89 characters: 24 neural characters (Johnson *et al.*, 1993) and 65 diverse morphological characters (Novacek *et al.*, 1988). Tree length = 153; consistency index = 0.680. This is in effect a combination of the trees of Figs. 5 and 6. Two of the set of 67 characters in Fig. 6 are also included in the set of 24 from Fig. 5; Character F in Table I appears as character 24 in the Novacek set, and character H as character 18. Thus, the total number of characters is 89. For the 24 neural characters, the score entered for a particular order (Table III) is the most primitive state appearing within that order; for the other 65 characters it is assumed that all members of the order show the same state, that given in Table 3.2 of Novacek *et al.* (1988). Several of the neural characters are multistate (characters C, E, L, M, N, O, Q, T, U, V); all of the nonneural characters are two-state. This "combined" tree does not have a polytomy at the base of the placental radiation: the only one on the tree occurs at the divergence of the well-defined archontan group from the insectivores and Glires-plus-Macroscelidea assemblage on one major branch of the placentals. On the other major branch, ferungulates including paenungulates are associated with Cetacea; edentates diverge early from this branch. Within the archontan group, Mega- and Microchiroptera are closest relatives; both are closer to Dermoptera than to Primates.

     This tree suggests five major supraordinal groupings of placentals: two on the first, and three on the second, limb of a basic bifurcation. The first limb dichotomizes into one subdivision consisting of the edentates, and another that diversifies into a large group of orders including the Cetacea and those making up the ferungulates as proposed by Simpson (1945). The second limb diverges into a trichotomy (the sole polytomy on the tree): the first subdivision is the Insectivora, the second is made up of Simpson's (1945) Glires plus the Macroscelidea, the third subdivision is the Archonta (as set forth by Novacek *et al.*, 1988).

The 29- and 94-Character Trees, with 5 Neural Traits Added from Pettigrew et al. (1989)

Recently, Pettigrew *et al.* (1989) presented an array of 24 neural characters, selected for their particular relevance to the phylogeny of bats and primates; this is of considerable interest here since few of the characters we have been using are directly relevant to this topic. One of our characters, character T, that does pertain was obtained from their earlier publication (Pettigrew, 1986); this character appears in the 1989 list as two characters, 14 and 15, which represent two different measures of the same phenomenon, the variations in projections from one retina to the ipsilateral and contralateral optic tecta (superior colliculi). Character 14 is listed as the "ratio of ipsilateral to contralateral eye input to superior colliculus" (Pettigrew *et al.*, 1989, p. 503) but no mention is made of how this ratio was measured. Character 15 is the surface representation of the superior colliculus occupied by the ipsilateral eye; this type of data is not available for many species and requires complex experiments to produce for each species. Another aspect of this character is relatively abundant in the published literature, namely, the relative density of ipsilateral projections in a given section through the colliculi. This is the measure we have used for our character T, which has a distribution of character states very much like those listed for characters 14 and 15 by Pettigrew *et al.* (1989). Excellent illustrations of the variable states of this character, as we measure them, are found in Huerta and Harting (1984), and details of the scoring of the states of character T are presented in Johnson *et al.* (in press). Primates and Megachiroptera share the most derived state with one another and with no other taxa. However, this unequivocal sharing of the derived state did not prevent PAUP 2.4.1 from placing megachiropterans with sirenians rather than primates on both the species tree of Fig. 3 and the orders tree of Fig. 5. This heightened our interest in other characters that might reinforce this commonality in brains of primates and Megachiroptera.

Of the remaining 22 characters used by Pettigrew *et al.* (1989), 5 proved amenable for use in our analyses. We are limited to characters that can be observed in normal specimens in a large number of available species. These 5 were their characters 4, 6, 18, 19, and 23 (we designate them P4, P6, P18, P19, and P23 for use in our analyses). Assessment of states of most of their other characters again requires lengthy and complicated experimental procedures for each species in turn; it would take countless person-years to assemble the requisite data for our comprehensive range of species.

The Five New Characters from Pettigrew et al. (1989)

Character P4

Character P4 is the presence of a derived clearly differentiated large-celled layer in the lateral geniculate nucleus. This character is similar but not identical in definition to their character 2, the laminar differentiation of the lateral geniculate nucleus. All lateral geniculate nuclei are in fact laminated, the lamination being visible in normal (as opposed to experimental) preparations in some species but not in others. In some cases this visibility may relate to phylogeny; in all it would be very difficult to score. But the presence of a large-celled layer is quite evident in any normal cell stain, and is seen in many or all members of several phylogenetic groupings (diprotodont marsupials, scandentians, primates, dermopterans, megachiropterans, sciurid rodents, carnivores, artiodactyls, hyracoids, and perissodactyls) and not in others (monotremes, edentates, insectivores, macroscelids, microchiropterans, murid rodents, and sirenians). As one illustrative case, in the past we and others (Johnson and Marsh, 1970; Pearson *et al.*, 1976) have reported *Pseudocheirus peregrinus* as different from all other diprotodont marsupials studied in its lack of lateral geniculate lamination; however, using this new criterion, close inspection reveals a laminar concentration of large cells along the border of the optic tract with the lateral geniculate nucleus, enabling us to score all of the diprotodonts as having the derived state.

Character P4 is an effective means of incorporating a wealth of data on distinctive organization of this dominant nuclear region of mammalian visual system into our analyses. It discriminates microchiropterans from megachiropterans, and shows the latter sharing a derived state with primates, scandentians, and dermopterans—and with groups from six other orders, which weakens its impact for supporting primate–megachiropteran affinities.

Character P6

Character P6 is the location of the magnocellular layer of the lateral geniculate immediately adjacent to the optic tract. Our survey shows that the magnocellular layer is next to the tract in all of the layer-possessing taxa except *Tupaia*. Pettigrew *et al.* scored *Tupaia* and all of the nonlayered species as primitive in this character. This character can be redefined as having the magnocellular layer located away from the tract, scoring *Tupaia* as derived and all other taxa as primitive. The effect in either case is to differentiate *Tupaia* from other taxa, as the only ones possessing the combination of 1 in character P4 and 0 in character P6 (or the only ones with 1 in character P6 in our alternative scoring scheme). We have used both scoring schemes; in the trees produced by PAUP 2.4.1 (as we use it) the two scoring methods yield identical results regarding placements of *Tupaia* and the whole archontan set, but turn up differences among the marsupials and ungulates (Figs. 11 and 12). (Simmons argues, in Chapter 1, that in this case all nonlayered species should be scored as "no data available" in this character; we disagree. Using either scoring method, it is a valid character distinguishing *Tupaia* and should be used, we believe.)

A separate problem arises when we come to use this character (and the preceding one) for scoring the order Scandentia as opposed to the species or genus *Tupaia*. Simmons (1979) reported an unlaminated lateral geniculate in another scandentian, *Ptilocercus*. Unfortunately, the photographs appearing in that article, of sections stained with the ineffective Klüver–Barrera method, show no cells whatever, so we cannot judge either lamination or large-celled regions. His verbal report is similarly unclear; it consists in two statements in the text (both on p. 77, Simmons, 1979): "LGN does not show a clear arrangement of cells and fibers sa [sic] seen generally in primates," and a curiously incomplete sentence: "On its medial edge, one or two regularly arranged rows of large, more deeply staining and stellate cells that may resemble those of the lateral layers of LGN (Layers 1 and 2) in *Tupaia* and primates." It will take a more adequate report to determine whether the lateral geniculate of *Ptilocercus* warrants scoring Scandentia differently from *Tupaia*. For now we score them similarly.

With regard to primates and megachiropterans, however, in one scoring (*Tupaia* = 0), they share the derived state, but again they share it with all of the other taxa sharing the derived state in character P4; with the other scoring (*Tupaia* = 1), Primates and Megachiroptera share the primitive state. Either way, there is little support for primate–megachiropteran affinity.

Character P18

Character P18 is what Pettigrew *et al.* (1989) describe as a reduced medial terminal nucleus (MTN) of the accessory optic tract, seen in primates and megachiropterans but not microchiropterans. Cooper, a coauthor of Pettigrew *et al.* (1989), reported soon after (Cooper *et al.*, 1990) extensive evidence that there are two parts to MTN, a dorsal (MTNd) and a ventral (MTNv); MTNd is present in all mammals studied, but MTNv is reduced or absent in some taxa, including several primates, megachiropterans, and dermopterans. In many species, experimental methods showing retinal terminals are necessary to detect MTNv; such methods can show presence but it is difficult to demonstrate complete absence as opposed to very few or very small terminals, hence the criterion of "reduced" rather than absent. In our normal cell-stained specimens, a cell group MTNv is recognizable in most species, but not in those described as having MTN reduced by Pettigrew *et al.* (1989) and by Cooper *et al.* (1990). Therefore, we scored all of our specimens according to whether we could or could not find MTNv in such normal specimens.

We followed Pettigrew *et al.* in calling the reduced MTNv the derived state, large and obvious MTNv being found in most mammals (especially marsupials) and therefore presumably primitive. However, in our specimens (and in the experimental studies of Campbell and Hayhow, 1971, 1972), the monotremes *Ornithorhynchus* and *Tachyglossus* show MTNd only, with no MTNv evident, implying that character polarity might be the reverse of what we propose here. Our scores agree with those of Pettigrew *et al.* (1989) for most individual taxa. Their macroscelid *Elephantulus myurus* and our *E. rufescens* show the primitive state (MTNv is present), as do their several microchiropterans and our *Myotis lucifugus,* their rodent squirrel *Petaurista petaurista* and our *Sciurus carolinensis,* and their scandentian *Tupaia* sp. (the species and source of their data are not clear) and our *Tupaia glis.* Derived states occur in: our sloth, *Choloepus,* and their sloth, *Bradypus,* among the edentates; their several megachiropterans and our *Pteropus giganteus;* their dermopteran *Cynocephalus variegatus* and our *C. volans;* and all of their primates except *Tarsius bancanus* and all of our primates except *Galago senegalensis. Galago* is the one case of disagreement: they report the derived state in this species, apparently from published data rather than experimental results or direct observation, although no reference is specifically identified as the source of this scoring; our specimen shows what appears to be a reasonably prominent MTNv. [Cooper *et al.* (1990) cite demonstrations of intraspecific variations in primate MTNv's.] In addition to the primate–dermopteran–megachiropteran group we also see the derived state in most insectivores, in our cetacean and sirenian specimens, the aforementioned monotremes, and the artiodactyl *Lama glama.* Presence is always more easily and securely demonstrated than absence, and for this reason identifications of the derived state are all somewhat tentative for this character.

More detailed study, such as quantitative measurement of volumes or cell numbers, probably requiring many experiments to identify cells of MTNv accurately, might reveal some characteristic of the reduction in the extent of MTNv that is phylogenetically shared among Megachiroptera, Dermoptera, and Primates and no other taxa. For now this character does not offer very strong evidence for affinity among these groups.

Character P19

Character P19 is listed by Pettigrew *et al.* (1989) as the ratio of inferior colliculus to superior colliculus. They point out that these two neighboring structures allow for ready comparative measurement (although they present no measures). They reported that the auditory inferior colliculus is relatively much larger in microbats; the visual superior colliculus is larger in megabats. This neural index of auditory versus visual specialization is another tempting candidate for phylogenetic discrimination, so we are measuring the relative sizes of the colliculi in all of our available specimens. The task is about half done, but we have measures for most of the species we have been using in our analytic programs thus far.

We measure the volumes of the colliculi on the left side by tracing outlines of their perimeters in sets of sections, ranging in size from 6 sections (cut sagittally or horizontally, where both colliculi appear together in the same sections) to 12 or more sections (cut coronally, where only one colliculus is seen in most individual sections). We follow the criteria of Paxinos and Watson (1986) in determining collicular boundaries. These outline tracings are entered into a three-dimensional reconstruction program (PC3D, Jandel Scientific, Corte Madera, Calif.) that also computes volumes of reconstructions. The measures obtained so far are presented in Table IV.

A derived measure from Table IV, the proportion of the whole collicular mass that is occupied by the auditory inferior colliculus, is also shown graphically in Fig. 8. (The advantage of using ratios or proportions, rather than absolute size, is that ratios provide automatic controls for body size and tissue shrinkage.) In Fig. 8 it can be seen that the proportion occupied by the inferior colliculus is extraordinarily high (indicating extreme auditory specialization) in the carnivorous microchiropterans, and quite low in our one frugivorous megachiropteran specimen. It is equally high in the echolocating cetacean specimen *Tursiops truncatus,* in the beaver *Castor canadensis* and porcupine *Erethizon dorsatum* among the rodents, and in the tenrec *Hemicentetes hemispinosus* (our two other tenrecs, *Setifer setosus* and *Tenrec ecaudatus,* have the next highest proportions). We were most interested to see in our *frugivorous* microchiropterans (*Brachyphylla cavernarum, Carollia perspicillata,* and *Monophyllus redmani*) that the proportions were moderate: the colliculi are about equal in size.

The lowest proportions (indicating visual specialization) are seen in the arboreal tree squirrel *Sciurus carolinensis* and tree shrew *Tupaia glis,* and the artiodactyl sheep *Ovis aries.* The startlingly low proportions seen in the monotremes are the result of extremely small and undeveloped inferior colliculi rather than expanded superior colliculi; otherwise we might make two characters from these data, one for expanded visual colliculi and the other for enlarged inferior colliculi. This may yet be possible if we can devise a metric for separating relatively large superior colliculi from relatively small inferior colliculi in determining the proportion or ratio.

We have scored as derived any inferior colliculus proportion of 0.64 or greater (inferior colliculus at least 1.8 times as large as the superior colliculus). This separates carnivorous microchiropterans from frugivorous megachiropterans, and also from frugivorous microchiropterans. The proportion for Megachiroptera is quite high, in the same range as that of the tree shrews and squirrels, artiodactyls, and monotremes. Primates show a quite normal (nondistinctive) range of proportions. We will be pursuing these measures for additional characters since many interesting features are beginning to appear (e.g., the very restricted range in all three marsupial orders).

Pettigrew *et al.* (1989) scored a larger inferior colliculus as derived according to their text, but then scored it as primitive in their Table 3. They pointed out that most outgroup nonmammalian vertebrates have smaller inferior colliculi; our monotreme data support this view. But then they claim in support of the opposite polarity that early mammals probably had large inferior colliculi. Our data support the larger inferior colliculi as the derived state, and this is the one entered in our analyses. Primates and Megachiroptera again in this case share the primitive state, but Microchiroptera is polymorphic. Frugivorous and carnivorous microchiropterans differ in the states of this character. In the scheme just described we would score Microchiroptera primitive

Table IV. Relative Sizes of Inferior Colliculi (IC) (Auditory Midbrain Tectum) and Superior Colliculi (SC) (Visual Midbrain Tectum)

Order, species	Proportion IC/(SC + IC)	Ratio IC/SC	Volume in mm³	
			SC	IC
Monotremata				
Ornithorhynchus anatinus	0.152857	0.180438	11.86	2.14
Tachyglossus aculeatus	0.286243	0.401037	11.57	4.64
Polyprotodonta				
Dasyurus viverrinus	0.446577	0.806938	16.9397	13.6693
Didelphis virginiana	0.496117	0.984589	10.1232	9.9672
Sminthopsis murina	0.518690	1.077666	1.4071	1.5164
Isoodon obesulus	0.521698	1.090732	10.9134	11.9036
Diprotodonta				
Vombatus ursinus	0.439836	0.785191	30.8089	24.1909
Pseudocheirus peregrinus	0.458789	0.847711	11.7806	9.9865
Macropus eugenii	0.467831	0.879104	16.7169	14.6959
Trichosurus vulpecula	0.478296	0.916797	16.4838	15.1123
Paucituberculata				
Lestoros inca	0.436905	0.775902	2.0853	1.6180
Caenolestes obscurus	0.468919	0.882953	1.7168	1.5159
Edentata				
Myrmecophaga tridactyla	0.446266	0.805922	38.3286	30.8899
Choloepus hoffmanni	0.508338	1.033918	17.4713	18.0639
Tamandua tetradactyla	0.512391	1.050827	22.9227	24.0878
Dasypus novemcinctus	0.536713	1.158490	14.2841	16.5480
Lagomorpha				
Lepus americanus	0.322126	0.475201	42.3963	20.1468
Oryctolagus cuniculus	0.338156	0.510930	35.6708	18.2253
Ochotona princeps	0.537296	1.161209	0.6899	0.8011
Insectivora				
Scalopus aquaticus	0.349737	0.537841	1.1986	0.6446
Blarina brevicauda	0.366263	0.577942	0.6622	0.3827
Condylura cristata	0.401615	0.671165	1.4657	0.9837
Erinaceus europaeus	0.487776	0.952273	3.4524	3.2876
Tenrec ecaudatus	0.598752	1.492229	5.1640	7.7059
Setifer setosus	0.632275	1.719429	2.0531	3.5302
Hemicentetes hemispinosus	0.685005	2.174657	1.2799	2.7834
Macroscelidea				
Elephantulus rufescens	0.391636	0.643755	12.8463	8.2699
Scandentia				
Tupaia glis	0.157366	0.186755	25.2150	4.7090
Primates				
Ateles geoffroyi	0.321166	0.473114	39.3799	18.6312
Saguinus sp.	0.355946	0.552666	7.8945	4.3630
Hylobates lar	0.367366	0.580694	40.9761	23.7946
Saimiri sciureus	0.382075	0.618321	37.6474	23.2782
Cercocebus torquatus	0.413860	0.706079	41.7045	29.4467
Macaca mulatta	0.422207	0.730724	20.2514	14.7982
Cebus capucinus	0.424835	0.738633	21.2495	15.6956

(continued)

Table IV. (*Continued*)

Order, species	Proportion IC/SC + IC	Ratio IC/SC	Volume in mm³	
			SC	IC
Lemur mongoz	0.440744	0.788090	14.1957	11.1875
Nycticebus coucang	0.465646	0.871419	9.4231	8.2114
Papio sphinx	0.482713	0.933166	24.1524	22.5382
Aotus trivirgatus	0.486851	0.948753	9.6608	9.1658
Microcebus murinus	0.489207	0.957743	3.9889	3.8204
Tarsius syrichta	0.491344	0.965966	6.8780	6.6439
Galago senegalensis	0.491375	0.966088	6.1749	5.9655
Presbytis entellus	0.523054	1.096675	26.8631	29.4601
Lemur catta	0.529088	1.123540	12.4777	14.0192
Pan troglodytes	0.577744	1.368234	30.6522	41.9394
Perodicticus potto	0.591947	1.450662	6.0223	8.7363
Dermoptera				
Cynocephalus volans	0.343039	0.522162	19.4878	10.1758
Microchiroptera				
Brachyphylla cavernarum	0.542507	1.185828	2.0511	2.4322
Carollia perspicillata	0.554811	1.246241	1.5410	1.9204
Monophyllus redmani	0.592247	1.452467	0.9853	1.4311
Noctilio albiventris	0.645607	1.821728	1.6641	3.0315
Rhinolophus ferrumequinum	0.679504	2.120172	1.0095	2.1404
Myotis lucifugus	0.692294	2.249858	0.4943	1.1121
Eptesicus fuscus	0.699854	2.331713	0.7423	1.7309
Desmodus rotundus	0.701760	2.353010	0.3433	0.8078
Saccopteryx bilineata	0.716557	2.528056	1.3237	3.3465
Hipposideros armiger	0.718138	2.547835	2.6729	6.8101
Vampyrum spectrum	0.726489	2.656165	4.3958	11.6760
Trachops cirrhosus	0.742597	2.884960	3.0606	8.8297
Rhinolophus hipposideros	0.747980	2.967940	0.4792	1.4222
Molossus molossus	0.777195	3.488237	0.7759	2.7067
Megachiroptera				
Pteropus giganteus	0.223310	0.287515	19.4594	5.5948
Rodentia				
Sciurus carolinensis	0.154104	0.182178	49.2775	8.9773
Dasyprocta leporina	0.365166	0.575215	39.6246	22.7927
Hydrochaeris hydrochaeris	0.367486	0.580993	126.0060	73.2087
Aplodontia rufa	0.457923	0.844756	6.7513	5.7032
Rattus norvegicus	0.476550	0.910405	12.8992	11.7435
Cavia procellus	0.567236	1.310728	8.0146	10.5049
Erethizon dorsatum	0.643176	1.802507	15.7221	28.3392
Castor canadensis	0.653800	1.888506	20.4047	38.5344
Cetacea				
Tursiops truncatus	0.794048	3.855512	204.834	789.74
Perissodactyla				
Equus burchelli	0.276459	0.382091	404.976	154.738
Artiodactyla				
Ovis aries	0.164417	0.196770	232.760	45.8002
Lama glama	0.219921	0.281922	472.513	133.212

(*continued*)

Table IV. (*Continued*)

Order, species	Proportion IC/SC + IC	Ratio IC/SC	Volume in mm^3	
			SC	IC
Sus scrofa	0.350035	0.538546	126.571	68.1644
Odocoileus virginiana	0.354910	0.550171	292.358	160.847
Tayassu tajacu	0.382677	0.619898	71.8426	44.5351
Carnivora				
Mustela erminea	0.403758	0.677172	5.0747	3.4364
Ursus maritimus	0.412176	0.701190	151.840	106.469
Canis familiaris	0.480907	0.926437	55.2209	51.1587
Procyon lotor	0.502959	1.011906	17.5362	17.7450
Felis catus	0.585498	1.412539	21.1449	29.8680
Pinnipedia				
Callorhinus ursinus	0.416596	0.714079	111.394	79.5442
Phoca vitulina	0.468413	0.881161	74.7987	65.9097
Hyracoidea				
Procavia capensis	0.372575	0.593818	20.7865	12.3434
Sirenia				
Trichechus manatus	0.594900	1.468530	134.224	197.112

for our ordinal analysis, owing to the primitive state in the frugivores. It can be argued that the frugivorous state is secondarily derived, leaving the ordinal score derived, and that is what we have done. Scoring primitive or derived for the order makes no difference in the trees of Figs. 13 and 14. At this time we have no other brain data from the microchiropteran frugivores, so they do not participate in the species analysis of Figs. 11 and 12.

Character P23

Character P23 is described by Pettigrew *et al.* (1989) as spinal cord with greatly enlarged dorsal roots and dorsal horn. In the caption of the figure illustrating this character, they qualify their character definition by restricting the locus of enlargement to the cervical spinal cord. They do not indicate how large roots or horns must be to qualify as "greatly enlarged."

We do not have specimens of dorsal roots, but many of the brains in our collections include the uppermost cervical cord. Since we are again dealing with quantifiable relative size, we have measured the dorsal horns in the available specimens of cervical spinal cords, at the level of the caudal pole of the cuneate-gracile nuclear complex, in the region of the first to third cervical segments (C1–C3).

In microchiropterans, a feature equal to dorsal horn width in distinctiveness is the extreme narrowness of the top of the intervening dorsal columns. Accordingly, we also measured these widths of dorsal columns (actually the angle intervening between the dorsal horns).

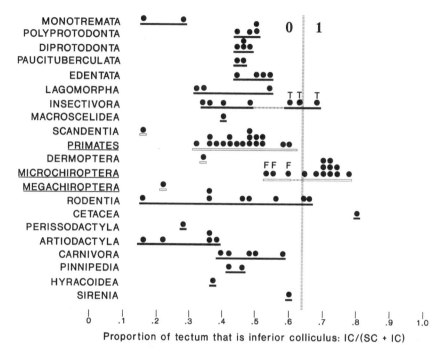

Fig. 8.   Based on character 19 of Pettigrew *et al.* (1989), the ratio of the sizes of the inferior and superior colliculi of the midbrain, this is a graphic illustration of the distribution, by orders, of the proportional sizes of inferior (auditory) and superior (visual) colliculi in the midbrains of 86 specimens representing 86 species in 21 ordinal groups (from Table IV). The value entered is the proportion of the total of both colliculi (the midbrain tectum) that is inferior colliculus (volume of the inferior colliculus divided by the sum of the volumes of the superior colliculus plus the inferior colliculus). We have listed Pinnipedia here as a separate order for easier localization of individual taxa; elsewhere in this chapter they are considered as within Carnivora. Dots indicate measures of individual specimens, bars show ranges seen within orders. Hollow bars are used for the archontan orders. Microchiropterans and our one cetacean specimen fall outside the range seen in all other mammals except for the tenrec *Hemicentetes* in Insectivora and the beaver *Castor* and the porcupine *Erethizon* in Rodentia. We therefore established the proportion of 0.64 as the boundary (vertical line) for scoring the derived state (1); this leaves all microchiropterans and cetaceans in the derived state and all but three of the remaining specimens in the primitive state. Megachiropterans and primates share the primitive state. An alternative character, scoring instead a relatively larger superior colliculus as a derived state, would at best associate megachiropterans with rodent tree squirrels, scandentian tree shrews, and artiodactyl sheep and llamas rather than with primates. IC, inferior (auditory) colliculus; SC, superior (visual) colliculus.

In our sample, the microchiropterans have very wide dorsal horns (gray matter) and narrow dorsal columns (white matter); the megachiropterans have wide dorsal columns and narrow dorsal horns. Our megachiropterans are larger than our microchiropterans, and it has long been known that spinal cords of larger animals have a relatively much greater volume of white matter than gray matter (Ariëns Kappers *et al.*, 1936–1960, Vol. I, p. 225). To measure horns and columns, while controlling for specimen size and tissue shrinkage, we use transverse sections and construct lines tangent to the borders of the substantia gelatinosa (a prominent and constant histological feature capping the dorsal horn) and intersecting the center of the central canal (Fig. 9). Then we measure the angles subtended by these lines and the midline to obtain measures of the right and left dorsal horns and dorsal columns. These measures are listed in Table V. As expected, animals within orders are arranged approximately according to body size.

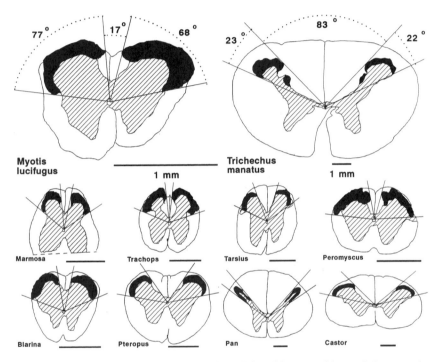

Fig. 9.   Top: Method used for measuring widths of dorsal horns and intervals between dorsal horns. Shown are tracings from transverse sections through the uppermost cervical spinal cord in two mammals, the small microchiropteran *Myotis lucifugus*, with wide dorsal horns and a narrow interhorn interval, and the large sirenian *Trichechus manatus*, with narrow horns and a wide interhorn interval. The substantia gelatinosa capping each dorsal horn is shown in black; the rest of the gray matter is shown as shaded. Lines were constructed tangent to the borders of the substantia gelatinosa and intersecting the center of the central canal; the angles described by these lines were measured to provide a quantitative expression of the widths of dorsal horns and interhorn intervals. In this illustration only the total interhorn interval is shown; separate measures were also taken for the right and left portions of this interval (on each side of the midline), and all three measures of the interval are listed in Table V.
Bottom: Using the same conventions (e.g., all scale bars = 1 mm), here are illustrated sections of the cervical spinal cord in eight diverse mammals. From left to right they are: two tiny non-microchiropterans, the polyprotodont marsupial *Marmosa murina* and the insectivore *Blarina brevicauda;* two bats, the microchiropteran *Trachops cirrhosus* and the megachiropteran *Pteropus giganteus;* two primates, the small *Tarsius syrichta* and the large chimpanzee *Pan troglodytes;* and two rodents, the small deer mouse *Peromyscus leucopus* and the large beaver *Castor canadensis*. The numerical measures for these examples are listed in Table V.

Table V. Widths of Interhorn Intervals in Dorsal Columns Arranged in Ascending Sequence by Orders[a]

Order, species	Total interhorn width L + RIH	Total dorsal horn width L + RDH	Proportional horn width (L + RDH)/(DH + IH)	Left dorsal horn width LDH	Left interhorn width LIH	Right interhorn width RIH	Right dorsal horn width RDH	Total bilateral width: horns + intervals DH + IH
Polyprotodonta								
Marmosa marina	16	85	0.8416	42	7	9	43	101
Isodon obesulus	32	55	0.6322	28	17	15	27	87
Antechinus flavipes	34	108	0.7606	59	18	16	49	142
Didelphis virginiana	53	52	0.4952	26	24	29	26	105
Dasyurus viverrinus	66	46	0.4107	25	35	31	21	112
Diprotodonta								
Petaurus breviceps	30	73	0.7087	43	16	14	30	103
Pseudocheirus peregrinus	38	63	0.6238	27	19	19	36	101
Macropus eugenii	42	83	0.6640	37	23	19	46	125
Trichosurus vulpecula	51	46	0.4742	23	24	27	23	97
Macropus rufus	77	61	0.4420	30	38	39	31	138
Vombatus ursinus	90	78	0.4643	36	45	45	42	168
Paucituberculata								
Lestoros inca	17	86	0.8350	41	11	6	45	103
Edentata								
Choloepus hoffmanni	76	88	0.5366	47	39	37	41	164
Myrmecophaga tridactyla	87	63	0.4200	32	43	44	31	150
Tamandua tetradactyla	92	76	0.4524	39	45	47	37	168
Lagomorpha								
Lepus americanus	60	62	0.5082	32	30	30	30	122
Insectivora								
Blarina brevicauda	24	102	0.8095	55	14	10	47	126
Scalopus aquaticus	37	84	0.6942	38	20	17	46	121
Tenrec ecaudatus	46	98	0.6806	45	21	25	53	144
Condylura cristata	51	91	0.6408	49	26	25	42	142
Erinaceus europaeus	55	126	0.6961	57	30	25	69	181
Hemicentetes hemispinosus	74	126	0.6300	65	43	31	61	200
Setifer setosus	75	139	0.6495	67	45	30	72	214

Scandentia								
Tupaia glis	39	66	0.6286	30	19	20	36	105
Primates								
Tarsus syrichta	32	69	0.6832	38	19	13	31	101
Saimiri sciureus	46	61	0.5701	31	23	23	30	107
Galago senegalensis	46	101	0.6871	52	23	23	49	147
Lemur mongoz	48	46	0.4894	22	25	23	24	94
Papio sphinx	57	57	0.5000	28	29	28	29	114
Macaca mulatta	60	51	0.4595	26	29	31	25	111
Saguinus sp.	64	70	0.5224	34	32	32	36	134
Cercocebus torquatus	66	31	0.3196	12	31	35	19	97
Perodicticus potto	67	60	0.4724	30	34	33	30	127
Presbytis entellus	68	53	0.4380	31	32	36	22	121
Nycticebus coucang	71	89	0.5562	45	35	36	44	160
Lemur catta	74	60	0.4478	30	36	38	30	134
Pan troglodytes	77	23	0.2300	11	38	39	12	100
Homo sapiens	80	34	0.2982	21	38	42	13	114
Dermoptera								
Cynocephalus volans	50	78	0.6094	39	29	21	39	128
Microchiroptera								
Rhinolophus ferrumequinum[b]	0	—	1.0000	—	0	0	—	—
Molossus molossus	2	54	0.9643	22	1	1	32	56
Trachops cirrhosus	13	126	0.9065	60	8	5	66	139
Brachyphylla cavernarum	16	126	0.8873	52	8	8	74	142
Myotis lucifugus	17	145	0.8951	77	9	8	68	162
Saccopteryx bilineata	18	70	0.7955	35	9	9	35	88
Carollia perspicillata	19	118	0.8613	64	8	11	54	137
Hipposideros armiger	20	102	0.8361	51	10	10	51	122
Monophyllus redmani	22	101	0.8211	51	10	12	50	123
Noctilio albiventris	24	81	0.7714	44	10	14	37	105
Megachiroptera								
Pteropus giganteus	75	87	0.5370	44	36	39	43	162

(continued)

Table V. (*Continued*)

Order, species	Total interhorn width L + RIH	Total dorsal horn width L + RDH	Proportional horn width (L + RDH)/ (DH + IH)	Left dorsal horn width LDH	Left interhorn width LIH	Right interhorn width RIH	Right dorsal horn width RDH	Total bilateral width: horns + intervals DH + IH
Rodentia								
Peromyscus leucopus	30	139	0.8225	68	14	16	71	169
Geomys bursarius	47	87	0.6493	48	22	25	39	134
Rattus norvegicus	52	74	0.5873	38	28	24	36	126
Ondatra zibethica	54	62	0.5345	30	27	27	32	116
Chinchilla laniger	63	106	0.6272	54	33	30	52	169
Glaucomys volans	70	82	0.5395	44	36	30	38	152
Myoxus glis	84	105	0.5556	55	40	34	50	189
Spermophilus tridecemlineatus	86	104	0.5474	53	43	44	51	190
Aplodontia rufa	88	81	0.4793	38	44	43	43	169
Hydrochaeris hydrochaeris	96	86	0.4725	42	46	44	44	182
Erethizon dorsatum	99	65	0.3963	33	51	50	32	164
Castor canadensis	101	58	0.3648	31	48	53	27	159
Perissodactyla								
Equus burchelli	86	88	0.5057	44	43	43	44	174
Artiodactyla								
Bos taurus	76	49	0.3920	19	34	42	30	125
Odocoileus virginiana	76	56	0.4242	28	42	34	28	132
Lama glama	81	77	0.4873	37	40	41	40	158
Tayassu tajacu	81	77	0.4873	36	40	41	41	158

Sus scrofa	83	85	0.5060	46	39	44	39	168
Ovis aries	87	62	0.4161	29	40	47	33	149
Carnivora								
Mustela erminea	37	51	0.5795	24	19	18	27	88
Cynictis penicillata	64	55	0.4622	30	31	33	25	119
Canis familiaris[c]	65	57	0.4672	26	33	32	31	122
Vulpes zerda	66	60	0.4762	32	30	36	28	126
Nasua nasua	71	71	0.5000	31	36	35	40	142
Canis familiaris[c]	73	65	0.4710	33	37	36	32	138
Ailurus fulgens	73	42	0.3652	21	35	38	21	115
Felis catus	75	38	0.3363	20	37	38	18	113
Potos flavus	82	53	0.3926	24	43	39	29	135
Procyon lotor	87	41	0.3203	20	41	46	21	128
Crocuta crocuta	90	37	0.2913	20	46	44	17	127
Vulpes vulpes	99	53	0.3487	24	51	48	29	152
Panthera leo	120	40	0.2500	22	61	59	18	160
Pinnipedia								
Callorhinus ursinus	63	75	0.5435	36	30	33	39	138
Phoca vitulina	104	44	0.2973	22	52	52	22	148
Hyracoidea								
Procavia capensis	80	92	0.5349	46	40	40	46	172
Sirenia								
Trichechus manatus	83	45	0.3516	23	40	43	22	128

[a]Entries (other than proportions) are in degrees of arc from the center of the central canal.

[b]Our specimen of *Rhinolophus ferrumequinum* was sectioned sagittally, making it difficult to measure widths of dorsal horns; but the angle of cut was such that the horns on each side were seen to contact one another at the midline septum.

[c]Two different specimens of *Canis familiaris* were measured.

The most distinctive of these measures, clearly separating the micro-chiropterans from nearly all other mammals, was the angular width between the dorsal horns, and this is illustrated graphically in Fig. 10. Only the tiny marsupials *Lestoros inca* (Paucituberculata) and *Marmosa murina* (Polyproto-donta) and the smallest insectivore *Blarina brevicauda* fall within the micro-chiropteran range. No other dimensions or ranges appear to have any phy-logenetic sequence. Forty-six of the eighty-nine species measured have dorsal horns as wide as do microchiropterans; and the proportion of the total width of horns plus columns that is occupied by horns shows not only the small marsupials and *Blarina*, but also small mice within the microchiropteran range. Megachiropterans and primates are well within the ranges of several other orders for any of the measures.

The microchiropteran range of the interhorn interval was scored as de-rived; primates and megachiropterans again share the primitive state. We still must determine whether this character is anything more than an index of body size.

The Effects of Adding the Five New Characters

In review, for our purposes there is one "good" character in the set used by Pettigrew *et al.* (1989), their character 14–15, which we have used as charac-ter T: there is a shared maximally derived state among only Primates and Megachiroptera, with less derived states in Dermoptera and Scandentia (but also in Carnivora). The five characters introduced more recently are less promising as phylogenetic indicators, particularly for indicating relationships between primates and megachiropterans. Characters P19, P23, and P6 (with our scoring of *Tupaia* derived) show primates and megachiropterans sharing the primitive state. Characters P4, P18, and P6 (with their scoring of *Tupaia* primitive) show primates and megachiropterans sharing derived states with many and diverse other taxa.

Adding these characters to the analysis that produced the species tree of Fig. 3 produced the trees seen in Figs. 11 and 12, respectively using the two alternative scorings for character P6. Compared with that of Fig. 3, the tree of Fig. 11 shows the following changes, which might be regarded as beneficial in terms of producing a more credible tree: the "unbelievable" placements of all but one of the edentates are eliminated as three of the four species are grouped near the base of the placental radiation, the polytomy involving the pinnipeds is resolved, and the scandentian *Tupaia* is associated with primates rather than insectivores. However, new "problems" appear in this tree: the perissodactyl *Equus* shares in a trichotomy with the most derived artiodactyls; the lagomorph *Oryctolagus* is inserted into the midst of the rodent group; and the dermopteran *Cynocephalus* is moved to a separate branch far from the archontans.

In the alternative tree of Fig. 12, compared with that of Fig. 11, the marsupials are rearranged without particular significance that we can detect,

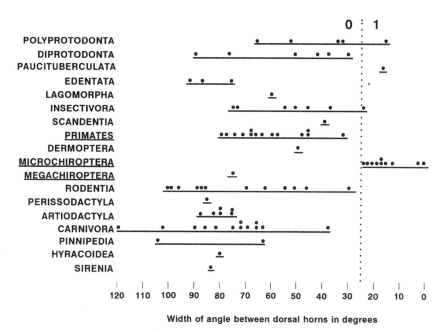

Fig. 10.   Based on character 23 of Pettigrew *et al.* (1989), the enlargement of the dorsal horns in the cervical spinal cord (and consequent narrowing of the dorsal columns intervening between the horns), this is a graphic illustration of the distribution, by orders, of the width of the inter-horn interval in cervical cords of 85 specimens representing 85 species in 18 ordinal groups (from Table V). As in Fig. 8, Pinnipedia are considered a separate order only for better localization of taxa in the figure, dots indicate measures of individual specimens, bars show ranges seen within orders. Microchiropterans fall outside the range seen in all other mammals except for the smallest marsupials, *Lestoros inca* in the Paucituberculata and *Marmosa murina* in the Polyprotodonta, and *Blarina brevicauda*, the smallest insectivore. We therefore established the angular width of 25° as the boundary (vertical line) for scoring the derived state (1); this leaves all microchiropterans in the derived state and all but three of the remaining specimens in the primitive state. Megachiropterans and primates are not distinctive in this measure: again they share the primitive state (0).

Equus regains its position as separate from the artiodactyls, the archontans are reunited with the move of *Cynocephalus* back to the limb leading to primates, but the rodents are fractured into at least two widely separated groups.

In neither of these species trees is there any change in the positions of primates relative to megachiropterans. Adding data from the new characters, using our methods of scoring and analysis, contributes no support for the hypothesis of affinity between these groups.

Orders trees based on strictly neural data, seen in Fig. 13, produced by the rules we used to produce the trees of Figs. 4 and 5, with data added from the five new characters, show no change in the positions of Primates, Megachiroptera, and Microchiroptera relative to one another.

The "grand" orders tree of Fig. 14 shows the effect of adding data from the 5 new characters to the 89-character tree of Fig. 7. The five major supra-

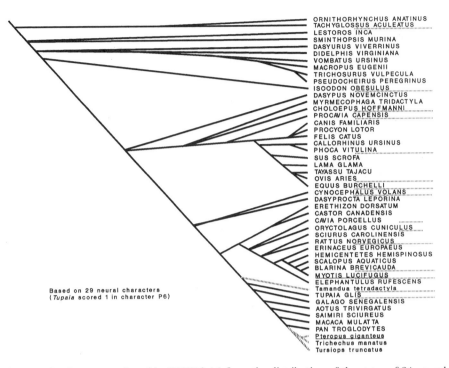

Fig. 11. Species tree produced by PAUP 2.4.1 from the distribution of the states of 24 neural characters in 49 taxa (from Fig. 3) with the addition of the states of 5 neural characters from Pettigrew *et al.* (1989); in character 6 of this set the distinctive state seen in *Tupaia* was scored derived and all other taxa were scored primitive. Tree length = 122; consistency index = 0.402. Dotted lines indicate boundaries of traditional ordinal groupings. In this tree, compared with that of Fig. 3, the "misplaced" edentates are more reasonably grouped (except for *Tamandua*); among the carnivores the polytomy leading to *Callorhinus* and *Phoca* is resolved; *Equus* appears as a most-derived artiodactyl; the dermopteran *Cynocephalus* is on an early-derived isolated branch from the second major limb of the placental dichotomy; *Oryctolagus* is here located between two groups of rodents; the microchiropteran *Myotis* is associated with the macroscelidean *Elephantulus* rather than with the insectivores; and the scandent *Tupaia* approaches the primates rather than the rodents and insectivores. More pertinent for this symposium, the addition of five characters originally formulated to support the association of Megachiroptera with Primates produces no change whatever in the positions of members of these groups.

ordinal groups are retained, but the polytomy at the base of the placental radiation is reintroduced.

Relations of Megachiroptera, Primates, and Archonta

While this manuscript was undergoing review, a report was published challenging the parallels in retinotectal connections seen in megachiropterans

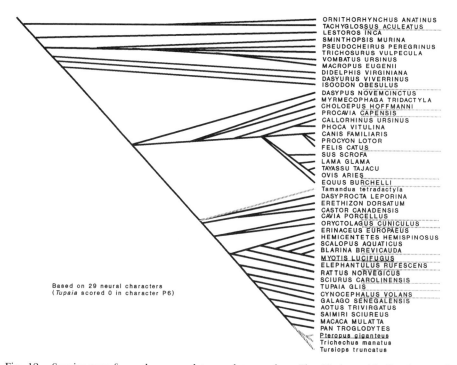

Fig. 12. Species tree from the same data used to produce Fig. 11, but with *Tupaia* scored primitive in character 6, and all other taxa that were scored derived in character 4 are again scored derived in character 6. This repetition of "shared-derived states" (as was done by Pettigrew *et al.*, 1989) for these taxa, which include primates, megachiropterans, and many other groups, produced the following alterations in the tree. Tree length = 127; consistency index = 0.386. Among the marsupials, in place of *Didelphis*, which now occupies a separate branch, *Sminthopsis* joins the branch leading to diprotodonts as the earliest diverging taxon, and the others in this group reverse their sequential order of divergence. *Equus* reverts to its own separate branch as in Fig. 3, departing from the most-derived artiodactyls. The rodents are fractured into two widely separated groups (*Dasyprocta, Erethizon, Castor,* and *Cavia* diverge before the insectivores; *Rattus* and *Sciurus* diverge later). The archontans are back together on this tree, with the dermopteran *Cynocephalus* now located between the scandent *Tupaia* and the primates. Again there is no change in the position of megachiropterans and primates.

compared with those of primates (Thiele *et al.*, 1991). The particular feature addressed in this report was that used as character 17 by Pettigrew *et al.* (1989), rather than those appearing as characters 14 and 15 and our character T.

This feature (character 17) is the sorting of cells in the retina such that all retinotectally projecting cells on the left side of each retina project to the left superior colliculus, while those of the right sides of each retina project to the right colliculus; the pattern is typical of all primates studied and was reported in three species of the megachiropteran genus *Pteropus* by Pettigrew (1986). This remarkable parallel, probably more than any other, generated the hy-

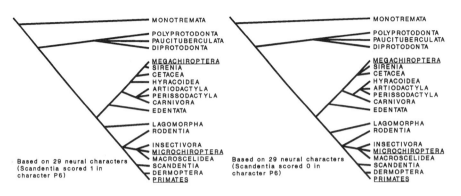

Fig. 13.   Orders trees computed by the PAUP 2.4.1 program, from the data used for generating the trees in Figs. 4 and 5, with the addition of data on the states of five additional neural characters. These data are listed in Table III. For the five added characters (characters P4, P6, P18, P19, P23 in Table III) the most primitive state occurring among members of traditional order was entered as the score for that order. No assumptions were made concerning any states of the five added characters: no entry was made unless a state was actually observed in a given taxon. The two trees were produced from alternative scorings in character P6, as in Figs. 11 and 12 for species. The first of these trees (Scandentia 1 in character P6) differs from that of Fig. 5 only in that Carnivora is more, and Hyracoidea is less, associated with Artiodactyla and Perissodactyla. Tree length = 83; consistency index = 0.530. The other tree (Scandentia 0 in character P6) differs from that in Fig. 5 in two respects: there is a polytomous, rather than sequential branching to Megachiroptera, Sirenia, and Cetacea; and the Hyracoidea–Artiodactyla–Perissodactyla group branches from the line leading to Carnivora rather than that leading to Megachiroptera–Sirenia– Cetacea. Tree length = 88; consistency index = 0.500. The addition of data from the five new characters produces no alteration in the positions of Megachiroptera, Microchiroptera, or Primates relative to one another.

pothesis of primate–megachiropteran affinities. This particular pattern of connections, discernable only with extensive experimentation, has as corollaries the equivalent ipsilateral and contralateral projections used as characters 14, 15, and T; these features, however, can exist without the precise detail of pattern involved in character 17.

Pettigrew *et al.* (1989) reported that *Rousettus aegyptiacus*, another megachiropteran species, shared the same derived states in characters 17, 14, and 15 with the *Pteropus* species and with primates, but presented no data. Thiele *et al.* (1991) showed data indicating that *R. aegyptiacus* has the primitive state in character 17, without directly addressing characters 14 and 15. Our scoring of the state of character T in *R. aegyptiacus* in the data presented by Thiele *et al.* (1991), and also in an earlier publication (Cotter, 1981), has the state as derived 2, and *Rousettus* sharing this state with *Tupaia* (based on data from Laemle, 1968) and *Cynocephalus variegatus* (based on data provided to us by J. D. Pettigrew); the more derived state 3 is seen in primates (Johnson *et al.*, 1993) and *Pteropus poliocephalus* (based on data provided to us by J. D. Pettigrew).

Thus, according to Pettigrew, primatelike character states are seen in *Rousettus* and *Pteropus,* suggesting megachiropteran–primate affinity. But ac-

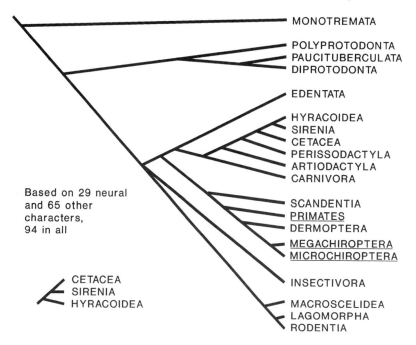

Fig. 14.   Tree computed by PAUP 2.4.1 from the data used to produce Fig. 7 plus the orders data from the five new characters used to generate the trees in Fig. 13. Tree length = 169; consistency index = 0.645. The only difference in trees using the alternative scorings for character P6 was in the relations among Cetacea, Sirenia, and Hyracoidea; scoring Scandentia as derived produced the whole tree depicted here; scoring Scandentia as primitive and several other orders with a shared derived state as in character P4 produced the relations shown in the inset at lower left. This tree differs from that of Fig. 7 in that a trichotomy is introduced whereby Insectivora, and the macroscelid–glirold group diverge at the base of the placental radiation. There is no change produced in any archontan relationships.

cording to Thiele *et al.,* the most detailed primatelike feature (the derived state in character 17) appears in *Pteropus* but not *Rousettus,* arguing for primate–megachiropteran homoplasy (unless, as they point out, *Rousettus* had undergone a secondary derivation, losing the primatelike state in this character). In our analysis, using character T, *Rousettus* shares a less derived state with other archontans (examples from Dermoptera and Scandentia), while *Pteropus* and primates share the most derived state; this could suggest homoplasy of the most derived state but a basic affinity of all of the archontans.

We have to note here that this character (T) is the only neural character from our set that groups any bats with any other archontans. None of our neural characters relates all bats with archontans. In every analysis using strictly neural characters, our sample microchiropteran is classed with insectivores, and our sample megachiropteran is placed with sirenians. The grouping of bats with primates, scandentians, and dermopterans occurs only when the analysis includes the nonneural data. Primates, *Tupaia,* and *Cynocephalus* are placed together using strictly neural characters, supporting a batless Archonta. We do not place too much emphasis on these results, since we have data from only two chiropteran species.

Neural Characters and Homoplasy

The parallel between *P. poliocephalus* and primates in the most derived states of characters T and 17 remains remarkable, particularly since accumulating evidence indicates that it is homoplasic (e.g., Bennett *et al.,* 1988; Baker *et al.,* 1991; Bailey *et al.,* 1992). Pettigrew has argued that homoplasy in neural characters is usually obvious in the distinctive structure of the various homoplasic features, and therefore neural structures that appear similar in all respects are most likely inherited and constitute a superior guide to phylogenetic relationships (Pettigrew *et al.,* 1989; Pettigrew, 1991). In our experience, the derived states of 5 (characters C, D, K, L, O) of the 15 characters we originally published appeared sufficiently similar that we needed the results of phylogenetic analysis to conclude that they represented homoplasies (Johnson *et al.,* 1984); the same is true of 5 of the additional characters (characters Q, T, U, V, X) we have used and now 4 of the 5 (characters P4, P18, P19, P23) of those introduced by Pettigrew *et al.* (1989) that we have been able to use in our analyses. In all, 14 of 29 neural traits show "obscure" homoplasy. Based on these data, rather than presumptive logic, we see no special property of neural characters that distinguishes them from other kinds of data in the likelihood of occurrence of obscure homoplasies.

We have retained homoplasic characters in all of our analyses, because our experience has indicated that despite homoplasies, the distribution of states of these characters adds useful information to the analyses (Kirsch and Johnson, 1983). For example, the presence of barrels in somatic sensory cortex (character K) occurs in several independent lineages, but in one of these, the diprotodont marsupials, it is a distinguishing feature of the related families Petauridae and Phalangeridae. This retention of apparently homoplasic characters is reflected in the tree lengths and consistency indices of trees generated by PAUP 2.4.1 (presented in the captions of Figs. 3–7 and 11–14). Trees are longer and consistency indices are smaller for the trees generated from our often homoplasic neural characters; much shorter trees with larger consistency indices are generated from nonneural characters (Fig. 6) since this character set, the only one of its kind available for our analysis, was culled specifically to eliminate homoplasic characters.

The Contribution of Neural Characters to Hypotheses about Interordinal Relationships, Including Those of Primates and Chiroptera

This is very much a progress report on an ongoing enterprise, and many tasks remain to be done before we can claim definitive suggestions, much less conclusions. For now we can say that the addition of brain traits changes the prior picture of mammalian relationships (such as that in Fig. 6) mainly by suggesting a sequential derivation of Edentata, Carnivora, Artiodactyla, Perissodactyla, Cetacea, Sirenia, and Hyracoidea along one branch of the placental radiation. The distal part of this branch resembles Simpson's cohort Ferungulata. [The embarrassing position of Perissodactyla on the tree of Fig. 6 is resolved by the addition of the brain traits; Novacek *et al.* (1988) found other means of resolution.]

The distribution of states of neural characters provides good support for the supraordinal groupings of the subclasses Prototheria, Metatheria, and Eutheria, and within Eutheria they are consistent with subdivision into the five major groupings of Edentata, Insectivora, Archonta, Glires-plus-Macroscelidea, and a large assemblage of Ferungulata-plus-Cetacea. They further suggest a closer relation of Edentata with the Ferungulata group, pointing toward a four-branched (rather than seven-branched) polytomy at the base of the placental radiation.

We tend not to trust the association of Archonta with the edentate–ferungulate assemblage suggested by Fig. 14. This mistrust may be arbitrary. We are tempted to weed our 29 characters, discarding those that superficially do not promise "good" sorting according to preconceptions about phylogeny. To resist this temptation, we resort to the caution that new ideas about relationships are not likely to emerge from data selected for their agreement with old ideas. On the other hand, if the old ideas are true (e.g., maybe there really was a polytomy at the beginning of the placental radiation), do we really want new ideas? We believe so: new ideas may be necessary to test and verify the old ones. In this vein we are grateful for the introduction of the hypothesis of primate–megachiropteran affinity even though our data so far do not offer it support.

Summary

All of our successive trees based on data from neural characters, including those with neural characters added to 65 nonneural traits, show placental resolutions not found in nonneural trees. The most recently generated trees, based on the most comprehensive set of neural and extraneural data, support five major supraordinal groupings among placentals: edentates, ferungulates (with cetaceans associated), insectivores, archontans, and a gliroid–macroscelidid group. Within the ferungulates, a sequential derivation of Carnivora, Artiodactyla, Perissodactyla, Cetacea, Sirenia, and Hyracoidea is suggested. Megachiropterans, microchiropterans, and primates are associated with dermopterans and scandentians in the archontan group. Addition of data from five new neural characters introduced by Pettigrew *et al.* (1989) supports the grouping into five major placental subdivisions; it does not support the hypothesis of special megachiropteran–primate relationships.

ACKNOWLEDGMENTS

These studies were supported by NSF Grants BSR 85-03687 and BNS 91-11592. We thank Dr. W. I. Welker for the use of the brain collection in the Department of Neurophysiology, University of Wisconsin-Madison, in obtaining a large portion of our data; Dr. R. C. Switzer for data from his collection at the Cole Neuroscience Laboratory, University of Tennessee, Knoxville; and Dr. Paul Pirlot for his generous donation to the Michigan State collection of many of the sections of brains of microchiropteran species used in these studies.

References

Ariëns Kappers, C. U., Huber, G. C., and Crosby, E. C. 1936–1960. *The Comparative Anatomy of the Nervous System of Vertebrates, including Man*, Vol. 1. Hafner, New York.

Bailey, W. J., Slighton, J. L., and Goodman, M. 1992. Rejection of the "flying primate" hypothesis by phylogenetic evidence from the ε-globin gene. *Science* **256**:86–89.

Baker, R. J., Novacek, M. J., and Simmons, N. B. 1991. On the monophyly of bats. *Syst. Zool.* **40**:216–241.

Bennett, S., Alexander, L. J., Crozier, R. H., and Mackinlay, A. G. 1988. Are megabats flying primates? Contrary evidence from a mitochondrial DNA sequence. *Aust. J. Biol. Sci.* **41**:327–332.

Bombardieri, R. A., Johnson, J. I., and Campos, G. B. 1975. Species differences in mechanosensory projections from the mouth to the ventrobasal thalamus. *J. Comp. Neurol.* **163**:41–64.

Campbell, C. B. G., and Hayhow, W. R. 1971. Primary optic pathways in the echidna, *Tachyglossus aculeatus:* An experimental degeneration study. *J. Comp. Neurol.* **143**:119–136.

Campbell, C. B. G., and Hayhow, W. R. 1972. Primary optic pathways in the duckbill platypus, *Ornithorhynchus anatinus:* An experimental degeneration study. *J. Comp. Neurol.* **145**:195–208.

Cooper, H. M., Baleydier, C., and Magnin, M. 1990. Macaque accessory optic system: I. Definition of the medial terminal nucleus. *J. Comp. Neurol.* **302**:394–404.

Cotter, J. R. 1981. Retinofugal projections of an echolocating megachiropteran. *Am. J. Anat.* **160**:159–174.

Huerta, M. F., and Harting, J. K. 1984. The mammalian superior colliculus: Studies of its morphology and connections, in: H. Vanegas (ed.), *Comparative Neurology of the Optic Tectum*, pp. 687–774. Plenum Press, New York.

Johnson, J. I. 1986. Mammalian evolution as seen in visual and other neural systems, in: J. D. Pettigrew, W. R. Levick, and K. J. Sanderson (eds.), *Visual Neuroscience: A Festschrift for P. O. Bishop*, pp. 196–207. Cambridge University Press, London.

Johnson, J. I., and Marsh, M. P. 1970. Laminated lateral geniculate in the nocturnal marsupial *Petaurus breviceps* (sugar glider). *Brain Res.* **15**:250–254.

Johnson, J. I., Kirsch, J. A. W., and Switzer, R. C. 1982a. Phylogeny through brain traits: Fifteen characters which adumbrate mammalian genealogy. *Brain Behav. Evol.* **20**:72–83.

Johnson, J. I., Kirsch, J. A. W., and Switzer, R. C. 1982b. Phylogeny through brain traits: The distribution of categorizing characters in contemporary mammals. *Brain Behav. Evol.* **20**:97–117.

Johnson, J. I., Kirsch, J. A. W., and Switzer, R. C. 1984. Phylogeny through brain traits: Evolution of neural characters. *Brain Behav. Evol.* **24**:169–176.

Johnson, J. I., Kirsch, J. A. W., Reep, R. L., and Switzer, R. C. 1993. Phylogeny through brain traits: More characters for the analysis of mammalian evolution. *Brain Behav. Evol.* (in press).

Kirsch, J. A. W. 1977. The classification of marsupials, in: D. Hunsaker (ed.), *The Biology of Marsupials*, pp. 1–50. Academic Press, New York.

Kirsch, J. A. W. 1983. Phylogeny through brain traits: Objectives and method. *Brain Behav. Evol.* **22:**53–59.

Kirsch, J. A. W., and Johnson, J. I. 1983. Phylogeny through brain traits: Trees generated by neural characters. *Brain Behav. Evol.* **22:**60–68.

Kirsch, J. A. W., Johnson, J. I., and Switzer, R. C. 1983. Phylogeny through brain traits: The mammalian family tree. *Brain Behav. Evol.* **22:**70–74.

Laemle, L. K. 1968. Retinal projections of *Tupaia glis. Brain Behav. Evol.* **1:**473–499.

Lillegraven, J. A. 1969. Latest Cretaceous mammals of upper part of Edmonton formation of Alberta, Canada, and review of marsupial–placental dichotomy in mammalian evolution. *Paleontol. Contrib. Univ. Kans.* **50:**1–122.

McKenna, M. C. 1969. The origin and early differentiation of therian mammals. *Ann. N.Y. Acad. Sci.* **167:**217–240.

McKenna, M. C. 1975. Toward a phylogenetic classification of the mammalia, in: W. P. Luckett and F. S. Szalay (eds.), *Phylogeny of the Primates*, pp. 21–46. Plenum Press, New York.

Novacek, M. J., Wyss, A. R., and McKenna, M. C. 1988. The major groups of eutherian mammals, in: M. J. Benton (ed.), *The Phylogeny and Classification of the Tetrapods*, pp. 31–71. Oxford University Press (Clarendon), London.

Paxinos, G., and Watson, C. 1986. *The Rat Brain in Stereotaxic Coordinates*, 2nd ed. Academic Press, New York.

Pearson, L. J., Sanderson, K. J., and Wells, R. T. 1976. Retinal projections in the ringtailed possum, *Pseudocheirus peregrinus. J. Comp. Neurol.* **170:**227–240.

Pettigrew, J. D. 1986. Flying primates? Megabats have the advanced pathway from eye to midbrain. *Science* **231:**1304–1306.

Pettigrew, J. D. 1991. Wings or brain? Convergent evolution in the origins of bats. *Syst. Zool.* **40:**199–216.

Pettigrew, J. D., Jamieson, B. G. M., Robson, S. K., Hall, L. S., McNally, K. I., and Cooper, H. M. 1989. Phylogenetic relations between microbats, megabats and primates (Mammalia: Chiroptera and Primates). *Philos. Trans. R. Soc. London Ser. B* **325:**489–559.

Shoshani, J. 1986. Mammalian phylogeny: Comparison of morphological and molecular results. *Mol. Biol. Evol.* **3:**222–242.

Simmons, R. M. T. 1979. The diencephalon of *Ptilocercus lowii. J. Hirnforsch.* **20:**69–92.

Simpson, G. G. 1945. The principles of classification and a classification of mammals. *Bull. Am. Mus. Nat. Hist.* **85:**1–350.

Switzer, R. C., Johnson, J. I., and Kirsch, J. A. W. 1980. Phylogeny through brain traits: The course of the dorsal lateral olfactory tract past the accessory olfactory formation as a palimpsest of mammalian descent. *Brain Behav. Evol.* **17:**339–363.

Swofford, D. L. 1984. Phylogenetic analysis using parsimony (PAUP). Illinois Natural History Survey, Champaign.

Thiele, A., Vogelsang, M., and Hoffmann, K. P. 1991. Pattern of retinotectal projection in the megachiropteran bat *Rousettus aegyptiacus. J. Comp. Neurol.* **314:**671–683.

Young, H. M., and Pettigrew, J. D. 1991. Cone photoreceptors lacking oil droplets in the retina of the echidna, *Tachyglossus aculeatus* (Monotremata). *Visual Neurosci.* **6:**409–20.

The Role of the Neurosciences in Primate Evolutionary Biology

10

Historical Commentary and Prospectus

TODD M. PREUSS

Introduction

If behavior is the leading edge of evolution, and if the brain is the principal organ of behavior, one might expect the neurosciences to occupy a central place in evolutionary biology. Obviously, this is not the case—at present. Yet the founders of modern physical anthropology and primatology included several individuals who also made significant contributions to neuroanatomy, particularly Grafton Elliot Smith, Wilfred E. Le Gros Clark, and Raymond Dart. (Examples of these contributions include Elliot Smith, 1897, 1910, 1919; Le Gros Clark, 1932, 1941, 1956; Dart, 1934). Such a confluence of professional interests was no accident: these individuals regarded the understanding of brain evolution as crucial for understanding primate phylogeny. As Elliot Smith (1924, p. 21) put it, the facts of brain evolution are "the cement to unite into one comprehensive story the accumulations of knowledge con-

TODD M. PREUSS • Department of Psychology, Vanderbilt University, Nashville, Tennessee 37240.

Primates and Their Relatives in Phylogenetic Perspective, edited by Ross D.E. MacPhee. Plenum Press, New York, 1993.

cerning the essential facts of Man's pedigree." In the works of Elliot Smith and Le Gros Clark, it was the increasing complexity of the brain, a result of life in the trees, that enabled primates to become behaviorally flexible or adaptable, and so escape the narrowing adaptations that beset terrestrial mammals (see especially Elliot Smith, 1924; Le Gros Clark, 1959).

However, following the collapse of the Le Gros Clark's "classical primatological synthesis" in the 1960s (Cartmill, 1982), in which neural evidence played a part (Campbell, 1966, 1980), the neurosciences came to play a greatly diminished role in the study of primate origins and relationships, a field dominated by comparative studies of skeletal anatomy and macromolecules. Students of primate brain evolution have focused almost exclusively on changes in brain *size,* rather than on discrete characteristics of neural organization that might serve to clarify phyletic relationships. Indeed, it has proven difficult to identify such discrete characteristics. As Falk (1982) has put it:

> Despite the voluminous amount of research in primate neuroanatomy that has transpired during the past 50 years, physical anthropologists are still not able to provide concrete hypotheses regarding specific qualitative (i.e., not size related) changes in the brain that occurred during primate, including human, evolution.

Moreover, since brain size varies as a single parameter (i.e., brain size can only increase or decrease), interpretations of brain evolution have typically been cast within the framework of a unitary phylogenetic scale, with taxa ranked by encephalization quotient (Jerison, 1973) or progression index (Stephan and Andy, 1969), for example. However, over the course of this century, evolutionary biologists have increasingly abandoned the scale as a metaphor for change, with its Victorian overtones of progress and improvement, in favor of a conception of evolution that emphasizes diversity and adaptive uniqueness, as represented by Darwin's (1859) metaphor of a branching tree. Thus, a fundamental difference in point of view has developed between workers who study the evolution of the brain and cognition and those who study the evolution of other biological systems (Boakes, 1984; Hodos and Campbell, 1969; Povinelli, 1993).*

The conceptual gap between the neurosciences and evolutionary primatology may now be closing. In recent years, connectional specializations of primate and other mammalian brains have been documented (e.g., Harting *et al.,* 1973; Kaas *et al.,* 1973; Krettek and Price, 1977; Kobler *et al.,* 1987; Nudo and Masterton, 1990) and the first cladistic analyses of neural characters have appeared (Johnson *et al.,* 1982; Kirsch and Johnson, 1983; Johnson and Kirsch, this volume). The lively response aroused by Pettigrew's interpretation of archontan relationships (Pettigrew, 1986, 1991; Pettigrew *et al.,* 1989), in which neural evidence figures prominently, is the most obvious sign that neuroscience

---

*In this chapter, references to neurobiologists should be understood as denoting specifically *mammalian* neuroscientists. Neuroscientists who study nonmammalian vertebrates typically have a more modern evolutionary orientation than those who study mammals, putting more emphasis on diversity and less on advancement.

has once again become a participant in the field of primate phylogeny. In the thick atmosphere of controversy, it is easy to lose sight of the fact that Pettigrew's work marks a significant reconciliation between neuroscientists and evolutionary biologists, since for the first time in many years they are arguing about common problems within a common conceptual framework.

The dramatic historical shifts in the status of neuroanatomy within primatology raise several questions. First, how did the brain come to occupy such an important position in the primatology of Elliot Smith and Le Gros Clark? Second, how did studies of brain evolution become focused on changes in size, to the virtual exclusion of changes in internal organization? Third, what changes within the neurosciences have made it possible for neuroscientists and primate biologists to speak the same language once again? In examining these issues, I will argue that Darwin's treatment of mental evolution provided a model for later interpretations of brain evolution in terms of a phylogenetic scale, rather than adaptive diversification. Further, I will suggest that the current reconciliation of neuroanatomy and evolutionary biology is a direct result of recent, fundamental changes in our understanding of brain organization, changes that make it possible to supplement or replace broad, scalar interpretations of neural evolution based mainly on brain volume, with branching or cladistic models that emphasize changes that have occurred in specific parts of the nervous system in particular mammalian groups.

The Question of Primate Neural Specializations

Darwin's Legacy: A Phylogenetic Scale of Mind

In the *Descent of Man* (1871), Charles Darwin placed the brain at the top of the agenda for human evolutionary studies. Darwin proposed that a large brain and prodigious intelligence were the fundamental human adaptations, allowing humans to compensate for physical weakness, notably the absence of large canines and other natural defenses. That is, the brain came to serve as a general functional mechanism, making possible the invention of artificial devices (tools) to substitute for specialized characteristics of musculoskeletal anatomy. What is more, Darwin argued that "there is no fundamental difference between man and the higher mammals in their mental faculties" (p. 35). He bolstered this point by attempting to identify "rudiments" of human mentality and morality in nonhumans. Human characteristics could then be explained as improvements of preexisting properties rather than as qualitative departures from ancestral traits. Darwin concluded (p. 105) that "the difference in mind between man and the higher animals, great as it is, is certainly one of degree and not of kind." The only exception that Darwin acknowledged, the only uniquely human characteristic, is language.

Darwin's insistence that humanity's close relatives share virtually all our moral and mental faculties is rather puzzling, for it is not required by his general theory of evolution. In the *Origin of Species* (1859), Darwin likened the results of evolution by natural selection to the branching of a tree, with independent lineages producing different adaptations. Darwin's tree metaphor emphasized uniqueness and diversity; yet his later discussion of mental evolution emphasized degrees of advancement along a single path. Why did Darwin not apply his general view of organic evolution to mental evolution? Why did he not propose a branching tree of minds? One can suggest several reasons for this.

One factor may have been Darwin's intellectual predilection for minute changes. It is true, as proponents of punctuational evolution have observed (Gould and Eldredge, 1977), that Darwin viewed evolutionary history as the gradual accumulation of very small modifications. Yet, even closely related taxa are sometimes separated by large morphological gaps, as Darwin recognized. Indeed, when pressed Darwin could provide plausible evolutionary accounts of the origin of new organs and "peculiar habits," as he put it, such as the evolution of wings and gliding in rodents (Darwin, 1859, pp. 179–180).

Another possible factor was Darwin's understanding of psychology, which was influenced by British empiricism. (Darwin's psychological ideas and their philosophical roots, as revealed in his notebooks, are discussed by Richards, 1987, pp. 105–107.) The principal tenet of empiricism is that complex ideas are the result of associations among simpler percepts. This process was considered the sole basis of intellectual activity in humans and animals, and some empiricists had by Darwin's era concluded in principle that there could be no qualitative distinction between human and animal thought (Richards, 1987, p. 107).

Perhaps the most important reason for Darwin's aversion to qualitative differences between humans and their close relatives is that during Darwin's day, gaps between humans and other animals—particularly intellectual or moral gaps—were seized upon as evidence against the theory of natural selection (Desmond, 1982; Richards, 1987, p. 107). To Darwin's contemporaries, gaps were considered signs of the miraculous, continuity the hallmark of natural process (Desmond, 1982). It was the perception of a gap between humans and lower animals that led Alfred Russel Wallace, the codiscoverer of natural selection, to posit supernatural intervention in human evolution (e.g., Richards, 1987, pp. 176–184). In this light, we can understand why Darwin was so concerned with identifying rudiments of human mentality in animals. We can also understand the significance of the famous exchange between Richard Owen and T. H. Huxley over the hippocampus minor, a ridge of cortical tissue that bulges from the wall of the lateral ventricle. Owen, searching for a difference to distinguish humans from beasts, and so refute Darwinian theory, argued that only humans possess a hippocampus minor. Unfortunately for Owen, a homologous structure had long before been demonstrated in great apes, as Huxley pointed out. (For accounts of this incident, see especially Desmond, 1982, and Elliot Smith, 1910.)

The vehemence and very public nature of this dispute over one possibly unique human characteristic—parodied in its day as the "great hippopotamus test"–illustrates how important it was to Darwin and his supporters to defend the anatomical similarity of human and nonhuman brains. It is difficult to imagine quite so heated an argument arising over some possibly unique aspect of human somatic anatomy, for instance, the longitudinal arch of the foot or the long flexor muscle of the thumb, or for that matter the absence of a baculum. There is simply no reason, under evolutionary theory, why humans could not have a unique structure of the brain or any other feature of anatomy. Yet the great hippopotamus affair set the tenor for later debates over brain evolution. As will become clear, the brain has been held to an unusually narrow standard of evolutionary continuity.

Classical Views of Primate Brain Evolution: Brodmann, Elliot Smith, and Le Gros Clark

While insisting that Owen had got it wrong, Darwin and his early supporters had little to say about what kinds of changes, other than enlargement, actually did occur in brain evolution. This is not surprising, as little was known about the histological organization or function of brain structures in the mid-1800s. During the late 1800s, however, new methods were developed for eliciting movements from brain structures with electrical stimulation, and new stains were used to study the cellular composition ("cytoarchitecture") and fiber plexuses ("myeloarchitecture") of the brain. Researchers used these techniques to try to identify anatomical and functional subdivisions—areas, that is—within the cerebral cortex, a structure that to most earlier workers had seemed quite homogeneous (Kemper and Galaburda, 1984; Phillips *et al.*, 1984). The idea that the cortex was divided into multiple anatomical divisions, a doctrine known as localizationism, was further reinforced by the clinical observations of Broca, Wernicke, Liepmann, and others, who demonstrated that lesions of specific brain regions may result in quite specific intellectual deficits (Jeannerod, 1985; Luria, 1973).

The early architectonic work of Brodmann (1909) was particularly influential in shaping ideas about cortical organization and evolution, owing to its broad comparative scope (he examined brains from many mammalian orders) and explicit phylogenetic conclusions. Although no less beholden to the phylogenetic scale than his contemporaries, Brodmann did not interpret brain evolution according to a narrow standard of continuity. Brodmann argued that cortical evolution is characterized by a process of expansion and differentiation, particularly of the parietal, temporal, and frontal lobes, by which anthropoid primates, and humans especially, came to possess more areas than other mammals. What is more, Brodmann identified a particular cortical territory, the granular frontal cortex, as distinctive of primates. This territory, which occupies the so-called "prefrontal" region anterior to the electrically excitable motor zone, is distinguished histologically by the presence of a distinct layer IV packed with very small cells; hence the term "granular." The motor region, by contrast, in which layer IV is poorly developed or absent, was characterized as "agranular." According to Brodmann, all primates possess granular frontal cortex, although the number of areas in the region is variable (Fig. 1). Humans have the most areas, making up a large portion of the frontal lobe. Monkeys have fewer areas than humans, and lemurs fewer still; in these animals the granular frontal region makes up a smaller part of the frontal lobe than in humans. In nonprimate mammals ("*bei den nächstniederen Mammaliern*," as Brodmann put it, using the language of the phylogenetic scale), granular frontal cortex is either negligible in extent or entirely absent (Brodmann, 1909, p. 203). Brodmann's evolutionary conclusions made sense from the standpoint of behavior, for the frontal lobe was commonly regarded at that time (as it is again today) as the locus of many higher-order intellectual functions (Stuss and Benson, 1986; Goldman-Rakic, 1987; Walsh, 1987). In fact, one neurologist declared that the evolutionary emergence of humans marked the "age of the frontal lobe" (Tilney, 1928, cited by Stuss and Benson, 1986, p. 1).

It was in this context that Elliot Smith and Le Gros Clark fashioned their primatological ideas, combining elements of localizationist neuroanatomy with Darwin's evolutionary psychology and Wood Jones's (1916) arboreal theory of primate origins (see especially Elliot Smith, 1924, and Le Gros Clark, 1959, and the historical essays by Cartmill, 1982, and Fleagle and Jungers, 1982). The commitment of early primates to an arboreal life was considered to promote visual acuity and tactile sensitivity, while permitting primates to retain a generalized musculoskeletal anatomy and behavioral flexibility. While other mammals were limited by their adaptive specializations, primates succeeded by avoiding narrow specializations, by being adaptable. The key to this adaptability was the brain, enlarged and differentiated to support keen senses, a quick wit, and an agile body (Elliot Smith, 1924, pp. 135–136). As Le Gros Clark (1959, p. 322) put it, "It is the increasing complexity of *general* organization (particularly in the higher functional levels of the brain) which gives to the Primates their distinctive capacity for a wider range of adjustments to any environmental change." In this fashion, Elliot Smith and Le Gros Clark extended Darwin's view of humans as brainy but otherwise unspecialized mammals to primates as a whole.

But how is one to reconcile the Darwinian idea that the brain acts as a general functional device with the localizationist doctrine that the brain is composed of discrete and functionally specialized units? It should be said that neither Elliot Smith nor Le Gros Clark explicitly challenged the localizationist doctrine, and both acknowledged that new cortical areas appear in evolution. In fact, Elliot Smith (1910, pp. 223, 226) advanced a model of how new areas could arise, by "budding off" from old areas, acquiring new connections and functions. This is very close to Brodmann's views. Yet at the same time, Elliot Smith denied that new structures appeared in the evolution of the human

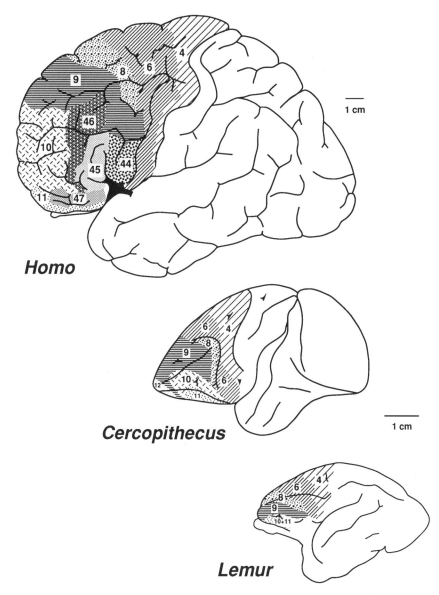

Fig. 1. The expansion and differentiation of primate frontal cortex, according to Brodmann (1909). Brodmann divided the frontal lobe into two zones, *Regio praecentralis*, which included areas 4, 6, and (in humans) 44, and *Regio frontalis*, encompassing the remaining, more anterior areas of the frontal lobe. The precentral region is motor in function; areas 4 and 6 correspond approximately to the primary motor area (MI) and premotor cortex, respectively. *Regio frontalis* corresponds to the granular frontal cortex, also known as prefrontal cortex or frontal association cortex. Brodmann regarded granular frontal cortex as a hallmark of primate anatomy. Moreover, he believed that the number of areas in this region increased during primate evolution, from prosimians through monkeys to humans.

brain. He twice recounted the hippocampus minor affair at length, each time drawing the conventional Darwinian conclusion, and dismissed several other claims of human cerebral specialization (Elliot Smith, 1910, 1924). It would seem that for Elliot Smith, new cortical areas did not deserve the status of new anatomical *structures*. Rather, he emphasized the unity of the cortex, in which new areas are not specializations in their own right, but servants in the cause of general functional capacity. The addition of areas in evolution, he wrote, does not interfere with the "fundamental purpose" of the cortex, "which continues to remain a unifying organ that acts as a whole . . ." (Elliot Smith, 1924, p. 26). Consistent with this approach, he construed the proliferation of cortical sensory areas in primate evolution not as a process of specialization, but rather as a release from specialization. In his view, it freed primates from the reliance on olfaction that marks primitive mammals, creating "a more equable balance of the representation of the senses" (Elliot Smith, 1924, p. 30).

Whereas Elliot Smith's conception of expansion and differentiation allowed for the evolution of areas with new functions (even as these areas were denied the status of new structures), Le Gros Clark's approach was if anything more conservative. He regarded brain evolution as a matter of refining or improving preexisting capacities, rather than adding new functions. For example, he wrote: ". . . the progressive elaboration and differentiation of the cortex in the evolving Primates has led to increasing powers of apprehending the nature of external stimuli, a greater capacity for a wider range of adjustments to any environmental change, and an enhancement of the neural mechanisms for effecting more delicately co-ordinated reactions." In this process, the cortical areas "undergo a progressive structural differentiation so that their extent and boundaries become more clearly definable by histological examination" (Le Gros Clark, 1959, pp. 228–229). Thus, generality of function is obtained not so much by creating a more equable balance of parts, as in Elliot Smith, but rather by the progressive unfolding of capacities latent within cortical tissue.

Lashley's Critique of Localizationism and the Rise of Allometry

While Elliot Smith and Le Gros Clark were attempting to reconcile localizationism with the doctrine of continuity—in essence by denying that new cortical areas are really new structures—an intellectual force was at work that promised to render the issue moot. Throughout the 19th century, two schools of thought had contended over the nature of cerebral organization: holism (globalism) and localizationism (Luria, 1973; Jeannerod, 1985). Localizationism, as discussed above, was fashionable at the beginning of the 20th century (although the influence of holism is apparent in Elliot Smith's work). By midcentury, there had been a decided shift toward holistic conceptions of cerebral organization. An important stimulus for this shift was the work of K.

S. Lashley, one of the principal figures in the history of American neuro-science. (For an appraisal of Lashley's career, see Orbach, 1982a,b.)

Lashley's major works, published in the period from about 1920 to 1950, explicitly challenged the localizationist tenets of clinical neurology and of architectonics. In experimental lesion studies of rats, Lashley (1929) concluded that retardation of performance on maze-learning tasks depends only on the amount of cortex removed, without respect to the location of the lesion. He concluded that, except for the primary sensorimotor areas, the cerebral cortex is relatively undifferentiated in structure and is functionally "equipotential." In a critical reappraisal of the architectonic method, Lashley and Clark (1946) were unable to identify many of the areas described by Brodmann and others, and concluded that architectonic workers had greatly overstated the extent of structural differentiation within the cortex.

Lashley's ideas cannot be dismissed as eccentric. A number of prominent clinicians of the same era espoused similar antilocalizationist views (Luria, 1973, pp. 23–26). The leading architectonic authorities of the day, Bonin and Bailey, while not adopting quite as extreme a position as Lashley, nevertheless recognized many fewer areas than Brodmann (e.g., Bonin, 1948; Bailey and Bonin, 1951). Moreover, Lashley's neurobiology was very much in step with behaviorism, the prevailing psychological system of the day. Behaviorism is a reductionist doctrine, which seeks to account for intelligent behavior with as few processes as possible (Boakes, 1984). Moreover, the mechanisms acceptable to behaviorists (like the empiricists of Darwin's era) were in general limited to associative processes forming links between stimuli or links between stimuli and responses. "Mentalistic" concepts like thinking and planning were regarded with suspicion as possibly unscientific; the most rigid behaviorists dismissed them entirely. Lashley himself was not a behaviorist of the same doctrinaire cast as, for example, Hull or Skinner (Orbach, 1982a,b; Boakes, 1984, pp. 234–235). Nevertheless, it should be apparent that Lashley's model of the cortex as undifferentiated and equipotential was more congenial to the behaviorist's ideal of explaining intellectual activity with a few simple processes, than was the localizationist model of the cortex as a mosaic of discrete and functionally specialized areas.*

Lashley had a strong influence on the study of brain evolution, an issue that he addressed directly (Lashley, 1949). Consistent with his premise that much of the cortex is structurally and functionally unspecialized, he argued there are no essential differences in the areal or connectional organization of the cortex across mammals; rather, species vary only in the total amount of cortex they possess. He wrote:

*It is interesting that in a brief comment in the *Descent of Man,* Darwin expressed a view of brain organization similar to Lashley's: "Little is known about the functions of the brain, but we can perceive that as the intellectual powers become highly developed, the various parts of the brain must be connected by the most intricate channels of intercommunication; and as a consequence each separate part would perhaps tend to become less well fitted to answer in a definite and uniform . . . manner to particular sensations or associations" (Darwin, 1871, p. 38).

The only neurological character for which a correlation with behavioral capacity in different animals is supported by significant evidence is the total mass of tissue, or rather, the index of cephalization, measured by the ratio of brain to body weight, which seems to represent the amount of brain tissue in excess of that required for transmitting impulses to and from the integrative centers. . . .

With this, Lashley paved the way for modern studies of brain–body allometry. Jerison (1961; 1973, p. 3) cited the foregoing passage as the starting point for his efforts to develop a quantitative theory of brain size, in which the allometrically adjusted encephalization quotient (EQ) takes the place of the index of cephalization. Many allometric studies have followed (major contributions and reviews include: Armstrong, 1983, 1985; Clutton-Brock and Harvey, 1980; Gould, 1975; Hofman, 1982; Martin, 1981; Martin and Harvey, 1985; Passingham, 1982; Stephan, 1972; Stephan and Andy, 1969; Stephan *et al.*, 1988). These studies offer various interpretations of the quantitative relationship between brain size and body size, and its biological explanation. However, all are founded on the same premise, derived from Lashley, that the organization of the brain (or some large part of the brain, such as the neocortex) does not vary among the taxa included in the regression, so brain and body size covary in the same fashion across taxa. Only if these conditions hold can empirical regression lines be interpreted in terms of biological scaling, and relative brain size be considered (possibly) to have an orderly, interpretable relationship to behavioral or cognitive capacity.

[This is not intended as a blanket condemnation of allometry. I regard as problematic specifically those studies that consider very distantly related taxa (i.e., "mouse-to-elephant" studies) and a heterogeneous collection of neural structures (commonly the entire brain or neocortex). Investigations restricted to circumscribed neural structures and closely related taxa (e.g., Armstrong, 1982; Frahm *et al.*, 1984) are less vulnerable to the criticisms offered here.]

The view that all mammalian brains or all primate brains are larger or smaller versions of the "same" brain—which may be termed the *invariance postulate* of allometry—has not gone unchallenged. Within anthropology, Holloway has been a consistent critic, arguing that animals with the same brain size may differ in neural organization, and therefore the comparison of cranial capacity is a comparison of unequal units (Holloway, 1966a,b; Holloway and Post, 1982). Within neuroanatomy, workers such as I. T. Diamond and his colleagues (e.g., Diamond and Hall, 1969; Harting *et al.*, 1972) continued to maintain that brain evolution was marked by the addition of areas and connectional changes. Nevertheless, as indicated by Falk's statement quoted at the beginning of this chapter, neuroanatomists were unable to cite cases of phyletic differences in mammalian cortical organization that were widely acknowledged as such by the neuroscientific or anthropological communities. As a result, Holloway's criticism of allometry had little force. In the neurosciences generally, Lashley's views became firmly entrenched. For example, strong challenges were mounted against Brodmann's claim for the uniqueness of primate prefrontal cortex (Rose and Woolsey, 1948; Akert, 1964), and it has become virtually axiomatic, particularly among researchers who work with rodents or carnivores, that homologues of the primate granular prefrontal areas are present in nonprimates (for reviews, see Markowitsch and Pritzel, 1979; Fuster, 1989; Preuss and Goldman-Rakic, 1991a,c). In this vein, Kolb (1984) writes:

> I conclude that in spite of the tremendous difference in the relative volume of frontal neocortex in humans versus the other species, and particularly rodents, there appears to be a remarkable unity in frontal cortex function across mammals. . . . The overall conclusion is that the frontal cortex of the rat may provide a good model to study frontal cortical control of behavioral processes in mammals, including humans.

The more general view, that there are no qualitative differences in neocortical organization between primates and nonprimates, or even among primates, has also been explicitly restated in recent years (e.g., Radinsky, 1975; Macphail, 1982, 1987; Kolb and Whishaw, 1985).

Foundations of a New Evolutionary Neurobiology

The Rebirth of Localizationism and the Documentation of Diversity

The past 20 years have witnessed a technical revolution in the neurosciences that has greatly altered our understanding of brain organization, particularly cortical organization. These technical and conceptual developments have profound implications for our understanding of brain evolution.

Perhaps the most important technical achievement in recent years has been the development of reliable and sensitive techniques for tracing the connections of neurons. These methods, which are based on the axonal transport of tracer substances such as tritiated amino acids and horseradish peroxidase (Cowan *et al.*, 1972; LaVail and LaVail, 1974), have revealed a wealth of cortical connections not detectable with older lesion-degeneration techniques. Another important development has been the use of microelectrodes to record the activity of a single neuron or small group of neurons, in preference to use of large electrodes, which record neural activity over a relatively large territory. The exploration of cortical regions beyond the classical primary sensory areas with microelectrodes has demonstrated a large number of previously unrecognized areal subdivisions (Fig. 2), each having a unique set of physiological characteristics (Allman and Kaas, 1971; Kaas, 1977, 1982, 1986, 1987a, 1989; Van Essen, 1985). Moreover, correlative anatomical studies have demonstrated that each area has a unique array of cortical and subcortical connections, and often distinctive cyto- and myeloarchitectonic features as well (Van Essen, 1985; Kaas, 1986, 1987b; Maunsell and Newsome, 1987; Livingstone and Hubel, 1988). Many new neurotransmitters and neuro-

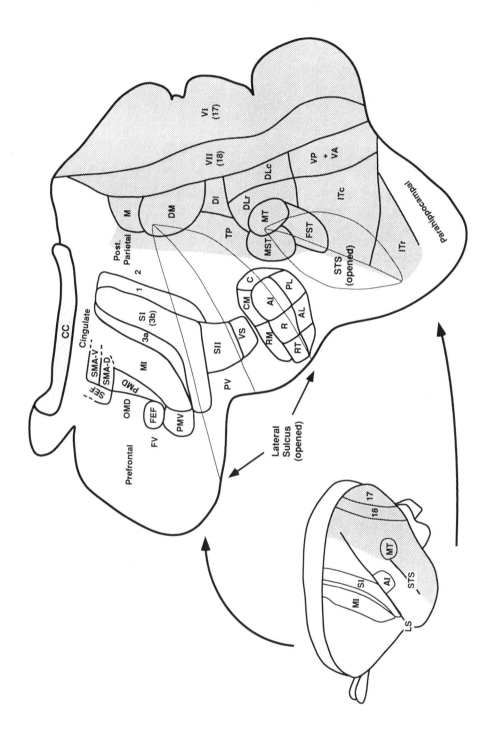

modulators have been discovered, and techniques devised to localize them; typically, these substances are found to be distributed selectively among nuclei and areas (McGeer *et al.*, 1987). Finally, methods have been developed to examine the metabolic activity of the brain during cognitive and behavioral activity (e.g., Roland, 1984). All of these methods confirm the high degree of structural subdivision and functional specialization within the cerebral cortex. It is probably not coincidental that at the same time that neuroscientists were returning to the classical, localizationist view of the brain as a mosaic of specialized structures, psychologists were coming to view the cognitive system as a collection of specialized information-processing devices, rather than as a general-purpose computer (Fodor, 1983; Gardner, 1983).

The advent of new methods of investigation, and the rebirth of localization, has led to new approaches to the study of brain evolution. Lashley's thesis that all mammals are fundamentally or qualitatively similar in organization is no longer tenable: important phyletic differences in the organization of neural systems have been documented. One example, now well known, is the difference between primates and most or all nonprimate mammals in the retinotopic organization of the superior colliculus (Kaas *et al.*, 1973; Lane *et al.*, 1973; see also Allman, 1977; Kaas and Preuss, 1993). In nonprimates, each of the bilaterally paired colliculi contains a complete representation of the contralateral visual hemifield, as well as a substantial portion of the ipsilateral hemifield. In primates, the colliculus represents only the contralateral visual hemifield. Among nonprimates, only megachiropterans (fruit bats) and dermopterans (colugos or flying lemurs) have been reported to share the primate condition (Pettigrew, 1986; Pettigrew and Cooper, 1986; Pettigrew *et al.*, 1989). This conclusion should be regarded with some caution, however, as the supporting data have yet to be published in detail. Furthermore, a recent report by Thiele *et al.* (1991) provides convincing evidence that at least one megachiropteran, *Rousettus aegyptiacus,* possesses the typical nonprimate condition.

In addition, it has become evident that the great variation in brain size that characterizes mammals is correlated with variation in the number of

---

Fig. 2. A modern view of cortical areal organization in owl monkeys (*Aotus* spp.), based on correlative electrophysiological, connectional, and architectonic studies. The figure at lower left shows the brain in dorsolateral perspective; the upper right figure represents the cortical mantle unfolded and flattened. Note the large region of cortex devoted to vision, shown here in gray; this is typical of primates. Currently, about 40 areas have been delineated in owl monkeys. This is certainly an underestimate of the total number of areas, as substantial parts of the cortical mantle [particularly the prefrontal, posterior parietal, superior temporal sulcal (STS), and cingulate regions] have not yet been systematically studied in owl monkeys. In this figure, *visual* areas include VI, VII, M, DM, DI, DLc, DLr, VP, VA, MT, MST, FST, ITc, and ITr. *Auditory* areas include AI, R, RT, PL, AL, C, CM, and Rm. *Somatosensory* areas include SI, 3a, 1, 2, SII, VS, and PV. *Motor* areas include MI, PMD, PMV, SMA-D, SMA-V, SEF, OMD, FEF, and FV. This figure is adapted from Kaas (1989), with additional auditory areas from Morel and Kaas (1992) and motor areas from Stepniewska *et al.* (1993).

cortical areas. Monotremes, marsupials, and small-brained eutherians possess only a very few (perhaps 10–20) neocortical areas (Kaas, 1987a, 1989). These areas include homologues of the primary and secondary sensory and motor areas found in larger-brained eutherians, such as carnivores and primates. In the latter groups, however, additional areas are present; anthropoid primates appear to possess on the order of 50–100 areas (Kaas, 1989; Preuss and Goldman-Rakic, 1991a,b). It is likely that many groups of mammals have evolved their own distinctive complements of cortical areas.

Recently it has been discovered that there are even variations in the basic cellular composition of mammalian cortex. Following Brodmann (1909), it is conventional to divide mammalian neocortex into six layers: the outermost stratum, layer I, consists predominantly of fibers and has few cells; layer II contains small cells; layer III contains larger, pyramid-shaped cells; layer IV forms a band of small, so-called "granule" cells in the middle of the cortex; and the deepest layers, V and VI, contain pyramidal cells and other large neurons. This pattern of cellular organization is present in many mammalian orders. However, the neocortex of at least one group—cetaceans—is markedly different, with large, pyramidlike cells in layer II and a paucity of small, granule cells in layer IV (Morgane *et al.*, 1985). (It should be noted that these findings were made not with a novel method, but with the venerable Golgi technique.) Morgane *et al.* interpret the distinctive features of cetaceans as retentions of a primitive mammalian organization. This seems unlikely, because the cortex of monotremes and marsupials resembles that of most eutherians, as does the cortex of ungulates, the closest relatives of cetaceans (see the reviews of Johnson, 1990, and Rowe, 1990). It is more probable that the unusual features of cellular organization in cetaceans are autapomorphies. The unusual cellular organization of cetaceran cortex suggests that its connectional organization may be unusual as well.

It should now be evident that mammalian brains are too diverse in organization to be ranked along a single scale of encephalization, if the goal is to make sensible inferences about the behavioral capacities of different taxa. The invariance postulate of allometry fails. However, with the technical means now available, it is possible for neuroscientists to document the diversity of neural systems in detail, identifying specific patterns of neural organization that are characteristic of particular phyletic groups. That is, we can now identify neural *characters,* assigning different, discrete *character states* to different taxa, and with this information construct cladograms and branching diagrams, just as one might with comparative information about the amino acid sequence of hemoglobin molecules or the course of the carotid artery through the middle ear. Thus, it is now possible to interpret brain evolution in terms of a branching tree.

While encouraged by the prospect of neurobiology moving into the mainstream of modern evolutionary biology, I must mention the practical difficulties involved in realizing this goal (some of which have been discussed previously; Kaas and Preuss, 1993). The main obstacle is the lack of comprehensive comparative studies; the only mammals that neuroscientists have studied in significant detail are primates (mainly macaque monkeys), domestic cats, and inbred rats. It is very important to recognize that the relatively broad comparative approach that has been adopted in the study of the retinocollicular system is the exception, not the rule. Moreover, neuroscientists as a group are still not alive to the possibility of significant differences across taxa. It is the rare investigator of rats or cats who does not conclude that the system under study is basically similar to that of primates, if the possibility of variation is considered at all. Often, evidence of differences presented in a Results section elicits little or no comment in the Discussion. When acknowledged, differences may be deemed "unexpected" and interpreted within the context of a phylogenetic scale (e.g., Berger *et al.,* 1991).

Some Candidate Primate Neural Specializations

Under these circumstances, one is usually unable to determine the character states of a given neural system in a large and diverse set of taxa, as is necessary for detailed phylogenetic analysis. Nevertheless, it is possible to identify a number of systems for which primates (or some subset of primates, such as anthropoids or catarrhines) have character states different from other mammals that have been investigated. The emphasis here will be on cortical organization, as the cortex has been the subject of particularly intensive investigation in recent years. Further, because the visual system has been considered in detail elsewhere (Kaas and Preus, 1993), I will focus primarily on other systems. This is not intended to be a comprehensive account of primate specializations; examples have been chosen primarily to illustrate the variety of ways in which cortical organization may be modified in evolution.

As discussed above, the addition of areas appears to be a frequent occurrence in cortical evolution. These added areas are most commonly modality-specific sensory areas (Kaas, 1987a,b). Primates are notable among mammals for the large number of visual areas they possess; recent estimates suggest that anthropoid primates, for example, possess on the order of 15–30 visual areas (Kaas, 1989). It is likely that many of these areas are unique to primates. The peculiar, bulbous appearance of the temporal lobe in primates, which reflects the large territory occupied by higher-order visual areas, can be seen in early euprimate endocasts from the Eocene (Radinsky, 1970; Allman, 1977).

The addition of areas has not been limited to sensory domains, however. Recall that Brodmann (1909) maintained that new areas appeared in the prefrontal region during primate evolution. Recently, Goldman-Rakic and I compared the frontal lobe organization of *Galago* and *Macaca,* using connectional and architectonic techniques (Preuss and Goldman-Rakic, 1991a,c). We were able to identify a number of areas shared by both taxa, specifically the frontal eyefield and surrounding areas, as well as additional areas present only in *Macaca* (Fig. 3). The latter areas are located anteroir to the frontal eyefield, along the deep principal sulcus of macaques. These areas may be specializations of anthropoid primates, or some group of anthropoids, such as catarrhines or cercopithecoids.

Brodmann's claim that new areas evolved in primate granular frontal cortex is not as controversial as his claim that the granular frontal cortex, in general, is unique to primates. As noted above, many contemporary workers believe that nonprimates possess homologues of granular frontal cortex, based on connectional similarities, although nonprimate mammals lack the well-developed granular layer present in primates. We have reconsidered this issue, concluding that Brodmann's claim has merit (Preuss and Goldman-Rakic, 1991a,c). Our review of recent research indicates that the connections of primate granular frontal cortex differ from its putative homologues in carnivores and rodents in several respects. In particular, primate prefrontal cortex has strong connections with higher-order parietal and temporal association cortex, whereas the predominant frontal connections in nonprimates are with limbic (specifically, cingulate and insular) areas. So, primates are different from nonprimates in connectional as well as cytoarchitectonic characteristics. These data do not definitively resolve the question of areal or regional homologies in favor of Brodmann, but they do highlight the anatomical distinctiveness of primate prefrontal cortex, which in turn suggests the possibility of unusual functional capacities.

There is reason to believe that the modifications that mark prefrontal cortex may be part of a larger connectional reorganization of primate higher-order association cortex. In primates, the prefrontal, parietal, and temporal association zones are all strongly interconnected, and each receives projections from a common thalamic nucleus, the medial pulvinar; this network of connections has no known counterpart in rodents or carnivores (Preuss, 1990; Preuss and Goldman-Rakic, 1991b,c). In fact, it is not clear that nonprimates possess a homologue of the medial pulvinar.

There are other examples of significant evolutionary rearrangements of connectivity. The somatosensory cortex provides a particularly well-studied example (Fig. 4). Mammals possess a pair of fields, the primary and secondary somatosensory areas (SI and SII), in which neurons respond strongly to light stimulation of the skin. These cutaneous inputs are relayed to the somatosensory cortex through the ventral posterior (VP) nucleus of the thalamus. SI and SII are also strongly linked by a system of corticocortical connections.

In most mammals that have been examined, including marsupials, carnivores, rodents, tree shrews, and strepsirhine primates (*Galago*), the VP nucleus projects to both SI and SII (see the review of Garraghty *et al.*, 1991). In anthropoid primates, by contrast, VP projects only to SI, although neurons in SII still respond to cutaneous stimulation, presumably by means of strong inputs from SI. This interpretation is supported by the comparative lesion

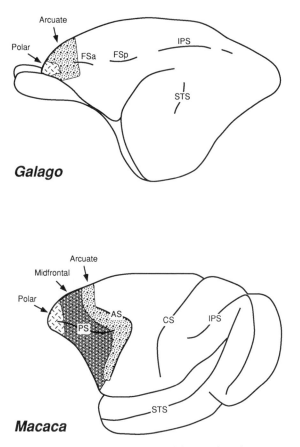

Fig. 3.   The granular prefrontal cortex in *Galago* and *Macaca,* based on recent connectional and architectonic studies (Preuss and Goldman-Rakic, 1991a,c). The granular frontal cortex is divided into several subregions (arcuate, midfrontal, polar), each of which consists of multiple areas. Galagos possess cortex homologous to the arcuate zone of macaques (the region that includes the frontal eye-field) and probably cortex corresponding to the polar zone. However, galagos apparently lack homologues of the midfrontal areas. This interpretation of galago–macaque differences is generally consistent with Brodmann's ideas and calls into question the widely held view that the organization of frontal cortex varies little across mammals. The figures are not to scale. AS, arcuate sulcus; CS, central sulcus; IPS, intraparietal sulcus; LS, lateral sulcus; FSa, FSp, anterior and posterior limbs of the frontal sulcus; PS, principal sulcus; STS, superior temporal sulcus.

studies reviewed by Garraghty *et al.* (1991). When SI is lesioned in anthropoid primates (macaques, marmosets), SII neurons no longer respond to cutaneous stimulation. Lesions of SI in nonanthropoid mammals (galagos, tree shrews, cats, rabbits), however, do not abolish cutaneous responsiveness in SII. On these grounds, Garraghty *et al.* suggest that anthropoid primates have a derived pattern of somatosensory information processing, supplementing or replacing a system of parallel processing with a serial system. The functional

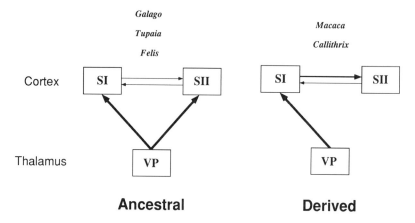

Fig. 4.   Evolutionary rewiring of the eutherian somatosensory system. In the presumed ancestral mammalian condition, the main somatosensory relay nucleus of the thalamus, the ventral posterior (VP) nucleus projects directly to both the primary (SI) and secondary (SII) somatosensory areas. Because SII receives a major projection from VP, it continues to respond to cutaneous stimulation after lesions of SI in cats, tree shrews, and bushbabies. In the derived anthropoid condition, shown on the right, SII receives its cutaneous inputs indirectly, through SI, rather than directly from VP. As a result, lesions of SI abolish the responsiveness of SII neurons to cutaneous stimulation in macaques and marmosets. Adapted from Garraghty *et al.* (1991).

significance of such a transformation is unclear, but it may be relevant to the use of the tactile sense in the active exploration of the environment (the so-called "haptic" sense) by anthropoid primates (Bishop, 1964).

Another example of connectional rearrangement, in this case involving the motor system, comes from the work of Berger and her colleagues (Berger *et al.*, 1991). They note striking differences between catarrhine primates (macaques and humans) and rodents in the innervation of the frontal lobe by dopamine-containing neurons of the midbrain. For example, the innervation of the primary motor cortex (area MI) is much denser in catarrhines than in rodents, whereas density differences in other regions such as the medial premotor and cingulate cortex are not so dramatic. Also, the laminar distribution of dopamine fibers in primary motor cortex varies: in rodents, fibers are restricted to the deepest cortical layer (VI), whereas in catarrhines they are distributed over all six cortical layers. Unfortunately, it is not possible at present to draw conclusions about the polarity of the rodent and catarrhine character states, because other taxa have not been studied in detail. However, these results suggest that mammals vary in the cortical organization of movement, a function in which dopamine projections may play an important role (Gaspar *et al.*, 1991).

Differences are also apparent in the biochemical properties of neurons located within motor cortex. In humans and Old World monkeys, the large layer V pyramidal cells of MI (Betz cells) stain strongly for the enzyme acetylcholine esterase (Okinaki *et al.*, 1961; Mrzljak and Goldman-Rakic,

1992). We have observed similar staining in platyrrhine monkeys (Stepniewska *et al.*, 1993). However, the Betz cells of strepsirhines (*Galago, Nycticebus*) stain little, if at all (Preuss, *et al.*, 1993). Again, the polarity of these character states is unclear, as is their functional significance.

On the Use of Neuroanatomical Information in Phylogenetic Analysis

The examples presented above demonstrate the existence of variation in many aspects of cortical organization. These examples probably represent just a small fraction of the variability that exists in the organization of mammalian brains: it is important to keep in mind just how few aspects of neural organization have been subjected to dedicated comparative analyses and just how few taxa have been studied in detail. It is likely that the mammalian nervous system holds a wealth of information, largely untapped at present, useful for phylogenetic reconstruction.

How are we to evaluate neural characters in relation to other kinds of information about evolutionary relationships and transformations? Pettigrew (1991) has argued rather strongly that neural systems are less subject to convergent evolution than features of musculoskeletal anatomy. Specifically, he suggests that neural systems are strongly canalized in development, enmeshed as it were in a dense "neural embroidery," which renders them resistant to change. When change does occur, it is usually not closely related to function. Instances of functional convergence in the nervous system do occur, although they are "usually readily recognizable" and can be distinguished from homologous similarities. Thus, according to Pettigrew, analyses of neural characters are less likely to be bedeviled by adaptive convergence than analyses of somatic characters, and therefore provide more reliable information about phyletic relationships.

I find these arguments unpersuasive. Indeed, the extent of neural diversity documented above belies the claim that the neural systems are particularly resistant to evolutionary change. Certainly, the cerebral cortex, which of all structures is unmatched for the complexity and density of its "neural embroidery," is not notably conservative in evolution.* Moreover, I do not share the view that the task of distinguishing homoplasy from homology in the nervous system is a simple or straightforward matter. For example, the

---

*Nor is the cortex as plastic in evolution as might be suggested by recent studies that demonstrate that cortex experimentally deprived of its normal input may come to respond to other, normally impotent stimuli (see Kaas, 1991, for a review of this field). These modifications typically consist of shifts or expansions of receptive field properties over a short distance within a single functional area (although more extensive changes may be induced by experimental manipulations early in development). Moreover, these studies do not demonstrate that the cortex is necessarily more plastic than subcortical structures; indeed, the changes observed in the cortex probably reflect in part changes occurring in subcortical structures (Garraghty and Kaas, 1991).

primary visual cortex of cats closely resembles that of (some) primates, in that inputs from each eye are strongly segregated into alternating bands known as ocular dominance columns. One cannot determine whether the ocular dominance columns of cats and primates are homoplastic or homologous merely by inspection; one must consider the phyletic distribution of ocular dominance columns. In fact, ocular dominance columns are absent in taxa thought to be closely related to primates, and therefore their presence in cats and primates suggests convergence (Kaas and Preuss, 1993). I have no doubt, however, that if carnivores and primates were thought to be sister groups, the presence of ocular dominance columns in each would be regarded as a homologous similarity and cited as additional, strong evidence of close relationship.

In my view, therefore, there is simply no compelling theoretical or empirical basis for concluding that neural characters as a class are more resistant to homoplasy than other types of characters. But the aspect of Pettigrew's argument that I find most problematic is the suggestion that many evolutionary changes in the nervous system are *functionless*. At present, very little is known about the relationship between structure and function in many parts of the nervous system. It seems unwise to conclude that a given evolutionary change is without functional significance merely because none is readily apparent. [These and related points are considered in a detailed response to Pettigrew by Baker *et al.* (1991).] This is not to say that *all* evolutionary changes must have functional and adaptive significance. Yet, to give short shrift to the possible functional significance of neural changes is to overlook what may be the greatest contribution that neuroscience has to offer evolutionary biology, namely, a unique perspective on the evolution of behavior.

A Paradigmatic Case: The Ventral Premotor Area

In order to illustrate the insights into behavioral function that may be gained from comparative neurobiology, I consider the specific case of the ventral premotor area (PMV). This area, which has also been called the arcuate premotor area (Dum and Strick, 1990) and inferior (or ventral) area 6 (e.g., Rizzolatti and Gentilucci, 1988; Preuss and Goldman-Rakic, 1991a), is located immediately anterior and ventrolateral to the primary motor cortex (MI), as illustrated in Fig. 5.

PMV has been most extensively studied in Old World monkeys (Rizzolatti and Gentilucci, 1988; Dum and Strick, 1990; Gentilucci and Rizzolatti, 1990), but there is evidence from connectional, physiological, and architectonic studies that a homologous area exists in New World monkeys and in strepsirhine primates (Nudo and Masterton, 1990; Preuss and Goldman-Rakic, 1991a,c; Preuss *et al.*, 1992; Stepniewska *et al.*, 1993). The particular suite of characteristics that distinguishes PMV from other motor areas has not been observed in nonprimate mammals and it is likely that nonprimates lack a homologous area. This point has been emphasized in a recent comparative study by

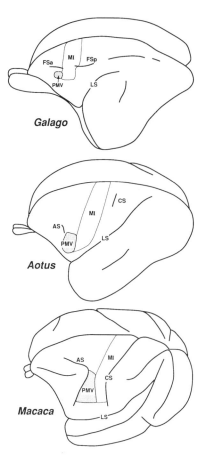

Fig. 5.   The location of the ventral premotor area (PMV) in strepsirhine and haplorhine primates. Area PMV is situated immediately anterior to MI, at about the level of the representation of face movements in MI. The figures are not to scale. Abbreviations as in Fig. 3.

Nudo and Masterton (1990). These workers mapped the cells of origin of the corticospinal tract in 22 mammalian species, representing seven orders. In all mammals, Nudo and Masterton found a dense concentration of corticospinal cells corresponding to the primary motor and somatosensory areas. However, in the strepsirhine and haplorhine primates they examined—and only in primates—they identified an additional patch of corticospinal neurons immediately anterior and ventrolateral to MI, corresponding to PMV. [Pettigrew *et al.* (1989) indicate that this anterolateral patch exists in megabats also, although the relevant studies have not yet been published.]

   While the comparative anatomical information alone would suffice to make PMV of interest to primatologists, its functional characteristics are especially intriguing. The behavioral role of PMV has been studied in detail in macaques by Rizzolatti and his colleagues, who consider it to play a role in visually guided reaching and prehension. [For recent reviews of this work, see Rizzolatti and Gentilucci (1988) and Gentilucci and Rizzolatti (1990); I present a simplified account here.] Neurons in PMV respond to visual stimuli, but more importantly, some neurons respond strongly only to stimuli that are located within reach. Many PMV cells also respond to tactile stimuli; receptive fields are located on the forelimbs and face. Visual and tactile receptive fields are often congruent. For example, a neuron may respond selectively to cutaneous stimulation of the hand (and no other part of the body) and also respond to the visual presentation of an object near the hand (and nowhere else).

PMV neurons are also active before and during movements, specifically, during various types of purposive, prehensive movements of the face and forelimbs. (Movements of other parts of the body are not represented in this area.) Neurons have been identified that fire selectively in association with the following classes of actions: reaching, bringing the hand to the mouth, holding, tearing, grasping with the hand, and grasping with the hand or mouth. The class of grasping neurons consists of three subtypes: neurons active in opposition of the thumb and index finger, those active during opposition of the thumb and other digits, and neurons active during whole-hand grasping.

It is important to point out that the activity of PMV neurons is related not simply to movement, but rather to *purposive acts*. This point is illustrated by the "grasping-with-the-hand-or-mouth" class of cells, which respond when an object is grasped, without regard to the body part used. Similarly, "reaching" neurons respond when the animal reaches toward a desired object with its forelimb but not when the animal makes a similar movement to push away an unwanted object.

Based on observations such as these, neurons in PMV have been said to represent a "vocabulary of motor acts" related to visually guided prehension (Rizzolatti and Gentilucci, 1988). Furthermore, PMV is but one component of an extensive, distributed network of interconnected cortical areas, some of which are known to have similar functional properties (Preuss and Goldman-Rakic, 1989). A substantial fraction of frontal and parietal cortex in macaques may be involved in the organization of visually guided reaching and grasping. Although behavioral studies of PMV and related areas have been carried out only in macaques, the anatomical similarity of macaques to New World monkeys and strepsirhines suggests that these areas perform similar functions in other primates. There may of course be variations in the representation of motor acts in PMV reflecting variations in motor repertoire. For example, whereas macaques commonly grasp objects by opposing the digits, strepsirhines and many New World monkeys typically use a whole-hand grasp (Bishop, 1964); one would expect to find the whole-hand grasp well represented within the PMV of these taxa.

Because reaching and grasping have long been deemed important components of primate behavior, PMV is precisely the sort of area we should expect to find in primates, if we suppose that the brain is a collection of specialized, "purpose-built" processors rather than a general-purpose computer. But PMV has additional significance, for its properties bear on the question of primate origins.

Modern interpretations of primate origins, based primarily on studies of skeletal anatomy and of the behavior of living taxa, focus on modifications of the limbs related to grasping with the hands and feet, particularly the replacement of claws with nails and the evolution of a widely divergent, powerfully grasping hallux (e.g., Cartmill, 1972, 1974; Covert, 1986; Dagosto, 1988; Szalay and Dagosto, 1988; Martin, 1990). These changes have been commonly regarded as adaptations for locomotion in the fine terminal branches of trees, the "fine-branch niche," as it has been called (Martin, 1990; Rasmussen, 1990; Sussman, 1991). However, Cartmill (1972, 1974) has added another element to this interpretation, proposing that ancestral euprimates were adapted for a particular style of predation in the fine terminal branches, using a visually guided hand strike to seize small prey. The neurobiological evidence regarding PMV—specifically its status as a primate specialization and its role in visually guided reaching and grasping—is consistent with Cartmill's visual predation hypothesis. Moreover, the neural evidence suggests an elaboration of the theory. PMV neurons respond selectively to grasping with the hand, but also bringing the hand to the mouth, and grasping with the mouth. This suggests that grasping with the mouth, by itself, and combined actions of the hand and mouth, were important components of early primate behavior. For instance, it may be the case that early primates were specialized for seizing prey with the hand and bringing it rapidly to the mouth for a lethal bite. Such an action sequence has been observed in extant strepsirhines by Charles-Dominique (1977).

I emphasize that although the evidence concerning PMV is consistent with the visual predation hypothesis, it does not exclude the possibility that early primates may be specialized for other types of visually guided prehension, such as might be involved in grasping and manipulating fruits and flowers, which are important elements in the diet of small strepsirhines (Rasmussen, 1990; Sussman, 1991). Nor does it constitute evidence against interpretations of primate origins that emphasize other aspects of anatomy and behavior, such as changes in the skeleton of the hindlimb related to grasping. My point is simply that neurobiology may provide insights into behavioral evolution in addition to those that have been gleaned from the study of somatic anatomy or naturalistic studies of behavior.

I chose the example of PMV because its status as a primate specialization is reasonably well established and because its role in prehension is relevant to a traditional field of primatological inquiry, namely, primate origins. The neuroscience literature contains hints of many additional primate neural specializations that bear on other aspects of behavior. For example, neurons in the temporal cortex of macaques respond selectively to the sight of particular body parts—hands or faces, usually (Rolls, 1984; Perrett *et al.*, 1987). In some instances, temporal neurons evidently respond preferentially to the sight of particular individuals and to particular social displays (Perrett *et al.*, 1987; Brothers *et al.*, 1990; Brothers and Ring, 1992). So, just as PMV is said to possess a vocabulary of motor acts, temporal cortex has been said to contain an "alphabet" of social signals (Brothers and Ring, 1992). Clearly, neuroscientific investigations may help us to understand how primates classify socially significant individuals and events, an issue that has recently attracted considerable attention among primate behaviorists (e.g., Byrne and Whitten, 1988; Cheney and Seyfarth, 1990). Finally, the areas that have been identified as likely specializations of the anthropoid frontal lobe (Preuss and Goldman-Rakic, 1989, 1991a,c) are believed to be involved in short-term or working memory, with particular subdivisions dedicated to different functional modalities (i.e., spatial memory, visual object memory, kinesthetic memory). These areas may therefore provide clues to the evolution of higher-order cognitive mechanisms, an aspect of primate biology that has received very little attention by modern anthropologists and psychologists. Moreover, if we allow that the cognitive systems of different taxa vary along multiple functional dimensions, rather than along a single phylogenetic scale, it becomes possible to go beyond the view of animal cognition as an inferior brand of human cognition, and conceive an evolutionary radiation of animal minds.

Summary and Conclusions

The study of brain evolution has for some time been divorced from the study of primate origins and relationships. Although Darwin, and later Elliot Smith and Le Gros Clark, stressed the importance of the brain in primate evolution, they emphasized quantitative changes—enlargement, improvement, and refinement—while denying discrete or qualitative changes. Because of this, and because of the reaction to cerebral localization led by Lashley, the field of brain evolution has for many years been dominated by allometric studies, in which taxa are ranked according to the relative size of the brain, or of some large component of the brain, such as the neocortex. The commitment to a view of the brain that emphasizes qualitative similarity across taxa, and a view of evolution that stresses advancement along a scale of progress, has created a conceptual and methodological gap between the field of brain–body allometry and other branches of primate and mammalian evolutionary biology. Workers in the latter disciplines have come increasingly to regard evolution as a process of diversification and adaptive specialization, and have adopted analytical methods appropriate to this view. These methods are used to identify monophyletic groups and to infer the ancestral (morphotypic) organization of those groups, based on discrete variations in somatic anatomy and molecular structure. However, brains are not amenable to this type of analysis if they vary only in size.

With the advent of improved techniques for investigating neural structure, it has become apparent that mammalian brains display a diversity of discrete character states, encompassing many facets of areal, connectional, and biochemical organization. This means that brain evolution can be approached with the same analytical techniques that have been applied to molecules, skeletal material, and soft tissue. In principle, one should be able to identify the derived characteristics of the brain that unite primates, or archontans, or any other group of mammals.

There is no reason to suppose, as has been claimed, that neural characters are unusually resistant to convergent evolution, and thus more useful for determining phylogenetic relationships than somatic or molecular characters. However, the comparative study of neural systems may provide uniquely valuable information about the evolution of behavior, including insights into the adaptive specializations of thought and action that distinguish various taxa. It should therefore be possible to construct branching trees of neural and cognitive systems.

ACKNOWLEDGMENTS

The author's work has been supported by the National Science Foundation, National Institutes of Health, and the James S. McDonnell Foundation. Some of the material in this chapter is derived from the author's dissertation (Preuss, 1990), which benefited from the critical evaluation of Alison Richard, Patricia Goldman-Rakic, Glenn Conroy, and Ford Ebner. The author also thanks Jon Kaas for his encouragement and comments on the present chapter, as well as Preston Garraghty, Nancy Handler, Pierre Lemelin, and Jon Marks.

References

Akert, K. 1964. Comparative anatomy of frontal cortex and thalamofrontal connections, in: J. M. Warren and K. Akert (eds.), *The Frontal Granular Cortex and Behavior*, pp. 372–396. McGraw–Hill, New York.

Allman, J. M. 1977. Evolution of the visual system in the early primates, in: J. M. Sprague and A. N. Epstein (eds.), *Progress in Psychology and Physiological Psychology*, pp. 1–53. Academic Press, New York.

Allman, J. M., and Kaas, J. H. 1971. A representation of the visual field in the caudal third of the middle temporal gyrus of the owl monkey (*Aotus trivirgatus*). *Brain Res.* **31**:85–105.

Armstrong, E. 1982. Mosaic evolution in the primate brain: Differences and similarities in the hominoid thalamus, in: E. Armstrong and D. Falk (eds.), *Primate Brain Evolution*, pp. 131–161. Plenum Press, New York.

Armstrong, E. 1983. Relative brain size and metabolism in mammals. *Science* **220**:1302–1304.

Armstrong, E. 1985. Allometric considerations of the adult mammalian brain, with special emphasis on primates, in: W. L. Jungers (ed.), *Size and Scaling in Primate Biology*, pp. 115–146. Plenum Press, New York.

Bailey, P., and Bonin, G. 1951. *The Isocortex of Man*. University of Illinois Press, Urbana.

Baker, R. J., Novacek, M. J., and Simmons, N. B. 1991. On the monophyly of bats. *Syst. Zool.* **40:**216–231.

Berger, B., Gaspar, P., and Verney, C. 1991. Dopaminergic innervation of the cerebral cortex: Unexpected differences between rodent and primate. *Trends Neurosci.* **14:**21–27.

Bishop, A. 1964. Use of the hand in lower primates, in: J. Buettner-Janusch (eds.), *Evolutionary and Genetic Biology of Primates*, pp. 133–225. Academic Press, New York.

Boakes, R. 1984. *From Darwin to Behaviourism*. Cambridge University Press, London.

Bonin, G. von. 1948. The frontal lobe of primates: Cytoarchitectural studies. *Res. Publ. Nerv. Ment. Dis.* **27:**67–83.

Brodmann, K. 1909. *Vergleichende Lokalisationslehre der Grosshirnrhinde*. Barth, Leipzig.

Brothers, L., and Ring, B. 1992. A neuroethological framework for the representation of minds. *J. Cogn. Neurosci.* **4:**107–118.

Brothers, L., Ring, B., and Kling, A. 1990. Response of neurons in the macaque amygdala to complex social stimuli. *Behav. Brain Res.* **41:**199–213.

Byrne, R. W., and Whitten, A. (eds.) 1988. *Machiavellian Intelligence: Social Expertise and the Evolution of Intellect in Monkeys, Apes, and Humans*. Oxford University Press (Clarendon), London.

Campbell, C. B. G. 1966. The relationships of the tree shrews: The evidence of the nervous system. *Evolution* **20:**276–281.

Campbell, C. B. G. 1980. The nervous system of the Tupaiidae: Its bearing on phyletic relationships, in: W. P. Luckett (ed.), *Comparative Biology and Evolutionary Relationships of Tree Shrews*, pp. 219–242. Plenum Press, New York.

Cartmill, M. 1972. Arboreal adaptations and the origin of the order Primates, in: R. Tuttle (ed.), *The Functional and Evolutionary Biology of Primates*, pp. 97–122. Aldine, Chicago.

Cartmill, M. 1974. Rethinking primate origins. *Science* **184:**436–443.

Cartmill, M. 1982. Basic primatology and prosimian evolution, in: F. Spencer (ed.), *A History of Physical Anthropology, 1930–1980*, pp. 147–186. Academic Press, New York.

Charles-Dominique, P. 1977. *Ecology and Behaviour of Nocturnal Primates*. Columbia University Press, New York.

Cheney, D. L., and Seyfarth, R. M. 1990. *How Monkeys See the World*. University of Chicago Press, Chicago.

Clutton-Brock, T. H., and Harvey, P. H. 1980. Primates, brains and ecology. *J. Zool.* **190:**309–323.

Covert, H. H. 1986. Biology of early Cenozoic primates, in: D. R. Swindler and J. Erwin (eds.), *Comparative Primate Biology*, Vol. 1, pp. 335–359. Liss, New York.

Cowan, W. M., Gottlieb, D. I., Hendrickson, A. E., Price, J. L., and Woolsey, T. A. 1972. The autoradiographic demonstration of axonal connections in the central nervous system. *Brain Res.* **37:**21–51.

Dagosto, M. 1988. Implications of postcranial evidence for the origin of euprimates. *J. Hum. Evol.* **17:**35–56.

Dart, R. A. 1934. The dual structure of the neopallium: Its history and significance. *J. Anat.* **69:**3–19.

Darwin, C. 1859. *On the Origin of Species*. John Murray, London. [Facsimile of first edition: Harvard University Press, Cambridge, Mass. 1984.]

Darwin, C. 1871. *The Descent of Man, and Selection in Relation to Sex*. John Murray, London. [Facsimile edition: Princeton University Press, Princeton, N.J., 1981.]

Diamond, I. T., and Hall, W. C. 1969. Evolution of neocortex. *Science* **164:**251–262.

Dum, R. P., and Strick, P. L. 1990. Premotor areas: Nodal points for parallel efferent systems involved in the central control of movement, in: D. R. Humphrey and H.-J. Freund (eds.), *Motor Control: Concepts and Issues*, pp. 383–397. Wiley, New York.

Elliot Smith, G. E. 1897. The origin of the corpus callosum; a comparative study of the hippocampal region of the cerebrum of Marsupalia and certain Chiroptera. *Trans. Linn. Soc. London 2nd Ser. Zool.* **7:**47–69.

Elliot Smith, G. 1910. Some problems relating to the evolution of the brain. *Lancet* **1910**-I:1–6; 147–153; 221–227.

Elliot Smith, G. E. 1919. A preliminary note on the morphology of the corpus striatum and the origin of the neopallium. *J. Anat.* **53:**271–291.

Elliot Smith, G. 1924. *The Evolution of Man: Essays.* Oxford University Press, London.

Falk, D. 1982. Primate neuroanatomy: An evolutionary perspective, in: F. Spencer (ed.), *A History of Physical Anthropology, 1930–1980,* pp. 75–103. Academic Press, New York.

Fleagle, J. G., and Jungers, W. L. 1982. Fifty years of higher primate phylogeny, in: F. Spencer (ed.), *A History of Physical Anthropology, 1930–1980,* pp. 187–230. Academic Press, New York.

Fodor, J. A. 1983. *The Modularity of Mind.* MIT/Bradford Books, Cambridge, Mass.

Frahm, H. D., Stephan, H., and Baron, G. 1984. Comparison of brain structure volumes in Insectivora and Primates. V. Area striata (AS). *J. Hirnforsch.* **25:**537–557.

Fuster, J. M. 1989. *The Prefrontal Cortex,* 2nd ed. Raven Press, New York.

Gardner, H. 1983. *Frames of Mind.* Basic Books, New York.

Garraghty, P. E., and Kaas, J. H. 1991. Functional reorganization in adult monkey thalamus after peripheral nerve injury. *NeuroReport* **2:**747–750.

Garraghty, P. E., Florence, S. L., Tenhula, W. N., and Kaas, J. H. 1991. Parallel thalamic activation of the first and second somatosensory areas in prosimian primates and tree shrews. *J. Comp. Neurol.* **311:**289–299.

Gaspar, P., Duyckaerts, C., Alvarez, C., Javoy-Agid, F., and Berger, B. 1991. Alterations of dopaminergic and noradrenergic innervations in motor cortex in Parkinson's disease. *Ann. Neurol.* **30:**365–374.

Gentilucci, M., and Rizzolatti, G. 1990. Cortical motor control of arm and hand movements, in: M. A. Goodale (ed.), *Vision and Action: The Control of Grasping,* pp. 147–162. Ablex, Norwood, N.J.

Goldman-Rakic, P. S. 1987. Circuitry of primate prefrontal cortex and the regulation of behavior by representational memory, in: V. B. Mountcastle, F. Plum, and S. R. Geiger (eds.), *Handbook of Physiology—The Nervous System,* Vol. 5, pp. 373–417. American Physiological Society, Bethesda.

Gould, S. J. 1975. Allometry in primates, with emphasis on scaling and the evolution of the brain, in: F. S. Szalay (ed.), *Contributions to Primate Paleobiology,* pp. 244–292. Karger, Basel.

Gould, S. J., and Eldredge, N. 1977. Punctuated equilibrium: The tempo and mode of evolution reconsidered. *Paleobiology* **3:**23–40.

Harting, J. K., Hall, W. C., and Diamond, I. T. 1972. Evolution of the pulvinar. *Brain Behav. Evol.* **6:**424–452.

Harting, J. K., Diamond, I. T., and Hall, W. C. 1973. Anterograde degeneration study of the cortical projections of the lateral geniculate and pulvinar nuclei in the tree shrew (*Tupaia glis*). *J. Comp. Neurol.* **150:**393–440.

Hodos, W., and Campbell, C. B. G. 1969. *Scala naturae:* Why there is no theory in comparative psychology. *Psychol. Rev.* **76:**337–350.

Hofman, M. A. 1982. Encephalization in mammals in relation to the size of the cerebral cortex. *Brain Behav. Evol.* **20:**84–96.

Holloway, R. L., Jr. 1966a. Cranial capacity, neural reorganization, and hominid evolution: A search for more suitable parameters. *Am. Anthropol.* **68:**103–121.

Holloway, R. L., Jr. 1966b. Cranial capacity and neuron number: A critique and proposal. *Am. J. Phys. Anthropol.* **25:**305–314.

Holloway, R. L., Jr., and Post, D. G. 1982. The relativity of relative brain measures and hominid mosaic evolution, in: E. Armstrong and D. Falk (eds.), *Primate Brain Evolution,* pp. 57–76. Plenum Press, New York.

Jeannerod, M. 1985. *The Brain Machine.* Harvard University Press, Cambridge, Mass.

Jerison, H. J. 1961. Quantitative analysis of evolution of the brain in mammals. *Science* **133:**1012–1024.

Jerison, H. J. 1973. *Evolution of the Brain and Intelligence*. Academic Press, New York.

Johnson, J. I. 1990. Comparative development of somatic sensory cortex, in: E. G. Jones and A. Peters (eds.), *Cerebral Cortex*. Vol. 8B, pp. 335–449. Plenum Press, New York.

Johnson, J. I., Switzer, R. C., and Kirsch, J. A. W. 1982. Phylogeny through brain traits: The distribution of categorizing characters in contemporary mammals. *Brain Behav. Evol.* **20:**97–117.

Kaas, J. H. 1977. Sensory representations in mammals, in: G. S. Stent (ed.), *Function and Formation of Neural Systems*, pp. 65–89. Dahlem Konferenzen, Berlin.

Kaas, J. H. 1982. The segregation of function in the nervous system: Why do sensory systems have so many subdivisions? in: W. P. Neff (ed.), *Contributions to Sensory Physiology*, pp. 201–240. Academic Press, New York.

Kaas, J. H. 1986. The structural basis for information processing in the primate visual system, in: J. D. Pettigrew, K. J. Sanderson, and W. R. Levick (eds.), *Visual Neuroscience*, pp. 315–340. Cambridge University Press, London.

Kaas, J. H. 1987a. The organization and evolution of neocortex, in: S. P. Wise (ed.), *Higher Brain Function: Recent Explorations of the Brain's Emergent Properties*, pp. 347–378. Wiley, New York.

Kaas, J. H. 1987b. The organization of neocortex in mammals: Implications for theories of brain function. *Annu. Rev. Psychol.* **38:**129–151.

Kaas, J. H. 1989. Why does the brain have so many visual areas? *J. Cogn. Neurosci.* **1:**121–135.

Kaas, J. H. 1991. Plasticity of sensory and motor maps in adult mammals. *Annu. Rev. Neurosci.* **14:**137–167.

Kaas, J. H., and Preuss, T. M. 1993. Archontan affinities as reflected in the visual system, in: F. S. Szalay, M. J. Novacek, and M. C. McKenna (eds.), *Mammalian Phylogeny*. Springer-Verlag, Berlin (in press).

Kaas, J. H., Harting, J. K., and Guillery, R. W. 1973. Representation of the complete retina in the contralateral superior colliculus of some mammals. *Brain Res.* **65:**343–346.

Kemper, T. L., and Galaburda, A. M. 1984. Principles of cytoarchitectonics, in: A Peters and E. G. Jones (eds.), *Cerebral Cortex*, Vol. 1, pp. 35–57. Plenum Press, New York.

Kirsch, J. A. W., and Johnson, J. I. 1983. Phylogeny through brain traits: Trees generated by neural characters. *Brain Behav. Evol.* **22:**60–69.

Kobler, J. B., Isbey, S. F., and Casseday, J. H. 1987. Auditory pathways to the frontal cortex of the mustache bat, *Pteronotus parnellii*. *Science* **236:**824–826.

Kolb, B., 1984. Functions of the frontal cortex of the rat: A comparative review. *Brain Res. Revs.* **8:**65–98.

Kolb, B., and Whishaw, I. Q. 1985. *Fundamentals of Human Neuropsychology*, 2nd ed. Freeman, San Francisco.

Krettek, J. E., and Price, J. L. 1977. Projections from the amygdaloid complex to the cerebral cortex and thalamus in the rat and cat. *J. Comp. Neurol.* **172:**687–722.

Lane, R. H., Allman, J. M., and Kaas, J. H. 1973. The visuotopic organization of the superior colliculus of the owl monkey (*Aotus trivirgatus*) and the bush baby (*Galago senegalensis*). *Brain Res.* **60:**335–349.

Lashley, K. S. 1929. *Brain Mechanisms and Intelligence*. University of Chicago Press, Chicago.

Lashley, K. S. 1949. Persistent problems in the evolution of mind. *Q. Rev. Biol.* **24:**28–42.

Lashley, K. S., and Clark, G. 1946. The cytoarchitecture of the cerebral cortex of Ateles: A critical examination of architectonic studies. *J. Comp. Neurol.* **85:**223–306.

LaVail, J. H., and LaVail, M. M. 1974. The retrograde intraaxonal transport of horseradish peroxidase in the chick visual system: A light and electron microscopic study. *J. Comp. Neurol.* **157:**303–358.

Le Gros Clark, W. E. 1932. The structure and connections of the thalamus. *Brain* **55:**406–470.

Le Gros Clark, W. E. 1941. Observations on the association fibre system of the visual cortex and the central representation of the retina. *J. Anat.* **75:**419–433.

Le Gros Clark, W. E. 1956. Observations on the structure and organization of olfactory receptors in the rabbit. *Yale J. Biol. Med.* **29:**83–95.

Le Gros Clark, W. E. 1959. *The Antecedents of Man.* Edinburgh University Press, Edinburgh.

Livingstone, M., and Hubel, D. 1988. Segregation of form, color, movement, and depth: Anatomy, physiology, and perception. *Science* **240:**740–749.

Luria, A. R. 1973. *The Working Brain.* Basic Books, New York.

McGeer, P. L., Eccles, J. C., and McGeer, E. G. 1987. *Molecular Neurobiology of the Mammalian Brain,* 2nd ed. Plenum Press, New York.

Macphail, E. M. 1982. *Brain and Intelligence in Vertebrates.* Oxford University Press, London.

Macphail, E. M. 1987. The comparative psychology of intelligence. *Behav. Brain Sci.* **10:**645–656.

Markowitsch, H. J., and Pritzel, M. 1979. The prefrontal cortex: Projection area of the thalamic mediodorsal nucleus? *Physiol. Psychol.* **7:**1–6.

Martin, R. D. 1981. Relative brain size and basal metabolic rate in terrestrial vertebrates. *Nature* **293:**57–60.

Martin, R. D. 1990. *Primate Origins and Evolution.* Princeton University Press, Princeton, N.J.

Martin, R. D., and Harvey, P. H. 1985. Brain size allometry. Ontogeny and phylogeny, in: W. L. Jungers (ed.), *Size and Scaling in Primate Biology,* pp. 147–173. Plenum Press, New York.

Maunsell, J. H. R., and Newsome, W. T. 1987. Visual processing in monkey extrastriate cortex. *Annu. Rev. Neurosci.* **10:**363–401.

Morel, A., and Kaas, J. H. 1992. Subdivisions and connections of auditory cortex in owl monkeys. *J. Comp. Neurol.* **318:**27–63.

Morgane, P. J., Jacobs, M. S., and Galaburda, A. 1985. Conservative features of neocortical evolution in dolphin brains. *Brain Behav. Evol.* **26:**176–184.

Mrzljak, L., and Goldman-Rakic, P. S. 1992. Acetylcholinesterase reactivity in cerebral cortex: Light and electron microscopic analysis in the adult monkey and human. *J. Comp. Neurol.* **364:**261–281.

Nudo, R. J., and Masterton, R. B. 1990. Descending pathways to the spinal cord. III: Sites of origin of the corticospinal tract. *J. Comp. Neurol.* **296:**559–583.

Okinaki, S., Yoshikawa, M., Uono, M., Muro, T., Mozai, T., Igata, A., Tanabe, H., Ueda, S., and Tomanaga, M. 1961. Distribution of cholinesterase activity in the human cerebral cortex. *Am. J. Phys. Med.* **40:**135–146.

Orbach, J. 1982a. At the Yerkes Laboratories of Primate Biology, Orange Park, Florida, in: J. Orbach (ed.), *Neuropsychology after Lashley,* pp. 21–51. Erlbaum, Hillside, N.J.

Orbach, J. 1982b. The legacy of *Brain Mechanisms and Intelligence,* in: J. Orbach (ed.), *Neuropsychology after Lashley,* pp. 1–20. Erlbaum, Hillside, N.J.

Passingham, R. E. 1982. *The Human Primate.* Freeman, San Francisco.

Perrett, D. I., Mistlin, A. J., and Chitty, A. J. 1987. Visual neurones responsive to faces. *Trends Neurosci.* **10:**358–364.

Pettigrew, J. D. 1986. Flying primates? Megabats have the advanced pathway from eye to midbrain. *Science* **231:**1304–1306.

Pettigrew, J. D. 1991. Wings or brain? Convergent evolution in the origins of bats. *Syst. Zool.* **40:**199–216.

Pettigrew, J. D., and Cooper, H. M. 1986. Aerial primates: Advanced visual pathways in megabats and flying lemurs. *Soc. Neurosci. Abstr.* **12:**1035.

Pettigrew, J. D., Jamieson, B. G. M., Robson, S.K., Hall, L. S., McAnally, K. I., and Cooper, H. M. 1989. Phylogenetic relations between microbats, megabats and primates (Mammalia: Chiroptera and Primates). *Philos. Trans. R. Soc. London Ser. B* **325:**489–559.

Phillips, C. G., Zeki, S., and Barlow, H. B. 1984. Localization of function in the cerebral cortex. Past, present, and future. *Brain* **107:**328–361.

Povinelli, D. J. 1993. Reconstructing the evolution of mind. *Am. Psychol.* in press.

Preuss, T. M. 1990. The granular frontal association cortex of the strepsirhine primate *Galago:* Comparative anatomy and evolutionary implications. Ph.D. dissertation, Yale University.

Preuss, T. M., and Goldman-Rakic, P. S. 1989. Connections of the ventral granular frontal cortex of macaques with perisylvian premotor and somatosensory areas: Anatomical evidence for somatic representation in primate frontal association cortex. *J. Comp. Neurol.* **282:**293–316.

Preuss, T. M., and Goldman-Rakic, P. S. 1991a. Ipsilateral cortical connections of granular frontal cortex in the strepsirhine primate *Galago*, with comparative comments on anthropoid primates. *J. Comp. Neurol.* **310:**507–549.

Preuss, T. M., and Goldman-Rakic, P. S. 1991b. Architectonics of the parietal and temporal association cortex in the strepsirhine primate *Galago* compared to the anthropoid primate *Macaca. J. Comp. Neurol.* **310:**475–506.

Preuss, T. M., and Goldman-Rakic, P. S. 1991c. Myelo- and cytoarchitecture of the granular frontal cortex and surrounding regions in the strepsirhine primate *Galago* and the anthropoid primate *Macaca. J. Comp. Neurol.* **310:**429–474.

Preuss, T. M., Stepniewska, I., and Kaas, J. H. 1992. Microstimulation studies of motor cortical organization in lorisid primates. *Eur. J. Neurosci (Suppl.)* **5:**174.

Radinsky, L. B. 1970. The fossil evidence of prosimian brain evolution, in: C. R. Noback and W. Montagna (eds.), *The Primate Brain,* pp. 209–224. Appleton–Century–Crofts, NewYork.

Radinsky, L. B. 1975. Primate brain evolution. *Am. Sci.* **63:**656–663.

Rasmussen, D. T. 1990. Primate origins: Lessons from a neotropical marsupial. *Am. J. Primatol.* **22:**263–277.

Richards, R. J. 1987. *Darwin and the Emergence of Evolutionary Theories of Mind and Behavior.* University of Chicago Press, Chicago.

Rizzolatti, G., and Gentilucci, M. 1988. Motor and visual–motor functions of the premotor cortex, in: P. Rakic and W. Singer (eds.), *Neurobiology of Neocortex.* pp. 269–295. Wiley, Chichester.

Roland, P. E. 1984. Metabolic measurements of the working frontal cortex in man. *Trends Neurosci.* **7:**430–435.

Rolls, E. T. 1984. Neurones in the cortex of the temporal lobe and in the amygdala of the monkey with responses selective for faces. *Hum. Neurobiol.* **3:**209–222.

Rose, J. E., and Woolsey, C. N. 1948. The orbitofrontal cortex and its connections with the mediodorsal nucleus in rabbit, sheep and cat. *Res. Publ. Nerv. Ment. Dis.* **27:**210–232.

Rowe, M. 1990. Organization of the cerebral cortex in monotremes and marsupials, in: E. G. Jones and A. Peters (eds.), *Cerebral Cortex,* Vol. 8B, pp. 263–334. Plenum Press, New York.

Stephan, H. 1972. Evolution of primate brains: A comparative anatomical investigation, in: R. Tuttle (ed.), *The Functional and Evolutionary Biology of Primates,* pp. 155–174. Aldine, Chicago.

Stephan, H., and Andy, O. J. 1969. Quantitative comparative neuroanatomy of primates: An attempt at a phylogenetic interpretation. *Ann. N.Y. Acad. Sci.* **167:**370–387.

Stephan, H., Baron, G., and Frahm, H. D. 1988. Comparative size of brains and brain components, in: H. D. Steklis and J. Erwin (eds.), *Comparative Primate Biology,* Vol. 4, pp. 1–38. Liss, New York.

Stepniewska, I., Preuss, T. M., and Kaas, J. H. 1993. Architectonics, somatotopic organization, and ipsilateral cortical connections of the primary motor area (M1) of owl monkeys. *J. Comp. Neurol.* **330:**238–271.

Stuss, D. T., and Benson, D. F. 1986. *The Frontal Lobes.* Raven Press, New York.

Sussman, R. W. 1991. Primate origins and the evolution of angiosperms. *Am. J. Primatol.* **23:**209–223.

Szalay, F. S., and Dagosto, M. 1988. Evolution of hallucial grasping in the primates. *J. Hum. Evol.* **17:**1–33.

Thiele, A., Vogelsang, M., and Hoffmann, K.-P. 1991. Pattern of retinotectal projection in the megachiropteran bat Rousettus aegyptiacus. *J. Comp. Neurol.* **314:**671–683.

Tilney, F. 1928. *The Brain, from Ape to Man,* 2 volumes. Harper & Row (Hoeber), New York.

Van Essen, D. C. 1985. Functional organization of primate visual cortex, in: A. Peters and E. G. Jones (eds.), *Cerebral Cortex,* Vol. 3, pp. 259–329. Plenum Press, New York.

Walsh, K. 1987. *Neuropsychology: A Clinical Approach,* 2nd ed. Churchill Livingstone, Edinburgh.

Wood Jones, F. 1916. *Arboreal Man.* Arnold, London.

Summary

11

ROSS D. E. MACPHEE

Introduction

In keeping with the title and intended theme of this book, this summary will be mainly devoted to reviewing current understanding of the interrelationships of primates and their relatives, as reflected in the investigations and discussions in preceding chapters. Inasmuch as knowledge cannot be divorced from the method by which it was gained, there will be some occasion to touch on methodological matters.

Homology and Justification

Homology is similarity *due to descent* (Patterson, 1982). This distinguishes homology from the other type of similarity that organisms can express, that is, convergence, or similarity that cannot be accounted for by an appeal to shared phylogeny. Finding reliable means for discriminating between these kinds of resemblances is one of the central problems of systematics, and in her chapter Simmons sets out the case for phylogeny being the only means of validating homologies. (Like Simmons, I am principally concerned with morphological homology.) Her central point, with which I agree, is that any hypothesis about a specific homology ("structure x in organism *1* is the homolog of structure y

ROSS D. E. MACPHEE • Department of Mammalogy, American Museum of Natural History, New York NY 10024.

Primates and Their Relatives in Phylogenetic Perspective, edited by Ross D.E. MacPhee. Plenum Press, New York, 1993.

in organism *2*") entails an equally specific hypothesis about phylogeny. This union of homology and phylogeny is both a great strength and a great weakness of systematic reasoning. The fact that two organisms possess a detailed structural similarity not found in other organisms is most simply and powerfully explained as a consequence of their sharing a common ancestor which also possessed the same structure (or a morphoclinal version of it). But the evidence for this inferred phylogenetic relationship is the similarity itself, which may or may not be combined with other resemblances that invite the same conclusion. Thus the very determinations of homology from which we try to infer phylogeny must themselves be validated by conclusions about phylogenetic relationships (the "problem of homology;" Wiley, 1981).

Cladistic methodology seeks to break the circularity inherent in the problem of homology by deductive methods, a reflection of the widespread but false notion that deductivism is the ideal form of knowledge (for an excellent critical evaluation, see Bryant [1989]). However, as Cartmill (1981) points out, in the absence of lawlike systematic principles suitable for use as minor premises in *modus tollens* argumentation (if A is true, B is true; but B is false; therefore A is false), the cladistic fallback position is to rely on *a priori* assumptions (samples: assume that a highly corroborated phylogeny is the correct one and adjudicate homology versus homoplasy according to the position of taxa and characters on the branching sequence; or, quasi-statistically, assume that evolution proceeds in a strictly parsimonious manner, so that phylogenetic hypotheses with the fewest steps can be regarded as the ones closest to reality). Such assumptions are open to numerous objections, but surely the crushingly definitive one is that "biological theory warrants no deductive inferences from the relationships of organisms to the distribution of various attributes among those organisms; and therefore *distributions of attributes cannot be used to falsify hypotheses of relationship*" (Cartmill, 1981, p. 90). That is, they cannot unless there are evolutionary laws from which distributions can be deduced; and there are none.

If phylogenetic statements—and the character distributions from which they are mainly inferred—cannot be falsified according to Popperian canons, what then can be done? From an epistemological standpoint, the problem with holding out any particular structural equivalency as a homology (or a particular branching sequence as a true phylogeny) is justifying one's belief in one's conclusion. An examination of the chapters of this book (or, indeed, any book like it on any group of real organisms) will reveal that most instantiations of homology (or homoplasy) have nothing to do with deduction and everything to do with justifying why a particular belief is held to be true. Justification of belief is an explicitly inductive procedure, whatever the preferred method of substantiating it (Pappas and Swain, 1978; Goldman, 1976; Dretske, 1971). Thus, in claiming that "phylogeny, not morphology, provides the crucial test of homology," Simmons (this volume, p. 4) overlooks the point that the only "test" provided by a stated phylogenetic diagram is whether it is consistent with something else held to be true, such as another such diagram (which in turn has to be consistent with something else, and so on, toward the infinite regress). Appealing to a cladogram does not solve the problem raised by "crossing synapomorphies," because distributions of characters have no lawlike bearing on relationships, as already noted.

At present, I see no way to defeat the problem of the vicious circle (and its relative, the infinite regress) inherent in homological determinations, at least in cases that concern morphology. But should this even concern us? It seems foolish to deny instances of knowledge just because one has only inductive reasons for regarding them as true, or because one can find no way of extracting equivalent knowledge from a *modus tollens* argument, or because one may occasionally and unwittingly accept worthless or untrue propositions as true. Only the last of these defects is a potentially serious problem in utilizing induction, but as far as I am aware the possibility that one might commit egregious error does not unduly affect how systematists do their everyday science. Instead, the real point is that we should turn our attention to formulating the soundest procedures for justification, and here, I think, is where cladistic methodology has already made its enduring contribution to systematic practice—in separating kinds of resemblance (primitive, derived, and homoplastic), in insisting that character polarities be justified and not merely asserted, in requiring that monophyly be the basis for the determination of relationships. These are not unimportant achievements, but they do not tell us everything that we want to know about evolution. It is a meager systematics indeed that cannot incorporate adaptational history, but in many quarters of systematic biology it is no longer considered pert to utilize information about the biological consequences of adaptation in classification. This is curious, because scenarios about adaptational history are no more or less implausible of reconstruction than are phylogenetic branching sequences. Fortunately, the tradition of exploring the connection between adaptation and morphology continues in evolutionary primatology (e.g., Cartmill, 1974; Sussman, 1991; Anthony and Kay, 1993; Preuss, this volume).

And what about molecular homology? As Novacek (1993, p. 17) points out in his review of recent work in higher-level mammalian phylogenetics, "molecular data potentially can recover all the heritable information [carried by the semaphoront], an attribute obviously not common to morphology or other phenotypic data." While molecular data are not free of interpretive uncertainties (e.g., alignment of sequence data, detectability of back mutations, information content of transitions versus transversions), it is reasonable to believe that, some day, systematics as a filing system will be exclusively predicated on genetic information. The justification for this is also clear: with enormous numbers of characters (sequence positions) to utilize in the mitochondrial and nuclear genomes, probabilistic statements about relationships based on sequence data may eventually reach such a level of mutual corroboration that it becomes difficult to imagine what evidence could be marshalled to defeat them—which is as good as it gets epistemologically (Pappas and Swain, 1978). Happily, a predictable consequence of this will be the rejuvenation of adaptational studies and the comparative method on which they are based, because a stable filing system will provide a platform for a more secure exploration of the *biological* consequences of phylogeny—in short, how evolution made us.

Relationships

Given the variety of methods, data, and philosophical positions represented in this volume, it is unsurprising that contributors failed to converge on a single solution to the problem of primate supraordinal relationships. On the other hand, there are some favorable signs of emerging consensus in some areas, especially in those instances where the indicators of morphology repeatedly appear to correspond to the indicators provided by molecular approaches. Table 1 presents a concordance of the systematic arguments of contributors who expressed specific conclusions on one or more of the following issues: (1) position of Scandentia; (2) monophyly of Chiroptera; (3) monophyly of Volitantia; and (4) content of Archonta. Issues collateral to these were also raised, and these will be identified in the appropriate places.

Position of Scandentia

The phylogenetic position of tree shrews has undergone several radical revisions in the past five decades (for earlier concepts, see papers in Luckett [1980]). Until the 1970s, the consensus was that tree shrews were either primates or the taxon most closely related to primates. The rhetorical strength of this argument derived its force from Le Gros Clark's (1971 and earlier editions) catalog of "evolutionary trends," a list of characteristic primate features which he felt were initiated in the primate ancestor and gradually modified in varying degrees in later primate lineages. In his view that ancestor was essentially tupaiidlike in its organization (or "total morphological pattern"), which in turn was a compelling reason for viewing tupaiids as the least evolved primates. As is now broadly acknowledged, the problem with Le Gros Clark's approach is that "total morphological pattern" mixes primitive and derived resemblances. Therefore it does not provide a basis for the identification of monophyletic groups, which should be justified on the basis of synapomorphous resemblances only. Most of the contributors to a review volume on tree shrew relationships, *Comparative Biology and Evolutionary Relationships of Tree Shrews* (Luckett, 1980), found few or no derived resemblances that justified placing tree shrews and primates in a sister-group relationship, and concluded by vacating the scandentian claim to primate status without necessarily providing a better placement (e.g., Cartmill and MacPhee, 1980).

In retrospect it can be seen that one problem with the range of comparisons carried out in the tree shrew volume was that most contributors tended to stress conditions in lipotyphlan insectivores and elephant shrews in their char-

**Table I. Hypotheses in This Volume About the Relationships
of Archontan Mammals**

Author(s)	Preferred Supraordinal Grouping That Includes Primates[a]	Chiroptera Monophyletic?	Volitantia Supported?	Notes
Szalay and Lucas	Pr, De, Sc, Ch	Yes	Yes	Microsyopidae, Mixodectidae, most "plesiadapiforms"[b]
Johnston and Kirsch	Pr, De, Sc, Ch[c]	Yes	Yes?	
Adkins and Honeycutt	Pr, De, Sc	Yes	No	Macroscelidea is sister-group of tree shrews?
Stanhope et al.	Pr, De, Sc	Yes	No	Rodentia and Lagomorpha (Glires) related to archontans? Chiroptera related to Carnivora?
Beard	Pr, De[d]	—	No	Most "plesiadapiforms" distributed among clades of a more inclusive Dermoptera
Wible and Martin	Pr, Sc?	—	—	Support for Archonta and internal groupings at ordinal level is weak, but evidence slightly favors primates and tree shews as close relatives
Thewissen and Babcock	—	Yes	Yes	
Luckett	—	Yes	?[e]	Glires (Rodentia + Lagomorpha) supported, but its relationship to any "archontans" considered unlikely

[a] As stated or illustrated in this book. Dashed lines indicate that no statement about the monophyly of the group in question was made. For possible sister-group relationships within concepts of Archonta, see individual papers. Key: Pr, Primates; De, Dermoptera; Sc, Scandentia; Ch, Chiroptera.

[b] Plesiadapiformes of some older classifications is paraphyletic.

[c] Johnson and Kirsch do not discuss optimal internal arrangement of taxa within Archonta. However, in their figure 14, on which this superordinal grouping is based, Dermoptera is the sister taxon of Chiroptera.

[d] Beard excludes bats and tree shrews from mirorder Primatomorpha (Primates + Dermoptera incl. Plesiadapiformes [in part]).

[e] According to Luckett, fetal membrane evidence neither supports nor denies Volitantia.

acter analyses—a reflection of the lingering hold of the precladistic notion of a "transition" between primates and insectivores and the assumption that insectivores could stand as "basal eutherians" (MacPhee and Novacek, 1993). There was, of course, no reason to exclude these taxa in the process of making decisions about character polarities, but, by the same token, there was no reason to exclude equally or more plausible outgroups such as colugos and bats, as recent history has shown. At present tree shrews are not considered by most observers to be the strongest candidates for the primate sister group, that position having being assumed by dermopterans (see below). However, the rehabilitation of scandentians as being at least reasonably close relatives of primates (and colugos) seems to be firmly established by the frequency with which the triad Scandentia-Dermoptera-Primates appears in diverse molecular investigations (e.g., Adkins and Honeycutt, 1991, this volume; Stanhope *et al.*, this volume) as well as recent morphological studies (e.g., Szalay and Lucas, this volume; Wible and Martin, this volume; see also Simmons, this volume).

Monophyly of Chiroptera

Within living Mammalia, powered flight is uniquely possessed by megachiropteran and microchiropteran bats. As far as can be determined from the existing fossil record, no extinct nonchiropteran mammals developed anything resembling the flight of bats, a point that is not diminished by the discovery that gliding adaptations were probably commoner among early Tertiary archontans than was once thought (Beard, 1993, this volume). For these reasons, powered flight, whether considered as a single complex or atomized into separate skeletal, muscular, and physiological adaptations, has usually been treated as essentially unassailable evidence of bat monophyly.

At the same time, it has also long been recognized that megabats and microbats are rather different in many details of anatomical structure not connected with the flight apparatus (Smith and Madkour, 1980). In most instances these differences, to the degree that they exist, can be interpreted as primitive characters retained by one bat clade and lost in the other, or as derived features that developed after the split between megabats and microbats. Most of the features in the second class do not occur in other mammals or, if they do (e.g., echolocation), have a different morphological basis and are therefore considered to be homoplastic. Pettigrew (1986, 1991a, b; Pettigrew *et al.*, 1989), however, challenged the bias for homoplasy by alleging that certain neural traits were exactly similar and unequivocally connected megabats and primates (and possibly also dermopterans), to the exclusion of all other eutherians—including microbats. If megabats are the sister group of primates, as the provocative "flying primate" hypothesis asserts, then flight must have evolved twice within Mammalia.

Pettigrew's hypothesis has been exhaustively treated, almost always with disconfirmatory results, in a number of papers (see in particular the bibliographies of Simmons, Stanhope *et al.*, Adkins and Honeycutt, this volume). Contributors to this volume recommended outright rejection of both of Pettigrew's chief contentions, namely that Chiroptera is diphyletic, and that Megachiroptera is the sister-taxon of Primates. Morphological evidence relevant to the refutation of diphyly includes Luckett's demonstration that the structure of fetal-membrane character complexes in bats is unique to them among eutherians, Thewissen and Babcock's evidence for identical patterning of propatagial musculature and innervation in megabats and microbats, Johnson and Kirsch's failure to find additional primate–megabat synapomorphies in neural characters, and the mtDNA and nuclear gene evidence assessed by Adkins and Honeycutt and Stanhope *et al.* Particularly telling is the comment by Stanhope *et al.* that, using the topology favored by Pettigrew and running their IRBP nuclear gene data set, the *least* parsimonious arrangement was one in which Megachiroptera was positioned as sister-group to Primates. A similar, highly unparsimonious result was found with their e-globin data set (see also Bailey *et al.*, 1992).

It is the strength of the comparative method that apparent contradictions caused by "crossing synapomorphies" are potentially resolvable with additional data. Here, the wing versus brain problem addressed by Pettigrew (1991a) is potentially resolved by the discovery (Thiele *et al.*, 1991) that *Rousettus* lacks the retinotectal pathways described for primates and *Pteropus*. How widespread the primitive pathway configuration is within megabats other than *Rousettus* has not been determined. However, the fact that megachiropterans are polymorphic at all goes against Pettigrew's argument that neural features are severely constrained compared to nonneural ones, and destroys the epistemologically privileged position he granted to neural characters (see Preuss, this volume).

Monophyly of Volitantia

While there was no disagreement on the monophyly of Chiroptera, contributors' responses to the monophyly of Volitantia (Chiroptera + Dermoptera) were quite mixed. Some authors enthusiastically endorsed Volitantia as a natural group, but others rejected it as highly unparsimonious compared to other solutions for the positioning of bats and colugos. At least at present, the only data sets that appear to support Volitantia as a monophyletic group are morphological (Novacek and Wyss, 1986; Novacek *et al.*, 1988; Szalay and Lucas, this volume), and even here there is no uniformity of opinion (Beard, this volume). By contrast, the recent molecular evidence seems to be running uniformly against this proposal (Adkins and Honeycutt, this volume; Stanhope *et al.*, this volume; Sarich, 1991).

The structural and functional antecedent to powered flight as practiced by bats is assumed to have been some form of gliding, and it has been recognized since the time of Leche (1886) that the patagia of colugos have some undeniable design features in common with the wings of bats that do not occur in other extant gliding mammals. But are these similarities due to homology or homoplasy? The jury is still out on this one. Thewissen and Babcock (this volume) argue that the unique propatagial muscle complex of megachiropterans and microchiropterans has an "adequate structural intermediate" in *Cynocephalus*. However, for this statement to have phylogenetic meaning, the colugo arrangement of propatagial musculature and innervation would have to be on the same morphocline as the equivalent (but more derived) version found in bats. This possibility is strongly contested by Beard (1993, this volume), who regards Dermoptera (including "plesiadapiforms") as the sister-group of Primates within his mirorder Primatomorpha. In his analysis of the morphological evidence, because primitive primatomorphs (Plesiadapidae and their relatives) lack any sign of having had a patagium, the last common ancestor of dermopterans must have been a nonglider. By contrast, taxa more closely related to galeopithecids (his Paromomyiformes) were patagiate—indicating that gliding was independently invented *within* Primatomorpha. Szalay and Lucas (this volume) come to an opposite phylogenetic conclusion about the connection between chiropterans and dermopterans. In their view, the fact that some "plesiadapiforms" had morphological attributes of gliders indicates that the last common ancestor of colugos and bats could have already possessed the necessary antecedent ("shared primitive") adaptations that provided the basis for the appearance of patagial development. As evidence for this they describe pertinent features of a partial skeleton of *Mixodectes* (considered to be an archontan by them but not by Beard). Among the problems with this line of argument from a cladistic standpoint is that shared primitive traits are not considered phylogenetically informative. In addition, hardly any of the traits signalled by Szalay and Lucas (p. 220) as diagnostic of the "protovolitantian" are known in *Mixodectes*, leading to the conclusion that this taxon is probably irrelevant to a decision concerning the validity of Volitantia.

Other contributors did not reach settled opinions. In Luckett's estimate, the fetal membrane evidence neither supports nor denies Volitantia, which parallels Wible and Martin's sense of the basicranial evidence.

Content of Archonta

Archonta has never been a clear-cut systematic concept, despite its now frequent use in the systematic literature. As the most inclusive taxon considered by most contributors to this volume, the content of this supraordinal taxon—or even its existence—was affected by other systematic decisions, such as the monophyly of Chiroptera. Most of these decisions have already been discussed in preceding sections, and here I will limit myself to a few summary observations.

The problem with Archonta is that it remains very poorly characterized (Novacek *et al.*, 1988). Szalay and Lucas (p. 221–222) have attempted to improve this situation by adding a host of new features described as "diagnostic attributes" of the "protoarchontan." Their list of diagnostic features includes only one feature of the skull and none of the teeth, despite the predominance of craniodental fossils in the archontan fossil record. This considerably amplifies the number of postcranial features marshalled to support Archonta, yet, as before (e.g., Szalay, 1977) the majority of characters used to define archontans remain those of the ankle region. A discussion of the diagnostic status of these traits and trait combinations is out of place here, but I will comment that it has proven acutely difficult to point to any cranial and dental characters that might stand as candidates for archontan synapomorphies. The single cranial trait identified by Szalay and Lucas ("entotympanic auditory bulla") is unlikely to do the job, because it appears to be increasingly likely that entotympanics are primitive for Eutheria as a whole (MacPhee and Novacek, 1993).

1. Chiroptera as outlier. Chiroptera is considered part of Archonta by Szalay and Lucas (this volume), but with the exception of Johnson and Kirsch (this volume) other contributors found little positive support for this position. The Szalay–Lucas position, as already discussed, is determined by their acceptance of a close Dermoptera–Primates connection and their assessment that bats "came from an ancestry that was dermopteran, i.e., a close cladistic relative of galeopithecids" (p. 220). In contrast to the prominence given neural characters by Pettigrew (1986, 1991a,b; Pettigrew *et al.*, 1989), Johnson and Kirsch's data set provides only limited endorsement for placing bats within Archonta. They emphasize that the only neural character in their current data set that unites either bat group with any other archontans is their character T (ratio of ipsilateral to contralateral retinal projections to superior colliculus). Even this is of limited effect. When strictly neural characters (including character T) are used to define relationships in a parsimony analysis, Primates, *Tupaia*, and *Cynocephalus* form a group, but *Myotis* joins insectivores and *Pteropus* ends up with sirenians. These results suggest that there is not enough character richness in the neural data set relative to the number of taxa being assessed. When neural and nonneural data are combined, as in the runs using 89 characters (fig. 7, p. 307) and 94 characters (fig. 14, p. 327), megachiropterans and microchiropterans unite and, moreover, associate with other archontans.

At present, the bulk of the evidence goes against retention of Chiroptera as an in-group of a monophyletic Archonta.

2. Menotyphla Redux? The most parsimonious resolution of Adkins and Honeycutt's (this volume) transversions-only data for mtDNA COII was a single tree, in which primates were sister to a clade containing galeopithecids and tupaiids. A somewhat surprising result was the appearance of macroscelideans within the Dermoptera-Scandentia clade—thus recreating the Menotyphla component of Gregory's (1910) original formulation of Archonta. This concept garners very little support from any other modern data set, morphological or molecular (Luckett, 1980).

3. The Glires Connection. Stanhope *et al.* (this volume) found that Rodentia and Lagomorpha consistently grouped close to Primates, Dermoptera, and Scandentia, raising (once again) the possibility that Glires and Archonta are closely related (see also results of Bailey *et al.*, 1992). In this context it is interesting that Adkins and Honeycutt (this volume) found that, for their

COII data set, *Mus* and *Rattus* are positioned between *Tarsius* and other primates when all of these taxa are included in the same analysis. The aberrant position of *Tarsius* is presumably a "long branch" problem occasioned by the lengthy separation of the tarsier lineage from that of most or all other primates (Beard and MacPhee, in press; MacPhee and Cartmill, 1986). However, independent of the positioning of *Tarsius*, something is pulling the rodents toward the primates in this data set—possibly a series of homologous substitutions unique to primates and rodents or, more generally, to rodents and archontans. These possibilities warrant additional exploration with more diverse data sets, both morphological and molecular (Sarich, 1991). Some interesting paleontological developments on this front can also be expected (M. C. McKenna, pers. comm.).

Prospects

As Stanhope *et al.* (this volume) point out, molecular systematics is still in its infancy, and we are in the steepest part of the learning curve regarding the possibilities and limits of molecular techniques. While it is predictable that such techniques will become increasingly important in systematic studies of all types, it is practically certain that morphological approaches will never be completely supplanted. If systematics is the study of anything and everything about organisms (Simpson, 1945), then phylogenetic reconstruction in the narrow sense cannot be the end-point of study. But it is just as clear that a secure, well-justified phylogeny is the best possible place to start in placing primates and their relatives in a truly evolutionary perspective.

References

Adkins, R. M., and Honeycutt, R. L. 1991. Molecular phylogeny of the superorder Archonta. *Proc. Natl. Acad. Sci. USA* **88:**10317–10321.

Anthony, M. R. L., and Kay, R. F. 1993. Tooth form and diet in ateline and alouattine primates: reflections on the comparative method. *Am. J. Sci.* **293:**356–382.

Bailey, W. J., Slightom, J. L., and Goodman, M. 1992. Rejection of the "flying primate" hypothesis by phylogenetic evidence from the e-globin gene. *Science* **256:**86–89.

Beard, K. C. 1993. Phylogenetic systematics of the Primatomorpha, with special reference to Dermoptera, in: F. S. Szalay, M. J. Novacek, and M. C. McKenna (eds.), *Mammal Phylogeny*, Vol. 2, *Placentals*, pp. 129–150. Springer-Verlag, New York.

Beard, K. C., and MacPhee, R. D. E. in press, Cranial anatomy of *Shoshonius* and the antiquity of Anthropoidea, in: J. G. Fleagle, R. F. Kay, and E. L. Simons (eds.), *Anthropoid Origins*. Plenum Press, New York.

Bryant, H. N. 1989. An evaluation of cladistic and character analyses as hypothetico-deductive procedures, and the consequences for character weighting. *Syst. Zool.* **38:**214–227.

Cartmill, M. 1974. Rethinking primate origins. *Science* **184:**436–443.

Cartmill, M. 1981. Hypothesis testing and phylogenetic reconstruction. *Zeitschr. Zool. Syst. Evolut.-forsch.* **21**:21–36.

Cartmill, M., and MacPhee, R. D. E. 1980. Tupaiid affinities: The evidence of the carotid arteries and cranial skeleton, in: W. P. Luckett (ed.), *Comparative Biology and Evolutionary Relationships of Tree Shrews*, pp. 95–132. Plenum Press, New York.

Dretske, F. 1971. Conclusive reasons. *Aust. J. Philos.* **49**:1–22.

Goldman, A. I. 1976. Discrimination and perceptual knowledge. *J. Philos.* **73**:771–791.

Gregory, W. K. 1910. The orders of mammals. *Bull. Am. Mus. Nat. Hist.* **27**:1–524.

Leche, W. 1886. Über die Säugethiergattung *Galeopithecus*. Eine morphologische Untersuchung. *K. Svenska Vet. Akad. Handl.* **21**:1–92.

Le Gros Clark, W. E. 1971. *The Antecedents of Man*. Edinburgh University Press, Edinburgh.

Luckett, W. P. (ed.) 1980. *Comparative Biology and Evolutionary Relationships of Tree Shrews*. Plenum Press, New York.

MacPhee, R. D. E., and Cartmill, M. 1986. Basicranial structures and primate systematics, in: D. R. Swindler and J. Erwin (eds.), *Comparative Primate Biology*, Vol. 1, pp. 219–275. Liss, New York.

MacPhee, R. D. E., and Novacek, M. J. 1993. Definition and relationships of Lipotyphla, in: F. S. Szalay, M. J. Novacek, and M. C. McKenna (eds.), *Mammal Phylogeny*, Vol. 2, *Placentals*, pp. 13–31. Springer-Verlag, New York.

Novacek, M. J. 1993. Reflections on higher mammalian phylogenetics. *J. Mamm. Evol.* **1**:3–30.

Novacek, M. J., and Wyss, A. R. 1986. Higher-level relationships of the recent eutherian orders: morphological evidence. *Cladistics* **2**:257–287.

Novacek, M. J., Wyss, A. R., and McKenna, M. C. 1988. The major groups of eutherian mammals, in: M. Benton (ed.), *The Phylogeny and Classification of Tetrapods*, Vol. 2, pp. 31–71. Systematics Association Special Volume No. 35B. Oxford University Press (Clarendon), London.

Pappas, G. S., and Swain, M. 1978. Introduction, in: G. S. Pappas and M. Swain (eds.), *Essays on Knowledge and Justification*, pp. 11–40. Cornell University Press, Ithaca.

Patterson, C. 1982. Morphological characters and homology, in: K. A. Joysey and E. A. Friday (eds.), *Problems in Phylogenetic Reconstruction*, pp. 21–74. Systematics Association Special Volume No. 21. Academic Press, New York.

Pettigrew, J. D. 1986. Flying primates? Megabats have advanced pathway from eye to midbrain. *Science* **231**:1304–1306.

Pettigrew, J. D. 1991a. Wings or brain? Convergent evolution in the origins of bats. *Syst. Zool.* **40**:199–216.

Pettigrew, J. D. 1991b. Primate relations as revealed in brain characters: the "flying primate" hypothesis. *Am. J. Phys. Anthrop.* **12 (Suppl.)**:142–143.

Pettigrew, J. D., Jamieson, B. G. M., Robson, S. K., Hall, L. S., McAnally, K. I., and Cooper, H. M. 1989. Phylogenetic relations between microbats, megabats, and primates (Mammalia: Chiroptera and Primates). *Philos. Trans. R. Soc. London Ser. B* **325**:489–559.

Sarich, V. M. 1991. A molecular perspective on primate origins. *Am. J. Phys. Anthrop.* **12 (Suppl. 1)**:157.

Simpson, G. G. 1945. The principles of classification and a classification of mammals. *Bull. Am. Mus. Nat. Hist.* **85**:1–350.

Smith, J. D., and Madkour, G. 1980. Penial morphology and the question of chiropteran phylogeny. *Proc. 5th Intern. Bat Res. Conf.*: 347–365.

Sussman, R. W. 1991. Primate origins and the evolution of angiosperms. *Am. J. Primatol.* **23**:209–223.

Szalay, F. S. 1977. Phylogenetic relationships and a classification of the eutherian Mammalia, in: M. K. Hecht, P. C. Goody, and B. M. Hecht (eds.), *Major Patterns in Vertebrate Evolution*, pp. 315–374. Plenum Press, New York.

Thiele, A., Vogelsang, M., and Hoffmann, K. P. 1991. Pattern of retinotectal projection in the megachiropteran bat *Rousettus aegypticus*. *J. Comp. Neurol.* **314**:671–683.

Wiley, E. O. 1981. *Phylogenetics: The Theory and Practice of Phylogenetic Analysis*. Wiley, New York.

Systematic Index

Author and Subject Index